Advances in Experimental Medicine and Biology

Volume 1419

Advances in Experimental Medicine and Biology provides a platform for scientific contributions in the main disciplines of the biomedicine and the life sciences. This series publishes thematic volumes on contemporary research in the areas of microbiology, immunology, neurosciences, biochemistry, biomedical engineering, genetics, physiology, and cancer research. Covering emerging topics and techniques in basic and clinical science, it brings together clinicians and researchers from various fields.

Advances in Experimental Medicine and Biology has been publishing exceptional works in the field for over 40 years, and is indexed in SCOPUS, Medline (PubMed), EMBASE, BIOSIS, Reaxys, EMBiology, the Chemical Abstracts Service (CAS), and Pathway Studio.

2021 Impact Factor: 3.650 (no longer indexed in SCIE as of 2022)

Zhanjun Zhang
Editor

Cognitive Aging and Brain Health

Springer

Editor
Zhanjun Zhang
State Key Laboratory of Cognitive
Neuroscience and Learning
Faculty of Psychology,
Beijing Normal University
Beijing, China

ISSN 0065-2598 ISSN 2214-8019 (electronic)
Advances in Experimental Medicine and Biology
ISBN 978-981-99-1626-9 ISBN 978-981-99-1627-6 (eBook)
https://doi.org/10.1007/978-981-99-1627-6

This Springer imprint is published by the registered company Springer Nature Singapore Pte Ltd.
The registered company address is: 152 Beach Road, #21-01/04 Gateway East, Singapore 189721, Singapore

Preface

The socio-economic development and advances in medical health around the world have enabled humanity to pursue longevity. China, as a major developing country with 1.4 billion people, will become an aging society soon with 14% of the citizens aged 65 and above. By 2060, one in three Chinese will be over 65 years old. Du Fu, a great poet in the Tang Dynasty, once lamented the brevity of life, "It is rare for a man to live to seventy." But living up to 70 years is not an exception any more nowadays.

Aging has become one of the most universal bodily and physical experiences for humanity. Life is like running on an extending track. As you keep going, it is inevitable that you will meet the issue of "aging" along the way. In the last stretch, aging and the accompanying physical decline and cognitive degeneration will limit our movement and thinking, undermining the quality of our life and our senses of happiness. By this stage, we will experience all kinds of "fatigue," which can have various meanings for different people: for some, fatigue might mean the pain caused by chronic physical diseases, and for others it represents the suffering of psychiatric disorders as terrifying as Alzheimer's Disease and other types of dementia. In China, 5.3% of the population aged 65 and above endure dementia. They should have enjoyed the remaining time of their life but end up leaving with fragmented memories.

China is not the only aging society. In fact, the entire world is aging fast. The United Nations (UN) predicts that by mid-century, the proportion of the global population aged 60 and above will equal or even exceed that of children aged 14 and under. Such a demographic shock, though unprecedented and historic, seems now irreversible and normal. These profound changes have urged individuals and groups alike to learn about "aging" in order to better cope with this issue.

It does not mean, however, we should indulge in negative emotions such as depression and fear. The aging of the body and the mind is inevitable in the foreseeable future, but effective interventions can still be found to delay it. Today, life sciences, behavioral sciences, and brain and cognitive sciences have developed by leaps and bounds, providing insights and inspirations into the deceleration of the aging process. Scientists in various fields around the world have been dedicated around the clock to improving the well-being and the dignity of people in their twilight years. The book is an epitome of their extensive and outstanding work.

This book brings home to us the process of aging from the perspectives of behavioral sciences as well as brain and cognitive sciences. It helps us dive

An Overview

1

Xin Li, Caishui Yang, and Jun Wang

Abstract

Nowadays, China has rapidly progressed into an aging society and is faced with huge challenges on public health. Aging is accompanied by the structural and functional alterations in the brain, which leads to the cognitive decline in the elderly and acts as the primary risk factor for dementia. However, the aging brain has not been well understood at a systemic level. This chapter presents the definition of brain health, the aging situation in China, an overview of the BABRI, the purpose of writing this book, and the introductions of the chapters, respectively, which will contribute to knowledge of the underlying mechanisms of healthy and pathological aging of the brain.

Keywords

Brain health · Active aging · Alzheimer's disease (AD) · Dementia · Beijing Aging Brain Rejuvenation Initiative (BABRI) · Aging brain

1.1 The Importance and Significance of Brain Health

1.1.1 Brain Health and Active Aging

Brain health is defined as "the ability to perform all the mental processes of cognition, including the ability to learn and judge, use language, and remember" [1]. A healthy brain performs normal cognitive functions, such as memory, emotions, and movements, which support our daily tasks and performance and affect the quality of our lives. Additionally, a healthy brain is also essential for pursuing longevity and health [2]. When discussing "pursuing longevity," healthy aging is a topic that could never be ignored, and more importantly, the current theme of "active aging" provides more inclusive information than "healthy aging" [3]. According to the definition provided by the World Health

X. Li · J. Wang (✉)
State Key Laboratory of Cognitive Neuroscience and Learning, Faculty of Psychology, Beijing Normal University, Beijing, China

Beijing Aging Brain Rejuvenation Initiative (BABRI) Centre, Beijing Normal University, Beijing, China
e-mail: lixin99@bnu.edu.cn; jun_wang@bnu.edu.cn

C. Yang
State Key Laboratory of Cognitive Neuroscience and Learning, Faculty of Psychology, Beijing Normal University, Beijing, China

School of Systems Science, Beijing Normal University, Beijing, China

Beijing Aging Brain Rejuvenation Initiative (BABRI) Centre, Beijing Normal University, Beijing, China
e-mail: psy_yang@bnu.edu.cn

Z. Zhang (ed.), *Cognitive Aging and Brain Health*, Advances in Experimental Medicine and Biology 1419, https://doi.org/10.1007/978-981-99-1627-6_1

Organization (WHO), "active aging" is "the process of optimizing opportunities for health, participation, and security to enhance the quality of life as people age" [4]. Here, "active" refers to "continuing participation in social, economic, cultural, spiritual, and civic affairs, more than the abilities to be physically active or to participate in the labor force," and the word "health" refers to health in the aspects of physical, mental, and social well-being [4]. When elderly individuals remain healthy, their subjective well-being level increases, and the costs of healthcare services and medical treatment can be saved [5]. To promote active aging, the impacts of the aging brain must be considered since a healthy brain is a fundamental principle for a healthy life.

1.1.2 The Impact of an Unhealthy Brain on Individuals

A healthy brain directly affects the physical and mental health of the individuals. With aging, the structure of the brain will change, such as a decrease in the volume of the whole brain [6], which in turn affects cognitive ability [7]. Moreover, the associations between cognitive deficits and psychiatric disorders [8] should be considered accordingly.

Individuals are affected by unhealthy brain, and the family members of elderly individuals and society are facing considerable challenges. Those who take care of individuals with Alzheimer's disease (AD) may experience a problem known as caregiver burden, which refers to unpleasant experiences with emotional, social, financial, physical, and spiritual functioning [9]. Moreover, the financial pressure experienced by the government and families should be given much more attention. According to the Alzheimer's association's report [10], the costs of AD and related dementia will reach 305 billion US dollars. Thus, society needs to build a public health system to provide special care services for elderly people.

1.2 The Current Aging Situation in China

1.2.1 Brief Overview of the Aging Population in China

The aging population in China is massively increasing. The estimated number of aging population is projected to be more than 400 million in 2050, which is 30% of the total population in China. According to the data in the 2019 Statistical Bulletin on the Development of Civil Affairs, 253.9 million residents in China are older than 60 years. Generally, the average time that the developed countries spend entering an aging society is 45 years. For instance, among the developed countries, France spent 130 years, Sweden spent 85 years, and Australia and the USA spent 79 years, respectively. However, China only spent 27 years, which is 18 years faster than the developed countries [11].

1.2.2 Challenges with the Aging Population

An aging population has an increased prevalence of dementia. Over six million Chinese elderly individuals suffered from dementia in 2010, and this number substantially increased over the next ten years and reached 10 million in 2020. Challenges are associated with the tremendous number of both the aging populations and dementia patients. The first challenge is the economic pressure faced by both the government and families. Elderly social welfare expenses in China, which were provided by the government, reached 45.3 billion RMB in 2019 [12]. The second challenge is the growing need for elderly special care services. Due to the transformation of Chinese society, families are no longer the primary care providers for elderly individuals. Therefore, the "empty nest" elderly require help from other elderly care services. However, these demands require more resources and attention from the current social welfare system in China. The third challenge is the poor awareness and acknowledgement of dementia in communities. Without a proper understanding and recognition, suitable services will be difficult to provide for the elderly. The fourth challenge is the lack of effective intervention and prevention programs [11]. Therefore, an evidence-based healthcare system is urgently needed in the society. Large-scale and longitudinal cohort studies can provide crucial research support in terms of the development of effective early prevention and intervention systems, which is a part of evidence-based healthcare system. To meet all these demands, the Beijing Aging Brain Rejuvenation Initiative (BABRI) was developed and initiated in 2008.

1.3 Beijing Aging Brain Rejuvenation Initiative (BABRI)

The BABRI has been launched and been conducted by Beijing Normal University since 2008, which included a data collection and prevention strategy design. Specifically, longitudinal information regarding aging, cognition, neuroimaging and biological data were collected accordingly, and prevention strategies have been designed for elderly individuals with normal cognition and cognitive impairment. Nevertheless, the BABRI is the first community-based cohort study focusing on primary and secondary prevention of dementia in China, and its longitudinal aging-related data collection makes early detection of dementia at communities possible. Moreover, the BABRI is also designed to help facilitate the development of the China's National Plan on aging and dementia.

The main goal of the BABRI is to identify risk markers of cognitive impairment and to depict developmental trajectory of AD and other types of dementia. The specific aims of the project are to (1) identify markers of different stages of cognitive impairment and dementia/AD, (2) understand the basic principles of cognitive aging and underlying brain mechanisms, (3) formulate and validate tools and norms for the cognitive screening of Chinese elderly individuals in community settings, and (4) test prevention strategies and potential alternative interventions for cognitive impairment and dementia at early stages [13].

All participants in the BABRI were native Chinese speakers, aged 50 years or above at the initial registration and baseline recruitment, and had normal or rectified normal vision, hearing, and speech functions. It is worth noting that unlike other cohort studies reported in the world, the BABRI has primarily recruited its participants from local communities, rather than memory clinics or hospitals, in Beijing since 2008. And after the BABRI was expanded into a multicenter project known as the BABRI-National Consortium (BABRI-NC) in the year of 2017, participants were also recruited from communities from other regions of China.

Older community residents were initially registered via brief telephone or face-to-face surveys on demographic information and medical history, followed by detailed assessments on health status, daily behaviors, and cognitive function, and would be revisited every two or three years over the 20 years of the project. During their participation in the BABRI project, the older individuals who met the inclusion criteria underwent magnetic resonance imaging (MRI) and positron emission tomography (PET) brain scans. Generally, the measurements included in the BABRI can be grouped into the following eight categories: (1) demographic characteristics; (2) medical records; (3) cognitive profiles; (4) lifestyles and habits; (5) emotional status; (6) biochemical markers; (7) genotypes; and (8) neuroimaging features.

By December 2020, 10,671 participants (60.28% female) joined the BABRI project. Among the older adults in Beijing, 2021 participants were followed up once or more at a two- or three-year interval. The BABRI is currently at its 15-year benchmark, which is more than halfway of the 20-year blueprint of the initiative. The research team of the BABRI expects to build a systematic framework for brain health, to apply the knowledge achieved regarding the underlying mechanisms of cognitive impairment and cognitive aging, and to identify the protective and risk factors of disease development and normal aging process.

1.4 Purpose of Writing This Book

The BABRI provides us the opportunities to offer better help to the aging population in China. Nevertheless, the BABRI aims to investigate the underlying mechanisms of the aging brain, but the acknowledgment of early prevention, and the intervention of aging and dementia still need more exploration. In order to achieve the goal of active aging, the proper usage of scientific methods to investigate the patterns and mechanisms is crucial for brain health.

Numerous studies have highlighted the exploration of the factors in active aging. Gholipour et al. [14] investigated the active aging model in Iran and have shown that policymaking, organizational structure, members component, control, financing, governance grants, services, and selected area to focus were the eight factors affecting management of active aging. Moreover, Steinmayr et al. [15] demonstrated active aging in the aspects of social networks, physical health, participation in society, lifelong learning, money, housing, and employment. Meanwhile, the exploration of successful brain aging is of utmost importance. Mora et al. [16] studied the mechanisms of brain aging from the aspect of brain plasticity, environmental enrichment, and lifestyle. Furthermore, Seinfeld and Sanchez-Vives [15] provided evidence on the effectiveness of neuroscientific information-based strategies to promote behavioral changes and lead to healthy aging.

Even though some accomplishments have been achieved in active aging and the underlying brain mechanism, more efforts are still needed to further improve the efficiency and to accelerate the progress of the research. A book that includes the fundamental knowledge and the recent findings in cognitive aging from the neurological perspective can attract the attention of the readers and provide an overview of brain health.

1.5 Introductions of the Chapters

The following information will be delivered and discussed in this book: (1) the definition of cognitive decline and how it is associated with aging; (2) cognitive aging models; (3) emotional and affective disorders in cognitive aging; (4) socioeconomic status and cognition; (5) the brain and biological mechanisms of aging patterns; (6) the early and late stage of abnormal aging; (7) the effectiveness and its imaging evidence on cognitive training; (8) the lifestyle intervention as the influential risk factors in cognitive aging.

In Chap. 2, the concept of cognitive aging, previous and present studies in this area, will be introduced. In Chap. 3, the decline in the different cognitive areas during the aging process, such as sensory perception, memory, language and reasoning, and space navigation, will be discussed. Behavioral models, educational, biological, and sociological theories in the field of cognitive aging will be presented in Chap. 4. In Chap. 5, affective and emotional disorders among elders, their impacts on cognitive aging, and the possible mechanisms underlying these findings will be introduced. In Chap. 6, a brief introduction on socioeconomic status and the measurement methods, and the relationship between socioeconomic status and cognitive aging will be discussed. In Chap. 7, an overview of the brain aging patterns which include the brain structure, function, and energy metabolism will be delivered. In Chap. 8, the neuroimaging evidence in brain aging will be introduced. In Chaps. 9 and 10, information regarding the neurobiological mechanisms of cognitive decline caused by brain aging, including the biomolecular markers of brain aging will be delivered. In Chap. 11, the syndrome, risk, and protective factors of cognitive impairment will be presented. In Chap. 12, dementia will be discussed as the late stage of abnormal aging from a neurological perspective. In Chap. 13, an overview of cognitive training will be delivered, and the imaging evidence of the effectiveness will be demonstrated. In Chap. 14, the impact of lifestyle on cognitive aging will be discussed.

1.6 Summary

In this chapter, the definition of brain health, the aging situation in China, an overview of the BABRI, the purpose of writing this book, and the introductions of the chapters are delivered, respectively. In the section of the definition of brain health, the importance of brain health is discussed from the aspect of active aging, as well as the meaning to the elderly individuals. The aging statistics and challenges brought by aging population are briefly introduced in the section of the aging situation in China. In the part of the BABRI overview, this project is introduced as a response strategy to the challenges of the Chinese aging society. In the last two sections, the purpose of writing this book and the introductions of the chapters are presented. The meanings of scientific research and active aging, and the importance of new concept in this field are discussed in the section of the purpose of writing this book.

References

1. Centers for Disease Control and Prevention (2009) Healthy ageing. What is a healthy brain? New research explores perceptions of cognitive health among diverse older adults
2. Wang et al (2020) What is brain health and why is it important? BMJ 371:m3683
3. Kalachea A, Kickbusch I (1997) A global strategy for healthy ageing. World Health 5(4):4–5
4. WHO (2002). Active ageing: a policy framework. Geneva
5. Steinmayr et al (2020) Gender differences in active ageing: findings from a new individual-level index for European countries. Soc Indic Res 151(2):691–721
6. Good et al (2001) A voxel-based morphometric study of ageing in 465 normal adult human brains. NeuroImage 14(1 Pt 1):21–36
7. Fletcher et al (2018) Brain volume change and cognitive trajectories in ageing. Neuropsychology 32(4):436–449
8. Reinlieb et al (2014) The patterns of cognitive and functional impairment in amnestic and non-amnestic mild cognitive impairment in geriatric depression. Am J Geriatr Psychiatry 22(12):1487–1495
9. Zarit et al (1986) Subjective burden of husbands and wives as caregivers: a longitudinal study. Gerontol 26(3):260–266
10. Alzheimer's Disease Facts and Figures (2020) Special reports. On the front lines: primary care physicians and Alzheimer's care in America by Alzheimer's associations
11. Zhang K (2018) Chinese elderly brain health report 2018, 1st edn. People's Medical Publishing House
12. Statistical Bulletin on the Development of Civil Affairs (2019) Ministry of Civil Affairs of the People's of Republic of China
13. Yang et al (2021) Early prevention of cognitive impairment in the community population: the Beijing aging brain rejuvenation initiative. Alzheimers Dement 2021(17):1610–1618
14. Gholipour et al (2020) Active ageing Management in Iran: designing a model. Asia Pac J Health Manage 15(2):1–10
15. Seinfeld S, Sanchez-Vives MV (2015) Healthy ageing promotion through neuroscientific information-based strategies. Int J Environ Res Public Health 12(10):12158–12170
16. Mora F (2013) Successful brain ageing: plasticity, environmental enrichment, and lifestyle. Dialogues Clin Neurosci 15(1):45–52

Part I

The Definition of Cognitive Aging

Cognitive Aging: How the Brain Ages?

2

Shaokun Zhao, Yumeng Li, Yuqing Shi, and Xin Li

Abstract

Cognitive aging refers to the cognitive changes or functional decline that comes with age. The relation between aging and functional declines involves various aspects of cognition, including memory, attention, processing speed, and executive function. In this chapter, we have introduced several dimensions about cognitive aging trajectories. Meanwhile, we have reviewed the history of the study of cognitive aging and expatiated two trends that are particularly noteworthy in the effort to elucidate the process of aging. One is that the differences between components of mental abilities have become gradually specified. The other one is a growing interest in the neural process, which relates changes in the brain structure to age-related changes in cognition. Lastly, as the basis of cognitive function, brain structures and functions change during aging, and these changes are reflected in a corresponding decline in cognitive function. We have discussed the patterns of reorganization of various structural and functional aging processes of the brain and their relationship with cognitive function.

Keywords

Cognitive aging · Cognitive domains · Brain aging

2.1 The Concept of Cognitive Aging

Across the human life span development, old age usually accompanies cognitive changes, generally referring to functional decline, which apparently affect the life independence of elderly individuals and restrict their fundamental abilities in their daily lives. This process is generally called cognitive aging. Initially, from the traditional perspective, cognitive aging is a monotonous and absolute development process that ignores the factors of individual differences and their complex interactions with other neurobiological and psychosocial changes [1]. However, researchers recognize the substantial heterogeneity in the specific "age" pattern [2, 3]. Therefore, cognitive aging is a dynamic process, with exceptions to aging-related cognitive decline.

Because neuropsychological test scores do not completely represent the performance of daily living activities in elderly individuals, aging

S. Zhao · Y. Li · Y. Shi · X. Li (✉)
State Key Laboratory of Cognitive Neuroscience and Learning, Faculty of Psychology, Beijing Normal University, Beijing, China

Beijing Aging Brain Rejuvenation Initiative (BABRI) Centre, Beijing Normal University, Beijing, China
e-mail: sk_zhao@mail.bnu.edu.cn; 202231061054@mail.bnu.edu.cn; 202021061037@mail.bnu.edu.cn; lixin99@bnu.edu.cn

Z. Zhang (ed.), *Cognitive Aging and Brain Health*, Advances in Experimental Medicine and Biology 1419, https://doi.org/10.1007/978-981-99-1627-6_2

should be understood from multiple perspectives. The experimental data tend to indicate that elderly individuals experience deficits in several domains of cognitive ability compared to younger adults [4] The relation between aging and functional declines involves various aspects of cognition, including memory, attention, processing speed, encoding, cognitive control, linguistic ability, expertise, and knowledge. Several dimensions will be introduced in detail for further descriptions of the distinctions among the organizations of cognition in terms of their trajectories of aging and change [5].

2.1.1 Attention

Attention refers to the ability to control the concentration on specific stimuli via executive function. Compared to the relatively simple attention task (usually known as digit span test that is measured by repeating a string of digits in a normal or reverse sequence), a more significant effect of aging is observed on executive attention measures, such as the Trial Making Test (A and B, especially B). Executive tests of planning and organizations require more mental flexibility, which is a critical component of attention that helps elderly individuals complete more complex activities [6, 7].

2.1.2 Memory

As people age, their memory performance exhibits an overall decline. Moreover, the effects of aging on several different types of memory have been interpreted independently of each other [8].

(a) Short-term memory and working memory (WM): Cognitive aging seems to affect working memory to a greater extent than short-term memory. The difference between them is that WM requires not only storage but also more processing components. Older adults experience more age-related impairments when the activities demand more processing procedures [9, 10].

(b) Explicit memory (declarative memory) versus implicit memory (automatic unconscious memory): Explicit memory refers to the collection of relevant facts and events requiring conscious and intentional retrieval (e.g., trying to recall what you did at a certain time). Whereas the implicit memory involves the influence on the current behavior without people's awareness (e.g., responding instantaneously to a familiar song). In the past 20 years, a notable difference has been identified between these two memory types: age-related implicit memory declines are much smaller than explicit memory declines [11, 12]. Thus, intentional and nonintentional memory retrieval are essential to the age-related cognitive changes. Therefore, assessment and measurement methods may take this factor into account.

(c) Episodic memory versus semantic memory: Both episodic memory and semantic memory are attributed to the declarative memory system [13]. On the one hand, episodic memory refers to the memory of the personal experience of events or scenes that happened at a certain time and place (e.g., remembering which friend you met in the park yesterday). On the other hand, semantic memory involves the meaning of words and concepts that are not related to personal experience (e.g., naming various types of animals or specific words). Similar to explicit and implicit memory, episodic and semantic memory are affected by cognitive aging to different degrees. Episodic memory may be significantly impaired as people age; however, semantic memory might hardly be affected by age-related cognitive decline [10, 14, 15], which is a potential question to be further discussed in detail.

(d) Prospective memory: Prospective memory refers to the memory of deferred behavior and involves two major types: time-based prospective memory and event-based prospective memory. Unlike other types of memory, the tasks for the encoding process of prospective memory have not yet performed. This period (from encoding to acting) is delayed, indicating that extra cognitive activities are needed, such as the simulation of the future or some conscious rehearsal of simulation-driven attention. These features make prospective memory different from other types of memory. Different theoretical perspectives have resulted in contradictory empirical findings, which made this memory type an attractive topic of discussion. One of the opinions is that the aging process should be accompanied by a sharp decline in prospective memory, which contrasts with the finding that the performance of older adults in prospective memory tasks is the same as that of younger adults. Another contradictory finding is that older adults performed significantly more poorly in other general memory tasks, such as free recall tasks, than younger adults [16, 17]. The uniqueness of prospective memory and the debates related to this type of memory make the future of research much more complex.

2.1.3 Executive Function

Executive function refers to "domain-general control processes that monitor and regulate other cognitive processes to guide the attainment of future goals" [18]. From a neuropsychological perspective, the frontal-executive hypothesis indicates that age-related declines mainly occur in the prefrontal cortex (PFC), which has been assumed to be responsible for executive function failures [19, 20]. From a cognitive perspective, executive function coexists with a wide range of abilities, such as problem solving, mental flexibility, concept formation, and self-monitoring [21]. Researchers have indicated that the processing efficiency and speed of executive function tend to decrease with aging [22].

2.1.4 Linguistic Ability

The acquisition of language skills accumulates over time. General linguistic ability remains stable as people age. However, some exceptions to the overall trend have also been observed, such as verbal fluency. This ability is not intact with aging. Verbal fluency refers to the capacity to generate words that belong to a specific category in a specific amount of time [23, 24]. It involves not only semantic knowledge of lexical items but also tracks prior responses and blocks intrusions from other semantic categories. Thus, verbal fluency shows declines with aging [25].

In the history of psychological aging, the field of cognitive aging has been described as [26] concentrating on "manifest changes or transformations that occur in human and animal behavior related to the length of life." Aging is not just related to the group of adults that become older. The field emphasizes the study of changes and transitions throughout nearly all of adulthood. Additionally, aging is not simply equivalent to a decline of function. Baltes and Willis [27] regard this process as a balance of trajectories because life cycle changes include processes that can be interpreted as growth and decline [27]. The developmental process is influenced by various circumstances, including social, cultural, ethnic, and individual factors.

2.2 Study of Cognitive Aging: Past and Present

Cognitive aging criteria vary between different areas, such as the pathological cognition and social perception. In this chapter, the term cognition will be defined as the process of mental function in healthy adults, and changes occurring in cognitive aging will be discussed.

Accumulating evidence has proven that cognitive performance varies in individuals of different ages, and researchers are still endeavoring to define the boundaries of knowledge and to develop more accurate and suitable theories and approaches to reveal the mystery of aging. Here, two trends are particularly noteworthy in the effort to elucidate this process. One possible explanation is that the differences between components of mental abilities have become gradually specified. The other explanation is a growing interest in the neural process, which relates changes in the brain structure to age-related changes in cognition.

The study of cognitive aging is accompanied by the rise of cognitive psychology after the mid-twentieth century. When behaviorism no longer dominated the psychology field and mental activities began to be viewed as observable events, cognition entered the realm of research, including attention, memory, and consciousness [28]. In the mid-1960s, increasing interest emerged in the mechanism underlying specific age-related cognitive domains. Previously, cognitive research was mostly descriptive and documentative rather than studying the explicit differences in intellectual function between the young and elderly individuals [29]. A series of research achievements was released in 1965. *Behavior, Ageing, and the Nervous System*, published by Welford and Birren [30], investigated the relationship between biological factors, the speed of behavior, and how behavior changes with age, which provided insights into cognitive aging research. The *Nature* journal received Schonfield's ideas, revealing that the ability to retrieve information from long-term memory declines with age, while recognition is maintained. These early studies were subsequently confirmed and supported by studies on attention resources, in which the reduced attentional resources of elderly individuals are compensated by a task environment with guided retrieval [31].

2.2.1 Support from the Development of Statistics

Improvements in statistics have facilitated with cognitive aging research, especially in studies of differences in the cognitive performance of individuals of different ages. As an ideal approach, structural models reveal and explain the hidden connections and mechanisms between age and certain cognitive functions. Salthouse [32] summarized four categories of cognitive aging in structural models: independence, mediation, shared influence, and hierarchical features. The independence model was first used to examine the effects of age on each cognitive variable. As the model was postulated to manipulate each variable, most single-variable studies only investigated separate and isolated age-related effects on the variables, and interrelations within variables were ignored. Since the 1990s, processing speed or working memory has served as the hypothesized mediator and has been frequently applied in a large number of studies with mediation models. Several variables are assumed to be mediated through more age-related or primitive variables [33]. Later, the shared influence structure was suggested and found suitable for the research data [34]. Instead of considering certain variables as a mediator, a common cause attributed to different types of variables was introduced, and the effects of age were postulated to manipulate only the common set of influences. The last pattern is the hierarchical structure, which is portrayed as age-related influences with an effect on any hierarchy level, but the effect decreases as the hierarchy level increases. The variable at the highest level of the hierarchy is explained by aging. Additionally, the loading of the variable displays a significant positive relation with the absolute value of the correlation of the variable with age [32].

2.2.2 Trend of Differentiating Specific Domains

The study of age-related effects on cognition is based on distinct cognitive fields, among which memory and attention are the two main ones. The research direction of cognitive aging has been revealed by the development of memory and attention studies in aging.

The early findings of memory and age are traced back to the beginning of the twentieth century when behaviorism was still the main focus of psychology research, and verbal learning paradigms were the popular measurements for assessing memory. In 1929, a study compared incidental memory for digit-symbol pairs in adults from different age groups [35]. Subsequent studies discovered that participants aged 12–17 years exhibited the best performance on the memory task, and adults in the 34–59 year-old age group exhibited better performance than those in the 60–82 year-old age group. These findings implied a gradual age-related difference in memory [36]. The strategy used in memory tasks was also assessed in later research, and older participants were less likely to use imagination or verbal mnemonic creations when attempting to remember paired associations [37]. Aging, however, does not lead to a global decline in memory. The concept of memory is not unitary and has been classified into specific aspects, including episodic and semantic memory, implicit and explicit memory, prospective memory, and encoding and retrieval processes, all of which show dissociations in normal aging. Many studies have focused on identifying and explaining the age-related changes in specific memory domains, and the findings indicate that some memory aspects change with aging, while the others are age-invariant. For instance, semantic memory is also referred to as crystallized intelligence, which is represented as the storage of factual knowledge in memory. It was initially presumed to be maintained or increased with age. Lars Nyberg [38] found that unlike episodic memory, age differences in semantic memory were not significant. Meanwhile, Park et al. [39] reported that although processes such as working memory, processing speed, and memory recall decline with age, as measured by vocabulary abilities, they increase across the life span.

However, further research called the stable maintenance of semantic memory into question. The research suggested that higher task demand might draw a different conclusion. In picture memory tasks, age effects are evident with the abstract pictures and the participants are required to integrate target and context information [40]. Nevertheless, some researchers claim that impairment in semantic memory retrieval does not represent an actual semantic deficit, but rather only the difficulties in specific information retrieval [41].

Instead of studying the overall attention process of aging, the research trend of aging-related changes in attention was stratified into different aspects, such as selective attention, inhibition, divided attention, and sustained attention. The early literature on the nature of attention was based on a study by Broadbent [42], focusing on the selection process. In 1965, Rabbitt compared the ability of participants to ignore irrelevant information when they were required to select targets, and the task seemed harder for older participants, implying that inhibition was also an aspect of attention [43]. Older adults have difficulty in divided attention, especially in tasks of higher complexity [44]. A review of a series of meta-analyses reveals that older adults show more decreases in the performance of dual tasks, which they are required to attend and process information from multiple sources [45]. Although these specific attention types are often recorded as highly interrelated, all aspects of attention are predicted to decline with aging due to age-related frontal lobe degradation [29]. The value of these age-related changes in certain areas is increasing and helps to confront and clarify the past doubts and future research directions on aging.

2.2.3 The Trend of Neuro Perspectives

Another notable trend is the advancement of neuroscience and imaging measurement techniques, including structural, resting, functional, and activation imaging. The perspectives of cognitive aging and neuroscience of aging were usually studied independently until the end of the twentieth century. An independent subject has emerged and is called as cognitive neuroscience of aging (CNA), which combines studies using behavior measures (e.g., memory and attention) and age-related neural decline (e.g., synaptic loss and cerebral atrophy). CNA focuses on changes in the brain that are affected by age-related effects on cognition. Cognitive aging is depicted as a consequence of neural aging at the structural and functional levels.

Evidence suggests that physiological changes occur in the brain structure and the volume of the brain decreases during aging; meanwhile, different parts and regions exhibit different rates of change. The results derived from a cross-sectional study of subjects ranging in age from 14 to 77 years show that the volume of grey matter may decrease in a linear manner, while the change in white matter follows a different trajectory. The age-related change in the volume of white matter exhibits an inverted U shape, which increases before the age of 38 and subsequently decreases in elderly individuals [46]. Compared with the temporal, parietal, and occipital cortices, the frontal cortex is the area with the highest rate of decrease [47]. Diffusion tensor imaging studies have shown more obvious age-related changes in white matter integrity in anterior regions accompanied by more vulnerable myelinated fibers. Meanwhile, the results derived from a volume-based imaging study also revealed the greatest age-related decrease in grey matter in frontal regions [47]. Another important debated topic is the domain of neuronal loss due to cognitive aging. Some evidence suggests that unlike individuals with Alzheimer's disease (AD), neuronal loss may not be entirely or primarily responsible for the condition of individuals with mild age-related cognitive decline, but not dementia. The findings implied that this mild cognitive decline may result from biochemical shifts in still-intact neural circuity and that these disruptions may involve selective vulnerability in the entorhinal cortex and other brain regions, such as the hippocampus [48].

Advances in functional neuroimaging provide technical support for studies of activated brain functions in different regions and functional connectivity during cognitive tasks, which are both attributed to changes in cognitive performance with aging. For instance, research comparing age-related effects on familiarity-related brain activity (the feeling that an event is familiar or not without the recovery of contextual details) to recollection-related activity (accompanied by specific contextual details) was tested in an event-related functional magnetic resonance imaging (fMRI) study with an episodic recognition task. Researchers found a dual dissociation within the medial temporal lobe that occurred with aging, with an increase in familiarity-related activity in the rhinal cortex and a decrease in recollection-related activity in the hippocampus. This result suggests a greater dependence on familiarity as compensation for deficits in recollection [49]. The dedifferentiation theory is also called cell dedifferentiation, which is defined as the process by which cells shift from a partially or terminally differentiated stage to a less differentiated stage. Concerning the functional activation among brain regions, this theory suggests that neural specialization or differentiation decreases with aging. This process leads to a less distinct transmission of information and less accurate mental representations of information resources [50]. Some researchers attribute this phenomenon of dedifferentiation to disruptions in the dopamine system. Additionally, the modulatory role of the prefrontal cortex is affected in older individuals, which leads to decreases in the discriminant network and stimulus response precisions [51]. In addition, the disconnection hypothesis proposes that certain deficits in communication between regions may lead to age-related cognitive decline [52]. The extent to which brain regions of older adults are interrelated is still being debated [53].

2.3 Brain Aging and Its Relationship with Cognitive Aging

The brain is the basis of cognitive function, and as people age, cognitive function declines and brain structures and functions change accordingly. We will discuss the patterns of reorganization of various brain structures and functions that occur during normal aging and their relationship with cognitive function.

2.3.1 Aging in the Gray Matter

With aging, the gray matter volume shows a trend of continuous decline [54, 55], and the thickness of the cerebral cortex decreases, especially in the brain areas related to perception function [56]. A previous study applied MRI techniques and morphological analysis methods based on voxels (voxel-based morphometry, VBM) to explore changes in the gray matter volume during the normal aging process. A significant decrease in the gray matter volume of the whole brain was observed. Frontal and temporal lobe atrophy were the most significant changes, followed by parietal lobe atrophy. However, the volume of the occipital lobe remained relatively stable [57].

The gray matter structural integrity in the brain is the foundation of normal cognitive function during the aging process. Brain aging is closely associated with cognitive decline, and previous studies have found that declines in working memory [58], attention [59], cognitive control [23, 60, 61], and processing speed [62] are all associated with a decrease in gray matter volume. The aforementioned evidence suggests that the structural reorganization of gray matter as a result of aging alters the performance of domain-general and domain-specific cognitive processes in the brain. Specifically, atrophy of the gray matter volume in the prefrontal lobe is associated with a decline in executive function [63, 64]. Atrophy of the middle frontal gyrus is mainly related to cognitive control and plays a crucial role in reorienting attention-based task requirements [65, 66]. Atrophy of the bilateral inferior frontal gyrus and superior temporal gyrus is mainly related to a decline in motor planning and pronunciation. Speech representation is related to executive processes [67]. Atrophy of the bilateral insula is mainly related to deficits of word recall and word generation. The structure of the hippocampus is closely related to memory, and its atrophy has been used to distinguish abnormal aging from normal aging [68].

2.3.2 Aging in White Matter

Similar to gray matter, the integrity of the structure of white matter changes with aging, which gradually increases until the age of 40 years, peaks at approximately 50 years, and gradually decreases after 60 years [69]. In normal aging, the white matter integrity decreases with aging, especially in the frontal lobe and parietal lobe [70, 71]. The subcortical white matter volume, which includes the amygdala, striatum, cerebellum and midline, cingulate cortex, and precuneus, also decreases with aging [72]. A large number of cross-sectional studies have identified significant age-related decreases in the composition and integrity of white matter fiber tracts, as manifested by a decrease in the fractional anisotropy (FA) and an increase in mean diffusivity (MD) of most fiber tracts [73]. The effect of aging on white matter integrity is region-specific, with more FA loss in the frontal lobe area, including the genu of corpus callosum and the posterior part of pericallosal white matter [74]. A previous study revealed a decreasing trend in the white matter connecting the frontal cortex [75], and FA showed a greater age-related decrease in the posterior region of the brain than in the anterior part of the brain [76]. The findings indicated a disproportionate change in anterior and posterior white matter. Other studies show that the effects of aging on the integrity of white matter are present not only in a transverse gradient but also in longitudinal gradients. Specifically, the upper white matter is more susceptible to aging than the lower white matter [77].

Structural integrity of white matter is the basis for normal cognitive function, and the cognitive decline related with normal aging is also associated with damaged structural integrity of white matter [78]. Previous studies have found that decreased white matter integrity in the prefrontal lobe leads to worse perceptual speed, quantitative reasoning [79], word fluency, visual motor function, and classification functioning [80]. A meta-analytical study of the relationship between white matter integrity damage and cognition [81] revealed that the damage of white matter around the ventricle [82–84] and subcortical white matter [85, 86] is associated with memory. Also, studies have found that the damage of white matter around the ventricle [83, 84, 87], and subcortical white matter [86] is related to executive function and processing speed.

2.3.3 Aging and Brain Functions

Task-related fMRI studies reveal unique patterns of changes in brain activity during normal aging. Compared with younger people, the brain activity of older people has areas of increased and decreased activation. Two main hypotheses have been proposed to explain the changes in activation of brain function in older people: dedifferentiation theory [88] and compensation theory. Dedifferentiation theory suggests that the loss of functional specificity in a brain region during task performance in elderly individuals is due to deficits in the dopamine regulatory system in the nervous system, which increases neural noise and further leads to the loss of cortical characterization specificity [89]. Compensation theory suggests that older people have higher activation than younger people to compensate for the loss of function in some brain regions. Furthermore, the theory is divided into three categories: (1) the pattern of hemispheric asymmetry is decreased, and the reduction is mainly observed in the frontal lobe [90, 91]; (2) the anterior brain compensates for posterior brain activity [92], where the activation of the posterior part of the brain decreases and the activation of the medial frontal cortex increases during the tasks of visual perception and episodic memory extraction; and (3) compensatory neural circuits [93], which include key neural circuits such as the prefrontal, parietal, precuneus, and posterior cingulate cortices, show increased activity.

A literature review summarized the variations in brain functions associated with cognitive reserve during aging [94]. High activation of the bilateral prefrontal lobes in semantic memory tasks was associated with better performance in older adults [95], and compared to the younger group, the level of frontal lobe activation was negatively correlated with the performance on the working memory task in the older group [96]. In the inspection time task, high activation of the right anterior cingulate gyrus is usually associated with better task performance [97]. The left inferior temporal lobe and the medial temporal lobe regions (the inferior temporal gyrus [ITG], the parahippocampal gyrus, and the fusiform gyrus) are negatively correlated with performance on the semantic memory task [95]. In the visual coding task, the activation of the right medial temporal gyrus, the right claustrum, the left thalamus, and the cerebellum are negatively correlated with performance [98]. In addition, a resting-state fMRI study has shown that a high cognitive reserve is associated with the activation intensity and betweenness centrality of the ITG in normal aging, as well as the local efficiency and clustering coefficient of bilateral precuneus and occipital lobe regions [99].

2.3.4 Aging and the Brain Network

During the normal aging process, changes in the overall brain functional connectivity mainly include the disconnection of large-scale networks and the reorganization of the whole-brain connectivity pattern. The functional connectivity of the default mode network (DMN), cerebellar network (CN), and salience network (SAN) decreases, while the functional connectivity strength of the somatosensory network (SSN) and subcortical network is increased [100–104]. Long-distance functional connectivity, local efficiency, global efficiency, SAN-visual network (VN) connections, and posterior connections of the DMN are reduced during normal aging. Short-distance functional connectivity, the DMN-CN connection, the VN-CN connection, the DMN-VN connection, the SAN-auditory network connection, and anterior DMN functional connectivity are enhanced in normal aging [100, 102, 105]. A study on brain network aging suggests that age exerts significant effects on both within network and between network connections; these networks include the SAN, dorsal attention network (DAN), and DMN.

The decrease in functional connectivity within the network increases the trend of dedifferentiation between networks. The decline in cognitive function in the normal aging process is associated with a decreased influence of the DMN on SAN and an increased influence of the right anterior insula on the left anterior insula. The internal functional connectivity of the DMN is reduced, while the functional connectivity of the DMN with other networks is increased [106]. The modularity of the brain refers to "a global measure of how well a network can be decomposed into a set of sparsely interconnected but densely interconnected modules" [107]. A study of age-related brain topological changes found that the modularity of older people decreases and local efficiency is low. These changes are due to the weaker connections within networks and individual brain regions that undergo different reorganization patterns in terms of functional characteristics. The global efficiency, local efficiency, and regional functional connectivity of the DMN are decreased, indicating that advanced cognitive functions decline, while the global efficiency, local efficiency, and functional connectivity of the sensorimotor network increase during the normal aging process. These results indicate the potential for a compensation mechanism of the sensorimotor area and cognitive decline. At the same time, the crucial functions of individual brain regions are transferred from central to non-central regions [107].

The cognitive function decline is associated with the reorganization of brain networks. Stern et al. [108] found that the negative activation of the two brain networks was positively correlated with the performance of working memory. The first network was composed of the left medial frontal gyrus (MFG), the bilateral inferior frontal gyrus (IFG)/precentral gyrus, and right IFG. The second network was composed of bilateral MFG/superior frontal gyrus (SFG). The negative activation in these two networks was stronger in young people than in old people. Studies have also found that working memory performance is positively correlated with negative activation of temporal lobe and limbic system networks (left hippocampus, left insula, right medial temporal gyrus/superior temporal gyrus, right cingulate, and cerebellum). A study of changes in resting state network of brain aging reveals that the elderly with a well-connected default network leads to better cognitive performance [109]. The findings of a Positron Emission Computed Tomography (PET) study also support this result, which indicated that not only the metabolic level of DMN inversely proportional to cognitive performance, so is the DAN [110].

2.4 The Significant Researches Related to Brain Aging

With the acceleration of the aging process, the proportion of the elderly population throughout human society is increasing significantly. With aging, elderly individuals always experience senile diseases. Various types of dementia, especially AD, are the main factors affecting the quality of life of elderly individuals.

According to the 2020 report from the Alzheimer's Association [111], more than five million people suffer from AD in the United States. In 2020, approximately 5.8 million Americans aged 65 years and older will develop to AD. By 2050, the number is expected to increase to approximately 14 million. The number of people with dementia will increase, but doctors announced that the health care industry is not yet ready to meet this demand. AD and other types of dementia will cost the United States $305 billion by 2020, and those costs could reach up to $1.1 trillion by 2050. AD is the sixth leading cause of death in the United States, with one-third of elderly individuals dying from AD or other dementia. These numbers caution us to support healthy aging, and serious attention should be given to the cognitive function and brain health of elderly individuals.

An increasing number of large longitudinal studies are currently in progress in several countries, such as the Australian Longitudinal Study of Ageing (ALSA) and the Beijing Aging Brain Rejuvenation Initiative (BABRI). The longitudinal design overcomes the limitations of cross-sectional studies, as the latter may confront the possible individual differences within people. Additionally, it can only serve as an indirect test of interrelationships among cognitive function and aging [112]. Studies concerning the aging brain would be another promising research field in the future. Imaging techniques with higher spatial and temporal resolution will provide a better understanding of the aging brain and nervous system. The function and microstructure of the brain (such as neurotransmitters and biomarkers) can be revealed by examining cognitive variables interrelated with aging and are useful to identify risk factors for cognitive decline and early-stage dementia. Meanwhile, implementing the research findings of interventional approaches into real life might help elderly individuals age successfully.

References

1. Craik F (1977) Age differences in human memory. In: Birren JE, Schaie KW (eds) Handbook of the psychology of aging. Van Nostrand Reinhold, New York
2. Perlmutter M (1988) Cognitive potential throughout life. In: Birren JE, Bengtson VL (eds) Emergent theories of aging. Springer Publishing Company, Berlin, pp 247–268
3. Salthouse TA (1990) Cognitive competence and expertise in aging. In: Birren JE, Schaie KW (eds) Handbook of the psychology of aging. Academic Press, Cambridge, MA, pp 310–319
4. Salthouse T (2012) Consequences of age-related cognitive declines. Annu Rev Psychol 63:201–226
5. Dixon RA (1998) The concept of gains in cognitive aging. In: Cognition, aging and self-reports. Psychology Press, London, pp 74–92
6. Carlson MC, Fried LP, Xue Q-L, Bandeen-Roche K, Zeger SL, Brandt J et al (1999) Association between executive attention and physical functional performance in community-dwelling older women. Gerontol B Psychol Sci Soc Sci 54(5):S262–S270
7. Swanberg MM, Tractenberg RE, Mohs R, Thal LJ, Cummings JL (2004) Executive dysfunction in Alzheimer disease. Arch Neurol 61(4):556–560
8. Willis SL, Schaie KW (2005) Cognitive trajectories in midlife and cognitive functioning in old age. pp 243–276
9. Bopp KL, Verhaeghen P (2005) Aging and verbal memory span: a meta-analysis. J Gerontol B Psychol Sci Soc Sci 60(5):P223–P233
10. Park DC, Lautenschlager G, Hedden T, Davidson NS, Smith AD, Smith PK et al (2002) Models of visuospatial and verbal memory across the adult life span. Psychol Aging 17(2):299
11. Light LL, La Voie D (1993) Direct and indirect measures of memory in old age. J Exp Psychol Learn Mem Cogn 18(6):1284–1297
12. Mitchell DB, Bruss PJ (2003) Age differences in implicit memory: conceptual, perceptual, or methodological? Psychol Aging 18(4):807
13. Tulving E (1972) Episodic and semantic memory 1:381–403
14. Laver GD, Burke DM (1993) Why do semantic priming effects increase in old age? A meta-analysis. Psychol Aging 8(1):34
15. Rönnlund M, Nyberg L, Bäckman L, Nilsson L-G (2005) Stability, growth, and decline in adult life span development of declarative memory: cross-sectional and longitudinal data from a population-based study. Psychol Aging 20(1):3
16. Einstein GO, Holland LJ, McDaniel MA, Guynn MJ (1992) Age-related deficits in prospective memory: the influence of task complexity. Psychol Aging 7(3):471
17. Einstein GO, McDaniel MA (1990) Normal aging and prospective memory. J Exp Psychol Learn Mem Cogn 16(4):717
18. Head D, Rodrigue KM, Kennedy KM, Raz N (2008) Neuroanatomical and cognitive mediators of age-related differences in episodic memory. Neuropsychology 22(4):491
19. Raz N (2000) Aging of the brain and its impact on cognitive performance: integration of structural and functional findings. In: Craik FIM, Salthouse TA (eds) The handbook of aging and cognition. Lawrence Erlbaum Associates Publishers, pp 1–90
20. West R (1996) An application of prefrontal cortex function theory to cognitive aging. Psychol Bull 120(2):272
21. Lezak M, Howieson D, Loring DJO (2012) Neuropsychological assessment, vol 10, 5th edn. Oxford University Press, New York, p 9780195395525
22. Bryan J, Luszcz MA, Pointer S (1999) Executive function and processing resources as predictors of adult age differences in the implementation of encoding strategies. Aging Neuropsychol Cognit 6(4):273–287
23. Salthouse TA (2010) Selective review of cognitive aging. J Int Neuropsychol Soc 16(5):754–760
24. Singh-Manoux A, Kivimaki M, Glymour MM, Elbaz A, Berr C, Ebmeier KP et al (2012) Timing of onset of cognitive decline: results from Whitehall II prospective cohort study. BMJ 344:d7622
25. Kemper S, Joan MM (2008) Dimensions of cognitive aging: executive function and verbal fluency. In: *In 2008*. Citeseer
26. Birren JE, Schroots JJ (2001) History of geropsychology
27. Baltes PB, Willis SL (1977) Toward psychological theories of aging and development. In: The handbooks of aging. Van Nostrand Reinhold, pp 128–154
28. Anderson JR (2000) *Cognitive psychology and its implications*. Worth Publishers
29. Anderson ND, Craik FI (2017) 50 years of cognitive aging theory. J Gerontol Series B 72(1):1–6
30. Welford AT, Birren JE (1965) Behavior, aging, and the nervous system: biological determinants of speed of behavior and its changes with age, vol 600. Thomas
31. Salthouse TA (2001) Structural models of the relations between age and measures of cognitive functioning. Intelligence 29(2):93–115
32. Bryan J, Luszcz MA (1996) Speed of information processing as a mediator between age and free-recall performance. Psychol Aging 11(1):3
33. Salthouse TA, Czaja SJ (2000) Structural constraints on process explanations in cognitive aging. Psychol Aging 15(1):44

34. Willoughby RR (1929) Incidental learning. J Educ Psychol 20(9):671
35. Kausler DH (2012) Experimental psychology, cognition, and human aging. Springer Science & Business Media
36. Hulicka IM, Grossman JL (1967) Age-group comparisons for the use of mediators in paired-associate learning. J Gerontol 22(1):46
37. Nyberg L, Bäckman L, Erngrund K, Olofsson U, Nilsson L-G (1996) Age differences in episodic memory, semantic memory, and priming: relationships to demographic, intellectual, and biological factors. J Gerontol Ser B Psychol Sci Soc Sci 51(4):P234–P240
38. Park D, Lautenschlarger G, Hedden T, Davidson N, Smith A, Smith P (2003) Models of visuospatial and verbal memory across the lifespan. Psychol Aging 17(2):299–320
39. Smith AD, Park DC, Cherry K, Berkovsky K (1990) Age differences in memory for concrete and abstract pictures. J Gerontol 45(5):P205–P210
40. Drag LL, Bieliauskas LA (2010) Contemporary review 2009: cognitive aging. J Geriatr Psychiatry Neurol 23(2):75–93
41. Donald B (1958) Perception and communication. Pergamon Press, New York
42. Rabbitt P (2019) 1 An age-decrement in the ability to ignore irrelevant information. In: Cognitive development and the ageing process: selected works of Patrick Rabbitt. p 5
43. McDowd JM, Craik FI (1988) Effects of aging and task difficulty on divided attention performance. J Exp Psychol Hum Percept Perform 14(2):267
44. Verhaeghen P, Cerella J (2002) Aging, executive control, and attention: a review of meta-analyses. Neurosci Biobehav Rev 26(7):849–857
45. Cabeza R, Nyberg L, Park DC (2016) Cognitive neuroscience of aging: linking cognitive and cerebral aging. Oxford University Press
46. Haug H (1983) Anatomical changes in aging brain: morphometric analysis of the human prosencephalon. Neurobiol Aging 12(4):336–338
47. Raz N, Gunning FM, Head D, Dupuis JH, McQuain J, Briggs SD et al (1997) Selective aging of the human cerebral cortex observed in vivo: differential vulnerability of the prefrontal gray matter. Cerebral Cortex (New York, NY: 1991) 7(3):268–282
48. Stern PC, Carstensen LL, Aging, N. R. C. C. o. F. D. f. C. R. o (2000) The aging mind: opportunities in cognitive research. National Academies Press (US)
49. Daselaar SM, Fleck MS, Dobbins IG, Madden DJ, Cabeza R (2006) Effects of healthy aging on hippocampal and rhinal memory functions: an event-related fMRI study. Cereb Cortex 16(12):1771–1782
50. Li S-C, Lindenberger U (1999) Cross-level unification: a computational exploration of the link between deterioration of neurotransmitter systems and dedifferentiation of cognitive abilities in old age. In: Cognitive neuroscience of memory. Hogrefe & Huber, pp 103–146
51. Li S-C, Lindenberger U, Sikström S (2001) Aging cognition: from neuromodulation to representation. Trends Cogn Sci 5(11):479–486
52. Bartzokis G, Sultzer D, Lu PH, Nuechterlein KH, Mintz J, Cummings JL (2004) Heterogeneous age-related breakdown of white matter structural integrity: implications for cortical "disconnection" in aging and Alzheimer's disease. Neurobiol Aging 25(7):843–851
53. Tucker-Drob EM, Salthouse TA (2008) Adult age trends in the relations among cognitive abilities. Psychol Aging 23(2):453
54. Driscoll I, Davatzikos C, An Y, Wu X, Shen D, Kraut M et al (2009) Longitudinal pattern of regional brain volume change differentiates normal aging from MCI. Neurology 72(22):1906–1913
55. Raz N, Lindenberger U, Rodrigue KM, Kennedy KM, Head D, Williamson A et al (2005) Regional brain changes in aging healthy adults: general trends, individual differences and modifiers. Cereb Cortex 15(11):1676–1689
56. Salat DH, Buckner RL, Snyder AZ, Greve DN, Desikan RS, Busa E, Morris JC, Dale AM, Fischl B (2004) Thinning of the cerebral cortex in aging. Cereb Cortex 14:721–730
57. Fjell AM, McEvoy L, Holland D, Dale AM, Walhovd KB, Alzheimer's Disease Neuroimaging Initiative (2014) What is normal in normal aging? Effects of aging, amyloid and Alzheimer's disease on the cerebral cortex and the hippocampus. Prog Neurobiol 117:20–40
58. Nyberg L, Lovden M, Riklund K, Lindenberger U, Backman L (2012) Memory aging and brain maintenance. Trends Cogn Sci 16(5):292–305
59. Geerligs L, Saliasi E, Maurits NM, Renken RJ, Lorist MM (2014) Brain mechanisms underlying the effects of aging on different aspects of selective attention. NeuroImage 91:52–62
60. Coppin AK, Shumway-Cook A, Saczynski JS, Patel KV, Ble A, Ferrucci L et al (2006) Association of executive function and performance of dual-task physical tests among older adults: analyses from the InChianti study. Age Ageing 35(6):619–624
61. Marchand WR, Lee JN, Suchy Y, Garn C, Johnson S, Wood N et al (2011) Age-related changes of the functional architecture of the cortico-basal ganglia circuitry during motor task execution. NeuroImage 55(1):194–203
62. Seidler RD, Bernard JA, Burutolu TB, Fling BW, Gordon MT, Gwin JT et al (2010) Motor control and aging: links to age-related brain structural, functional, and biochemical effects. Neurosci Biobehav Rev 34(5):721–733
63. Lemaitre H, Goldman AL, Sambataro F, Verchinski BA, Meyer-Lindenberg A, Weinberger DR et al (2012) Normal age-related brain morphometric changes: nonuniformity across cortical thickness, surface area and gray matter volume? Neurobiol Aging 33(3):617. e611–617.e619

64. Manard M, Bahri MA, Salmon E, Collette F (2016) Relationship between grey matter integrity and executive abilities in aging. Brain Res 1642:562–580
65. Rossi AF, Pessoa L, Desimone R, Ungerleider LG (2009) The prefrontal cortex and the executive control of attention. Exp Brain Res 192(3):489
66. Japee S, Holiday K, Satyshur MD, Mukai I, Ungerleider LG (2015) A role of right middle frontal gyrus in reorienting of attention: a case study. Front Syst Neurosci 9:23
67. Hickok G (2009) The functional neuroanatomy of language. Phys Life Rev 6(3):121–143
68. Fjell AM, McEvoy L, Holland D, Dale AM, Walhovd KB (2014) What is normal in normal aging? Effects of aging, amyloid and Alzheimer's disease on the cerebral cortex and the hippocampus. Prog Neurobiol 117:20–40
69. Liu H, Wang L, Geng Z, Zhu Q, Song Z, Chang R et al (2016) A voxel-based morphometric study of age-and sex-related changes in white matter volume in the normal aging brain. Neuropsychiatr Dis Treat 12:453
70. Resnick SM, Pham DL, Kraut MA, Zonderman AB, Davatzikos C (2003) Longitudinal magnetic resonance imaging studies of older adults: a shrinking brain. J Neurosci 23(8):3295–3301
71. Scahill RI, Frost C, Jenkins R, Whitwell JL, Rossor MN, Fox NC (2003) A longitudinal study of brain volume changes in normal aging using serial registered magnetic resonance imaging. Arch Neurol 60(7):989–994
72. Draganski B, Ashburner J, Hutton C, Kherif F, Frackowiak RS, Helms G et al (2011) Regional specificity of MRI contrast parameter changes in normal ageing revealed by voxel-based quantification (VBQ). NeuroImage 55(4):1423–1434
73. Cox SR, Ritchie SJ, Tucker-Drob EM, Liewald DC, Hagenaars SP, Davies G et al (2016) Ageing and brain white matter structure in 3,513 UK biobank participants. Nat Commun 7:13629
74. Sullivan EV, Pfefferbaum A (2003) Diffusion tensor imaging in normal aging and neuropsychiatric disorders. Eur J Radiol 45(3):244–255
75. Davis SW, Dennis NA, Buchler NG, White LE, Madden DJ, Cabeza R (2009) Assessing the effects of age on long white matter tracts using diffusion tensor tractography. NeuroImage 46(2):530–541
76. Salat DH, Tuch DS, Greve DN, van der Kouwe AJ, Hevelone ND, Zaleta AK et al (2005) Age-related alterations in white matter microstructure measured by diffusion tensor imaging. Neurobiol Aging 26(8):1215–1227
77. Zahr NM, Rohlfing T, Pfefferbaum A, Sullivan EV (2009) Problem solving, working memory, and motor correlates of association and commissural fiber bundles in normal aging: a quantitative fiber tracking study. NeuroImage 44(3):1050–1062
78. Bennett IJ, Madden DJ (2014) Disconnected aging: cerebral white matter integrity and age-related differences in cognition. Neuroscience 276:187–205
79. Caserta MT, Bannon Y, Fernandez F, Giunta B, Schoenberg MR, Tan J (2009) Normal brain aging: clinical, immunological, neuropsychological, and neuroimaging features. Int Rev Neurobiol 84:1–19
80. Bartres-Faz D, Clemente I, Junque C (2001) White matter changes and cognitive performance in aging. Rev Neurol 33(4):347–353
81. Bolandzadeh N, Davis JC, Tam R, Handy TC, Liu-Ambrose T (2012) The association between cognitive function and white matter lesion location in older adults: a systematic review. BMC Neurol 12(1):126
82. Burns JM, Church JA, Johnson DK, Xiong C, Marcus D, Fotenos AF et al (2005) White matter lesions are prevalent but differentially related with cognition in aging and early Alzheimer disease. Arch Neurol 62(12):1870–1876
83. Kim JH, Hwang KJ, Kim J-H, Lee YH, Rhee HY, Park K-C (2011) Regional white matter hyperintensities in normal aging, single domain amnestic mild cognitive impairment, and mild Alzheimer's disease. J Clin Neurosci 18(8):1101–1106
84. Cees De Groot J, De Leeuw FE, Oudkerk M, Van Gijn J, Hofman A, Jolles J et al (2000) Cerebral white matter lesions and cognitive function: the Rotterdam scan study. Ann Neurol 47(2):145–151
85. Silbert L, Nelson C, Howieson D, Moore M, Kaye J (2008) Impact of white matter hyperintensity volume progression on rate of cognitive and motor decline. Neurology 71(2):108–113
86. Baune BT, Roesler A, Knecht S, Berger K (2009) Single and combined effects of cerebral white matter lesions and lacunar infarctions on cognitive function in an elderly population. J Gerontol Ser A 64(1):118–124
87. Ishii H, Meguro K, Yamaguchi S, Ishikawa H, Yamadori A (2007) Prevalence and cognitive performances of vascular cognitive impairment no dementia in Japan: the Osaki–Tajiri Project. Eur J Neurol 14(6):609–616
88. Wig GS, Grafton ST, Demos KE, Wolford GL, Petersen SE, Kelley WM (2008) Medial temporal lobe BOLD activity at rest predicts individual differences in memory ability in healthy young adults. Proc Natl Acad Sci U S A 105(47):18555–18560
89. Li SC, Lindenberger U, Sikstrom S (2001) Aging cognition: from neuromodulation to representation. Trends Cogn Sci 5(11):479–486
90. Cabeza R (2002) Hemispheric asymmetry reduction in older adults: the HAROLD model. Psychol Aging 17(1):85–100
91. Berlingeri M, Danelli L, Bottini G, Sberna M, Paulesu E (2013) Reassessing the HAROLD model: is the hemispheric asymmetry reduction in older adults a special case of compensatory-related utilisation of neural circuits? Exp Brain Res 224(3):393–410
92. Davis SW, Dennis NA, Daselaar SM, Fleck MS, Cabeza R (2008) Que PASA? The posterior-anterior shift in aging. Cereb Cortex 18(5):1201–1209

93. Schneider-Garces NJ, Gordon BA, Brumback-Peltz CR, Shin E, Lee Y, Sutton BP et al (2010) Span, CRUNCH, and beyond: working memory capacity and the aging brain. J Cogn Neurosci 22(4):655–669
94. Anthony M, Lin F (2018) A systematic review for functional neuroimaging studies of cognitive reserve across the cognitive aging spectrum. Arch Clin Neuropsychol 33(8):937–948
95. Springer MV, McIntosh AR, Winocur G, Grady CL (2005) The relation between brain activity during memory tasks and years of education in young and older adults. Neuropsychology 19(2):181
96. Steffener J, Reuben A, Rakitin BC, Stern Y (2011) Supporting performance in the face of age-related neural changes: testing mechanistic roles of cognitive reserve. Brain Imaging Behav 5(3):212–221
97. Waiter GD, Fox HC, Murray AD, Starr JM, Staff RT, Bourne VJ et al (2008) Is retaining the youthful functional anatomy underlying speed of information processing a signature of successful cognitive ageing? An event-related fMRI study of inspection time performance. NeuroImage 41(2):581–595
98. Solé-Padullés C, Bartrés-Faz D, Junqué C, Vendrell P, Rami L, Clemente IC et al (2009) Brain structure and function related to cognitive reserve variables in normal aging, mild cognitive impairment and Alzheimer's disease. Neurobiol Aging 30(7):1114–1124
99. Marques P, Moreira P, Magalhães R, Costa P, Santos N, Zihl J et al (2016) The functional connectome of cognitive reserve. Hum Brain Mapp 37(9):3310–3322
100. Onoda K, Ishihara M, Yamaguchi S (2012) Decreased functional connectivity by aging is associated with cognitive decline. J Cogn Neurosci 24(11):2186–2198
101. Wang L, Laviolette P, O'Keefe K, Putcha D, Bakkour A, Van Dijk KR et al (2010) Intrinsic connectivity between the hippocampus and posteromedial cortex predicts memory performance in cognitively intact older individuals. NeuroImage 51(2):910–917
102. Geerligs L, Renken RJ, Saliasi E, Maurits NM, Lorist MM (2015) A brain-wide study of age-related changes in functional connectivity. Cereb Cortex 25(7):1987–1999
103. Sala-Llonch R, Bartres-Faz D, Junque C (2015) Reorganization of brain networks in aging: a review of functional connectivity studies. Front Psychol 6:663
104. Tomasi D, Volkow ND (2012) Aging and functional brain networks. Mol Psychiatry, 17(5):471, 549–458
105. Achard S, Bullmore E (2007) Efficiency and cost of economical brain functional networks. PLoS Comput Biol 3(2):e17
106. Tsvetanov KA, Henson RN, Tyler LK, Razi A, Geerligs L, Ham TE et al (2016) Extrinsic and intrinsic brain network connectivity maintains cognition across the lifespan despite accelerated decay of regional brain activation. J Neurosci 36(11):3115–3126
107. Song J, Birn RM, Boly M, Meier TB, Nair VA, Meyerand ME et al (2014) Age-related reorganizational changes in modularity and functional connectivity of human brain networks. Brain Connect 4(9):662–676
108. Stern Y, Zarahn E, Habeck C, Holtzer R, Rakitin BC, Kumar A et al (2008) A common neural network for cognitive reserve in verbal and object working memory in young but not old. Cereb Cortex 18(4):959–967
109. Arenaza-Urquijo EM, Landeau B, La Joie R, Mevel K, Mézenge F, Perrotin A et al (2013) Relationships between years of education and gray matter volume, metabolism and functional connectivity in healthy elders. NeuroImage 83:450–457
110. Bennett DA, Wilson RS, Boyle PA, Buchman AS, Schneider JA (2012) Relation of neuropathology to cognition in persons without cognitive impairment. Ann Neurol 72(4):599–609
111. Alzheimer's Association (2020) 2020 Alzheimer's disease facts and figures
112. Deary IJ, Whiteman MC, Starr JM, Whalley LJ, Fox HC (2004) The impact of childhood intelligence on later life: following up the Scottish mental surveys of 1932 and 1947. J Pers Soc Psychol 86(1):130

Part II

Aging and Cognition

Cognitive Decline Associated with Aging

3

Yiru Yang, Dandan Wang, Wenjie Hou, and He Li

Abstract

Cognitive decline is one of the most distinct signs of aging, and age-related cognitive decline is a heterogeneous issue varying in different cognitive domains and has significant differences among older adults. Identifying characteristics of cognitive aging is the basis of cognitive disease for early-detection and healthy aging promotion. In the current chapter, age-related decline of main cognitive domains, including sensory perception, memory, attention, executive function, language, reasoning, and space navigation ability are introduced respectively. From these aspects of cognition, we focus on the age-related effects, age-related cognitive diseases, and possible mechanisms of cognitive aging.

Keywords

Cognitive decline · Sensory perception · Memory · Attention · Executive function · Language · Reasoning · Space navigation

3.1 Introduction

Faced with the "silver wave" worldwide, a crucial challenge is to ensure a concurrent improvement in the mental health practices that may enable the maintenance of cognitive and intellectual integrity in old age [1]. An understanding of the nature of human cognition and the regularity of age-related cognitive development and decline roughly corresponds to the history of scientific psychology. From the life-course perspective, aging is a continuum with a changing trajectory that starts from a high and stable capacity in middle life and then decreases with individual variability to substantial loss toward the end of life [2]. Studies that focus on the underlying mechanism of age-related cognitive decline are of great

Y. Yang
State Key Laboratory of Cognitive Neuroscience and Learning, Faculty of Psychology, Beijing Normal University, Beijing, China

Beijing Aging Brain Rejuvenation Initiative (BABRI) Centre, Beijing Normal University, Beijing, China

School of Nursing and Rehabilitation, Cheeloo College of Medicine, Shandong University, Jinan, China

D. Wang · W. Hou
State Key Laboratory of Cognitive Neuroscience and Learning, Faculty of Psychology, Beijing Normal University, Beijing, China

Beijing Aging Brain Rejuvenation Initiative (BABRI) Centre, Beijing Normal University, Beijing, China
e-mail: 201931061030@mail.bnu.edu.cn

H. Li (✉)
Institute of Basic Research in Clinical Medicine, China Academy of Chinese Medical Sciences, Beijing, China

Beijing Aging Brain Rejuvenation Initiative (BABRI) Centre, Beijing Normal University, Beijing, China

Z. Zhang (ed.), *Cognitive Aging and Brain Health*, Advances in Experimental Medicine and Biology 1419, https://doi.org/10.1007/978-981-99-1627-6_3

value for disease prevention, improving the quality of life of the elderly and achieving healthy aging. Cognitive decline commonly occurs with aging, as revealed by life-span developmental studies of significant cognitive changes. For instance, fluid intelligence and crystallized intelligence were conceptualized to explain human cognition. Within these concepts, fluid intelligence is an innate, neurologically based and age-related capacity, whereas crystallized intelligence is acquired, automated, and stable over time [3]. Numerous studies focused on various cognitive domains covering both aforementioned concepts of intelligence, which include memory, language, attention, executive function, and other functions [4]. The verified decline trajectories and the underlying mechanism explain human aging, which is the process of losing and maintaining cognitive function.

3.2 Age-Related Decline of Sensory Perception

Sensory perception is the reflection in the human brain of objective things that directly affect the sense organs. It is the beginning of cognitive activities and the basis of all complex human psychological activities, which generally include vision, hearing, balance, olfaction, and time and space perception [5]. Our common sense of understanding of the world depends on the combined processing of multiple sensory input information, which is the result of the synergistic effect of multiple sensors [6].

As people age, their sensory perception ability will gradually decline. After the age of 50 or 60, a decrease in the individuals' sensitivity to stimuli and an increase in the sensory threshold generally occur [5, 7]. Age-related sensory perception loss is significantly associated with cognitive decline or impairment and exerts a considerable effect on public health and quality of life, imposing a heavy burden on families and society [8]. Therefore, the elderly must understand the influencing factors contributing to various degenerative changes in sensory perception and adapt to the changes in sensory perception during the aging process.

3.2.1 Age-Related Vision Impairment

Vision is one of the most crucial human senses. With aging, a decrease in normal eye tissue function occurs, and the incidence of eye pathology increases. At present, the five major causes of vision impairment in elderly individuals are presbyopia, age-related macular degeneration, cataracts, glaucoma, and diabetic retinopathy [9].

A visual impairment in elderly individuals can seriously affect their daily activities. It can lead to reduced contrast sensitivity, inaccuracy in visual-motor coordination, gait instability, and color discrimination [8], which is related to an increased incidence of mental disorders, such as depression and anxiety. Meanwhile, visual impairment is also significantly associated with cognitive decline in multiple domains, such as episodic memory, language, attention, spatial processing, and executive function, and with an increased risk of dementia [9, 10].

A variety of neuroimaging studies support the conclusion that visual impairment is an important component of the functional impairment in individuals with Alzheimer's disease (AD) [8]. Older people with visual impairment show faster atrophy of the occipital cortex, superior temporal cortex, and posterior cingulate cortex (PCC). Atrophy in these areas accelerates with cognitive aging. In diffusion tensor imaging (DTI) data, increased diffusion in the lateral occipitotemporal cortex and PCC [11] reduces the speed of information processing and transmission. Decreased functional connectivity of the lateral occipital cortex was also observed in functional magnetic resonance imaging (fMRI) research. Studies have shown that hypoactivation in the posterior visual association areas measured using fMRI is strongly associated with impairments in patients with AD. Regional cerebral blood flow measurements obtained using positron emission tomography (PET) showed that dorsal stream visual motion areas were more vulnerable than ventral object-oriented areas [8, 12].

Visual impairment among the elderly is a common and major health problem worldwide. Active screening for vision loss, earlier detection of visual impairment, and the treatment of related problems in elderly individuals are very important.

3.2.2 Age-Related Hearing Loss and Balance Impairment

Hearing is the other most crucial sense in addition to vision. Age-related hearing loss (ARHL) is a progressive bilateral symmetrical sensorineural hearing loss [7]. The causes of ARHL development include the damage caused by physical and environmental factors, increased vulnerability to alterable lifestyle behaviors, and genetic predisposition. All these factors persist throughout life. The possible conceptual model for the association between hearing impairment and cognitive and physical impairments in the elderly is shown in Fig. 3.1 [14]. The prevalence of ARHL is high in the elderly, with approximately one-third of adults over the age of 65 showing varying degrees of hearing impairment, compared with 83% of those over the age of 70 [15].

ARHL is associated with faster atrophy in the gray matter volume of the primary auditory cortex and the right temporal lobe [16], which might mediate the relationship between ARHL and cognitive decline. Additionally, studies of white matter structure showed that the decreased fractional anisotropy (FA) values of the inferior fronto-occipital fasciculus (IFOF), inferior longitudinal fasciculus (ILF), superior longitudinal fasciculus (SLF), and FA value of the IFOF showed a significant negative correlation with the average hearing threshold. Meanwhile, fMRI studies showed increased functional connectivity between the dorsolateral prefrontal cortex (dlPFC), temporoparietal cortex, and vision areas in patients with a hearing impairment. This result indicates a compensatory mechanism between cognitive abilities. The cognitive load caused by hearing loss might lead to a reduced availability of resources for the other cognitive tasks, the depletion of cognitive reserve, and an earlier presentation of dementia symptoms [8, 17].

ARHL is related to many health and social issues, including affective disorders, balance impairments, altered self-motion abilities,

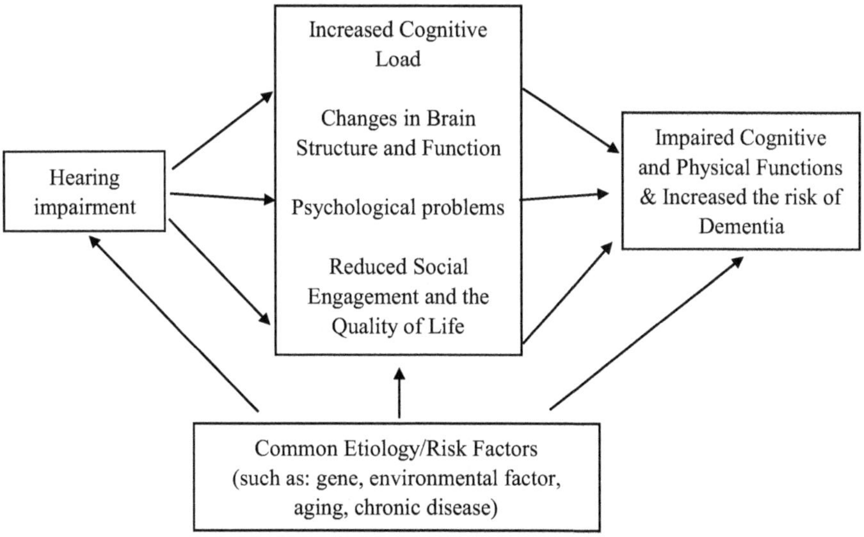

Fig. 3.1 The conceptual model of possible associations between hearing impairment, impaired cognitive and physical function in older adults, adapted from [13]. Reprinted with permission from prof. Frank Lin

decreased socioeconomic status, and decreased subjective well-being. A 30–40% rate of accelerated cognitive decline and increased earlier mortality was discovered in individuals with ARHL [7, 16].

3.2.3 Age-Related Olfaction Loss and Anosmia/Taste

Olfaction is the detection of odorous gaseous substances. Multiple factors cause age-related olfactory loss (AROL), including nasal congestion, olfactory epithelial damage caused by environmental disruption, decreased levels of mucosal metabolic enzymes, loss of odor perception by olfactory receptor cells, and changes in neurotransmitters and neuroregulatory systems. AROL is a common sensory impairment, and the prevalence of olfactory dysfunction in the general population is approximately 3.8% to 5.8%. Among people over the age of 65, this proportion increases to 13.9% [18].

People with an olfactory impairment exhibit a faster decline in almost all cognitive domains, including visuospatial ability, perceptual speed, semantic memory, and especially the episodic memory domain. Olfactory performance is associated with changes in gray matter volume. For instance, an olfactory impairment is associated with lower volumes of the fusiform and middle temporal cortices, which include the entorhinal cortex and hippocampus with structural MRI studies [19]. The medial temporal lobe includes structures such as the hippocampus that are vital to memory processing. This structure might be an important link between age-related olfactory impairment and accelerated cognitive decline. In addition, when patients with an olfactory impairment process olfactory information, less brain activity is detected in olfactory structures, including the superior and inferior frontal and perisylvian regions, left orbital pole, piriform, amygdala, and periamygdaloid cortex. However, the activation of the frontoparietal areas was increased, suggesting that individuals with an olfactory impairment might activate more executive control of cognitive resources to compensate for the decline or loss of olfactory functioning [20]. PET imaging of the brain shows decreased glucose metabolism in the hippocampus, entorhinal cortex, and thalamus. Additionally, a significant correlation between olfactory dysfunction and dopaminergic denervation of the nigrostriatal system in the elderly was observed [21].

The research evidence indicates that olfactory and taste dysfunction might be an effective potential biomarker for mild cognitive impairment (MCI), AD, Lewy body dementia, and Parkinson's disease (PD). Meanwhile, it also exerts a substantial negative effect on physical health, nutritional status, quality of life, and daily safety, imposing a large economic burden on families, society, and the country [18].

3.2.4 Age-Related Loss of Time and Space Perception

Time perception is a human perception experience of the continuity and sequence of objective things [22]. A precise time perception ability is extremely important for the normal life of people and is the essence of many aspects of our lives, including duration judgment, sound localization and speech recognition, motion perception and motor coordination [23]. Generally, with increasing age, the elderly start underestimating the duration in the replication time task, which can be explained by an alteration in the working memory ability of the elderly. In the task of generating time, the elderly will overestimate the duration, which can be explained by the slower internal clock of the elderly that is related to a slower processing speed [24].

A time perception impairment was observed among the individuals with neurodegenerative diseases. For example, in patients with AD, time distortions may lead to impaired episodic memory, and conversely, impaired episodic memory in patients with AD may also result in a loss of the ability to retrieve temporal information. The relationship between time perception decline and memory impairment in individuals with AD may be explained by the involvement of the hippocampus since this brain region supports time perception and memory. Additionally, it is an early target for the neuropathological processes of AD [25].

Space perception is the ability to recognize the spatial relationship of objects in the objective world, which includes shape perception, size perception, depth and distance perception, and orientation perception [5, 26]. Age-related space perception loss is also common in the elderly and can lead to spatial cognitive impairment, including deficits in spatial memory, spatial orientation and visualization, navigation, and place learning. This type of perception loss also severely affects the normal life of older people [27]. This result supports the research on spatial memory tasks based on fMRI. Age affects neural activity, and an interaction between age and parietal lobe activation has also been observed [28]. During the memory encoding process, increased activities are observed in the right hemisphere of young people, while the activities in the left hemisphere of 64–79-year-old people decrease. Therefore, aging individuals may shift from the coordinate processing of spatial information into classification processing, which supports the hypothesis that the coordinate representation of spatial information may be damaged during normal aging [29]. Aging disrupts the coordinated processing of spatial relationships, and the age-related loss of right parietal lobe activity might represent an accelerated loss of right hemisphere function. The findings may be consistent with the hypothesis of right hemisphere aging. The hypothesis assumed that some structures in the right hemisphere might undergo a faster aging process than the corresponding structures in the left hemisphere [30].

3.2.5 Summary

Aging exerts a global effect on sensory and perception functions in elderly individuals. The aging process is manifested in many systems, including deficits in vision, hearing, olfaction, and time and space perception [31]. Age-related decreases in sensory function and perception are significantly associated with faster brain atrophy in the relevant areas and accelerating deficits in cognitive ability. These changes increase the risk of age-related cognitive impairment [32] and significantly affect the quality of life of patients. Therefore, prompt and effective treatment of sensory and perception deficits might have a profound effect on the patient's quality of life and may even indirectly increase the patient's life expectancy [31].

3.3 Age-Related Memory Decline

Aging is an inevitable stage in life. In the process of aging, from perception, attention, and processing speed to executive function, memory, language, and other cognitive functions would suffer various degrees of decline based on behavioral measures [33]. Undoubtedly, memory performance is one of the most severely impaired areas, both during normal aging and during the development of neurodegenerative diseases [34].

Numerous studies have shown that different memory systems and processes are affected inconsistently with age. Generally, declarative long-term memory and working memory are more substantially impaired than implicit, nondeclarative long-term memory [35, 36]. Furthermore, in declarative long-term memory, episodic memory is more affected than semantic memory in older adults; in particular, more severe deficits are observed in episodic recall tasks than in recognition tasks [37, 38]. However, due to interindividual variability, some older adults were still able to maintain high levels of cognitive performance, although many older adults showed evidence of a decline with aging [39]. Potential explanations for this finding are ages at which the declines occur differ, and the degrees of decline vary accordingly. Altogether, a consensus on the best methods for characterizing age-related changes in memory has not been achieved. The purpose of this chapter is to summarize the development of different memory

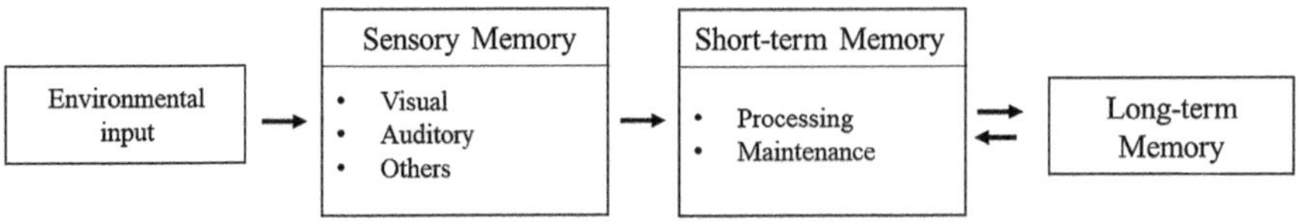

Fig. 3.2 The classification of memory. Information flows from the sensory modalities to the short-term memory for processing and maintenance via rehearsal and then transfers to the long-term memory, which depends on the amount and quality of processing in the short-term memory

subtypes with age and their underlying neural mechanisms.

Here, we introduce one of the classifications of memories based on the degree of their persistence (see Fig. 3.2). This model includes sensory memory, short-term memory (including working memory), and long-term memory [40, 41].

3.3.1 Sensory Memory

Sensory memory should be studies as an early response, which might be involved in ensuing higher-order perceptual operations [42]. Although sensory memories are proposed to exist for all senses, most studies have focused on vision and audition, which are known as iconic memory (visual sensory memory, VSM) and echoic memory (auditory sensory memory), respectively [43].

3.3.1.1 Iconic Memory

Regarding iconic memory, two different experimental procedures are commonly used to study the storage capacity and duration of visual sensory memory, which is the delayed partial sampling procedure (partial report techniques) and the direct measurement procedure. The storage capacity of visual sensory memory decreases with age, while the rate of decline remained unchanged. A decrease in the holding capacity of the nervous system was observed with increasing age [44]. However, inconsistent conclusions have been obtained for the visual persistence of visual sensory memory. On the one hand, the iconic memory of young adults persists longer time than that of older adults [45]. On the other hand, researchers found that older adults showed significantly higher levels of persistence than younger adults [46].

fMRI and PET might not be the appropriate tools to study VSM because of its highly transient nature [47]. Researchers have suggested that VSM always involves the continuous firing of relevant neurons. It may involve reverberatory circuits of neurons of some type [48]. Additionally, the activation trace in the primary visual cortex (area V1) might be the cortical source of VSM. As visible persistence might be due to the persistence of reverberating activation beyond stimulus offset [49], the persistence of the synchronized activity of neurons in the primary visual cortex might be a candidate for the neural mechanism of VSM [47].

3.3.1.2 Echoic Memory

In the exploration of echoic memory, a recent study obtained clear evidence for the faster decay of auditory sensory memory traces in aging brains [42], consistent with the diminished MMN (mismatch negativity) of subsequent increases in the interstimulus interval (ISI) in previous studies [50–52]. Namely, human auditory cortices exhibit an age-dependent decrease in auditory sensory memory, which might be associated with age-related deficits in the encoding of sound properties [52]. A previous study investigated the neural circuits of sensory memory in patients with unilateral brain damage to dorsolateral prefrontal cortex, posterior association cortex, or the hippocampus; the result showed that in comparison with control subjects, temporal-parietal and dorsolateral prefrontal patients had impaired auditory discrimination and reduced MMNs, implying that a temporal-prefrontal neocortical network critical for the transient storage of auditory stimuli. They further pointed out that auditory traces may be located in the posterior association cortex and actively maintained through the effect of the dorsolateral prefrontal cortex [53].

In addition, the accuracy of auditory memory was reported to be higher than that of visual memory [54] and that the accuracy of these two types of memory was lower in older adults than in young adults [55]. Regarding neural mechanisms, visual sensory memory activated the visual center of posterior visual areas, namely, the occipital lobe (BA17/18/19) and bilateral superior parietal gyrus (BA 7); auditory sensory memory activated the auditory center of the superior temporal gyrus (BA 22). However, their activated areas are the bilateral left middle frontal gyrus (BA 9) and BA 7, with lower activation of BA 7 in auditory sensory memory than in visual sensory memory [56].

3.3.2 Short-Term Memory and Working Memory

An important distinction between short-term memory and working memory is that working memory emphasizes the active manipulation and use of information, whereas short-term memory is merely focused on its maintenance or storage [40].

3.3.2.1 Short-Term Memory

Researchers used "free recall of words without any distracting tasks" to illustrate the effects of age on short-term memory. The results revealed a relatively stable performance level across the age cohorts without any age deficit [57], which were consistent with other studies using various short-term memory tasks [58–62]. The findings suggested that the ability to store and retrieve information in the brain over a brief period might be relatively preserved in old age [62, 63].

3.3.2.2 Working Memory

Performance on working memory (WM) tasks continuously decreases with age, including processing speed, accuracy, storage capacity, etc. [61, 64, 65]. Resting neuroimaging studies showed strong negative correlations. For instance, one study indicated that greater white matter degeneration was associated with lower functional activity during WM tasks. Additionally, dlPFC and white matter hyperintensity (WMH) volumes were associated with decreased activity in the posterior parietal and anterior cingulate cortices during WM task performance. Based on these results, the disruption of white matter, especially within the dlPFC, is considered as the mechanism of age-related changes in WM function [35, 66].

Furthermore, there are task-related activation differences on the neural activation WM tasks between older and younger people in the frontal cortex [65, 67, 68]. For example, researchers used PET as a research tool to report the result of PFC activity in the verbal task and the spatial task. The result showed that for the elders, a reduced left PFC activity and an increased right dlPFC activity were shown in the verbal task. In the spatial task, the older age group showed greater left activation for spatial task and greater right activation with only a weak tendency toward left activation for verbal task, while the youngers showed greater right activation for spatial task and greater left activation for verbal task [65]. The dlPFC activity pattern in older adults activity might indicate a compensatory mechanism, which reflects a response to increased task demands [69], or retrieval effort [64], or dysfunction [70].

In addition, older adults exhibit diminished left hippocampal activation compared with younger adults [64]. A previous study reported an equivalent performance of older adults and young adults on a single item of WM task, whereas in the combination (location plus object) task, a marked decrease was observed. This decrease in performance was accompanied by decreased activity in the left anterior hippocampus and anteromedial PFC (right BA 10) compared with younger adults [71], suggesting that a disruption in hippocampal-PFC circuitry may underlie age-related deficits in the combination condition. Moreover, several fMRI studies have also shown that working memory maintenance facilitates the encoding of these items into long-term memory by activating rehearsal processes in the hippocampus [72] or the parahippocampal cortex [73].

3.3.3 Long-Term Memory

Long-term memory is a vast resource of knowledge and a recorder of prior events. It refers to the ability to recall information from the past [74]. Since no consensus has been achieved for the taxonomy of long-term memory, we will sample some representative categories to illustrate this type of memory. Additionally, a summary of the effect of aging and the potential neural mechanisms will be presented.

3.3.3.1 Episodic Memory

Episodic memory decreases in normal aging [61, 75, 76]. A recent study including middle-aged and older adults reported significant decreases in both immediate and delayed memory with aging [77] .

Resting neuroimaging studies showed a reduced hippocampal volume in older adults with decreasing episodic memory performance, consistent with DTI measures of fractional anisotropy in the anterior corpus callosum [78]. Reduced episodic long-term memory performance was also observed in normal aging and was related to the reduced integrity of a distributed network that depends on white matter pathways supporting episodic memory [77].

Episodic memory encoding studies typically include the following topics: intentional encoding studies, incidental encoding studies, and subsequent memory studies. Task-related activation recorded in neuroimaging studies showed increased activation in the left prefrontal cortex of both younger and older groups during incidental episodic encoding tasks with blocked designs. Additionally, right prefrontal activation in elderly subjects was associated with the most substantial decrease in memory performance [78–83]. However, recent subsequent memory studies using event-related fMRI showed increased age-related activity in the dorsolateral and orbitoprefrontal cortices [80, 84, 85]. The inconsistent results may be due to the use of different experimental designs. Moreover, large numbers of fMRI studies also reported an age-related reduction in medial temporal lobe (MTL) activity. Aging may be accompanied by reduced neural activity and diminished connectivity between PFC and MTL areas [86]. Meanwhile, older adults also exhibit decreased activities in the posterior parietal region, which may be associated with deficits in the encoding of visuospatial information and visual regions compared to younger adults [87, 88].

Research regarding aging and imaging during episodic memory retrieval are divided into three main categories: recognition, recall, and contextual memory. Additionally, older adults showed larger age deficits when the difficulty of cognitive tasks increased (i.e., context> recall>recognition tasks). Task-related activation recorded in neuroimaging studies suggested that older adults showed decreased activation in occipital and parietal regions coupled with increased activation in PFC regions during incidental episodic memory retrieval. Several studies documented age-related decreases in right PFC activity [82, 87, 89–91] and age-related increases in MTL activity [87, 92–94]. In addition, researchers also observed a dissociation between weaker hippocampal activity and stronger parahippocampal gyrus activity in older adults [57, 94].

In addition, WMH volume of the dorsal PFC was associated with decreased activity in medial temporal and anterior cingulate regions during episodic retrieval tasks. Thus, in many older individuals, a disruption of white matter integrity might be one of the mechanisms of PFC dysfunction, which manifests as lower functional brain activity and reduced episodic memory performance [35, 66].

3.3.3.2 Autobiographical Memory

Autobiographical memory is a type of episodic memory that is important and unique. Compared with other kinds of episodic memories, autobiographical memory seems more "emotional" and "personally meaningful" than other forms of episodic memories, which is a lack of the relevance to human cognitive and emotional life [95].

Age-related changes in autobiographical memory at the behavioral level showed that memories of the past generated by older adults contained less episodic, perceptual, and contextual details. They also contained more generic content, fewer vivid recollections and correct details, and more false information regarding simulated crimes than the memories of younger adults [96–98]. On the positive side, both older and younger adults provide similar amounts of contextual information in autobiographical free recall [99]. Additionally, autobiographical recollections of older adults' judgments are considered more interesting and informative than those of younger adults [100].

Studies using fMRI or other research techniques to investigate the brain function of normal adults suggest that the left prefrontal cortex may be particularly important for autobiographical memory [79, 95]. Imaging studies conducted with normally functioning adults also revealed an important role for the hippocampus in autobiographical recall [101]. Functional neuroimaging studies suggested that normally aging adults showed increased activity in the hippocampus, parietal cortex, and occipital cortex during autobiographical memory recollection [93, 102]. Decreased activity in the medial temporal lobes and precuneus during the construction of autobiographical events was also reported in previous studies [103, 104]. Furthermore, a study showed reduced activation of the prefrontal cortex and reduced coupling between the ventrolateral prefrontal cortex and hippocampus during the task. Additionally, the elaboration phases of autobiographical memory retrieval were noticed [105, 106].

3.3.3.3 Semantic Memory

Semantic memory (SM) includes not only memories of facts and concepts but also high-level knowledge structures such as schemas and scripts within this concept [40]. Behavioral evidence indicates that knowledge-based semantic memory is relatively preserved in healthy older adults [57, 75, 104, 107–109]. However, results from other studies consistently show an increase in semantic memory performance through the early age of 60, followed by a gradual decline [110, 111]. Meanwhile, severe age-related deficits occur in certain semantic memory tests, such as vocabulary and general knowledge tests [57, 110].

The perirhinal cortex, parahippocampal cortex, and inferior temporal gyrus are associated with the accuracy of recalling recent and remote semantic memory questions. This result supports the hypothesis that the medial temporal lobe plays an important role in semantic memory [112]. In addition, several imaging studies have observed similar brain activity during SM tasks in young and older adults, although older adults may show increased frontal activity [70, 113].

3.3.3.4 Implicit Memory

The effects of aging on implicit memory were studied with the following tasks: (1) priming tasks, which include word stem completion priming [114] and repetition priming [115]; and (2) learning tasks, which include sequence learning [116, 117] and probabilistic categorical learning [118]. Age differences in the priming tasks are small and approximately equivalent [114, 115]. In the learning tasks, the differences were still intact, but the rate of learning was slower in elderly adults [117, 118]. A neuroimaging study showed a higher level of prefrontal cortex and caudate activation in young adults and higher level of parietal cortex activation in older adults. Thus, older adults appear to be able to recruit additional resources within the parietal cortex to learn the probabilistic associations as well as young adults [118], which may explain the equivalent performance of younger and older adults on implicit memory tasks.

Regarding the neural mechanism underlying the effect of aging on implicit memory, studies show that priming is mediated by frontal regions [119, 120], and sequence and categorical learning are attributed to changes in a network of brain regions, including the dorsolateral prefrontal cortex and the striatum [116]. The findings might extend previous frontal-centric views of implicit memory aging, including changes in the subcortical structure.

3.3.3.5 Procedural Memory

Procedural memory involves memory of cognitive and motor skills, such as reading and driving [40]. Procedural memories are frequently assessed using performance-based tasks and change relatively little from early to late adulthood [121–123]. Furthermore, adults with mild AD are also able to perform procedural memory tasks [123, 124].

3.3.4 Summary

In summary, large numbers of studies have focused on cross-sectional comparisons; however, a longitudinal study is more feasible to reveal the trajectory of memory function throughout the life span development. In longitudinal studies, we can investigate the age-related decline in memory processes continuously across the adult life span, and steeper age deficits in certain types of memory only occur in the later stages. Nevertheless, individual differences must also be considered as the contributing factor along with other factors that may help slow the rate of cognitive decline and decrease the risk of dementia associated with aging.

3.4 Age-Related Decline of Attention and Executive Function

Executive function is the most advanced cognitive activity in the human brain and is responsible for controlling and coordinating various specific cognitive processes. It has always been a hot topic in neuropsychology, cognitive psychology, and cognitive neuroscience [125]. An increasing number of studies show that executive function is not a cognitive system with a single structure. Namely, it involves multidimensional structures, including inhibitory control, attentional control, working memory, and cognitive flexibility [126]. All those components are affected by aging.

3.4.1 Inhibitory Control

Inhibitory control refers to the ability to suppress disturbing information or previously activated cognitive processes to focus attention on relevant task requirements [127, 128].

The inhibitory control deficits theory was proposed by Hasher and Zack [129] and was the most classic theory explaining the deficit in inhibitory control in aging adults. The theory proposes that the decline in this ability is the main cause of cognitive aging. With aging, the efficiency of the inhibitory process in older adults is reduced, which is mainly reflected in three aspects: (1) a decrease in suppressing irrelevant information into working memory, (2) deletion of unrelated messages to current tasks, and (3) successful inhibition of the dominant response.

With the development of advanced technology, Westerners [130] proposed the executive decline hypothesis of cognitive aging, which is also known as the frontal lobe hypothesis and based on considerable neuropsychological research. The hypothesis states that cognitive decline in elderly individuals is related to a decrease in the executive function of the frontal lobe, in particular the prefrontal lobe. Moreover, scholars found that inhibitory control deficits in the elderly are a direct manifestation of a loss of integrity in the frontal cortex [127]. The degenerating frontal lobe fails to inhibit irrelevant stimuli; therefore, older adults suffer from inhibitory control deficits [128].

In summary, inhibitory control, which refers to the ability to successfully suppress the dominant behavior and irrelevant stimuli, is an important mechanism of cognitive function and is related to the frontal lobe. Two important theories have been proposed. One proposes that decreases in inhibitory control are attributed to cognitive aging, and the other assumes that a loss of function in the frontal lobe is an obvious indicator of an inhibitory control deficit.

3.4.2 Attentional Control

Attentional control is the ability to maintain goal-oriented behavior by sustaining information-processing activity over time in terms of distraction, temporarily stopping the activity to respond to other information, and coordinating the course of concurrent activities [131].

Current models of attention indicate that attentional control is responsible for inhibitory functions. A distributed network of structures within the brain, including the anterior cingulate cortex, prefrontal cortex, parietal cortex, extrastriate cortex, superior colliculus, thalamus, and basal ganglia, supports attentional function [132, 133]. Aging impairs the brain's ability to implement attentional control, allowing greater activation of irrelevant representations and actions capable of decreasing the efficiency of working memory processes [133]. Specifically, age-related changes in the responsiveness of the prefrontal and parietal cortex are related to attentional control. In addition, attentional control of executive function also decreases significantly during the early stages of AD [134].

In summary, attentional control involves the interaction between perceiving environmental cues and the allocation of perceptual processing resources, which involves many brain areas and is easily affected by aging.

3.4.3 Cognitive Flexibility

Cognitive flexibility refers to the ability to switch between different tasks. It is a central function of executive control and is related to the functions of various prefrontal cortical structures [135, 136]. Cognitive flexibility has been divided into proactive control and reactive control. Proactive control refers to the active maintenance of goal-relevant information and prevention of interference. Reactive control refers to the reactivation of task goals and reservations of interference as needed [136]. The former is reflected in sustained lateral prefrontal cortex activation and associated with decreased transient anterior cingulate cortex activation. Additionally, the latter is reflected in the transient activation of the lateral prefrontal cortex and the anterior cingulate cortex [137].

Other terms used interchangeably in describing cognitive flexibility include mental flexibility, mental set shifting, cognitive shifting, task switching/shifting, and attention switching/shifting. Despite some disagreement in the literature related to the appropriate definition of the term operationally, one commonality is that cognitive flexibility is a component of executive functioning.

Several studies using task-switching paradigms have documented the complexities of the network involved in human cognitive flexibility. Activation of the dorsolateral PFC has been shown during the resolution of interference of irrelevant task sets [138]. In addition, numerous studies using animal models have reported age-related deficits in flexible decision-making processes that are presumed to depend on prefrontal cortex function [139, 140].

Older adults tend to exhibit less flexible performance on these types of tasks than younger adults because of a decreased ability to acquire and update relevant information [135].

In conclusion, cognitive flexibility has been described as the mental ability to shift one's thoughts between two different concepts, which is related to the prefrontal cortex in both humans and animals.

3.4.4 Working Memory

Memory will decline with aging, particularly working memory (WM), which refers to an executive memory process that allows transitional information to be held and manipulated temporarily in memory stores before being forgotten or encoded in long-term memory [141, 142].

Additionally, WM deficits are a recognized feature of AD [142]. Although the mechanism underlying the age-related WM decline remains unclear, the information obtained from previous studies has been employed to improve the theoretical understanding of the processes. On the one hand, neural atrophy occurs in the prefrontal and parietal regions [111]. A subtle loss of white matter integrity has been proposed to lead to the disruption of working memory in particular because it relies on the dynamic and reiterative activity of corticocortical pathways [143]. On the other hand, changes in the activation of brain regions such as the prefrontal cortex are one presumed contributor to WM decline in the healthy aging population [144].

In particular, neuroimaging studies show that younger adults exhibit more intense bilateral frontal activity performing WM tasks than older adults [141]. In addition, the decrease in activity of the cortical area in the right frontal region may explain age-related decline in working memory [144]. Meanwhile, working memory deficits are a recognized feature of AD, which is commonly ascribed to a central executive impairment and assumed to be related to frontal lobe dysfunction [142].

In summary, working memory is an essential component of executive function and an important manifestation of early AD. Changes in neural atrophy and brain activation are the causes of age-related working memory deficits in elderly adults.

3.4.5 Attention

In addition to executive function, many studies in the aging literature have also explored the effects of attention on the performance of elderly adults [145, 146]. Attention is a multidimensional concept that includes three functionally different dimensions: sustained, selective, and divided attention [146, 147]. The first describes a fundamental component of attention characterized by the subject's readiness to detect rarely and unpredictably occurring signals over prolonged periods of time. The second refers to the ability to select some stimuli and ignore other stimuli. The last is the ability to divide attention between two or more tasks.

Previous research on attention in the visual modality indicates that although the ability to process task-relevant information remains largely intact, the ability to suppress task-irrelevant information decreases with aging [148]. Additionally, scholars documented a longer time to a slower reaction time in older adults than in younger adults, particularly in trials in which a single salient distractor was present [149]. These results suggested age-related behavioral deficits in attention.

The prefrontal lobe (PFC) is an important neural basis of our attention capacity. Proper PFC function is crucial for the inhibition of distracting stimuli and irrelevant thoughts, such as proactive interference. Patients who present lesions in the PFC are easily distracted [150, 151]. PFC lesions lead to the inability to sustain attention over a long period [152]. Lesions in the dorsolateral PFC impair the ability to shift an attentional set to more prominent stimuli [153]. In addition, divided attention is also affected by PFC lesions, and these attentional deficits are associated with lesions in the left superior PFC [150]. Additionally, researchers have reported a decrease in the cerebral cortex thickness during aging, and atrophy of the PFC is more severe in elderly adults than in the whole brain [154, 155]. Therefore, the attention ability is vulnerable to aging.

In general, attention is also susceptible to aging, and it is impaired in different dimensions. Most importantly, during aging, changes in both behavioral performance and brain areas have been observed.

3.5 Age-Related Decline of Language and Reasoning

3.5.1 Language

Language is an important method by which humans express themselves and communicate. However, with the aging of individual physiological structures and functions, the language cognitive ability also shows an aging trend [33]. The aging process typically involves cognitive decline associated with brain atrophy [156], loss of neuronal synaptic connections, and neuropathic signs associated with dementia [157]. A longitudinal population-based study by Cullum et al. [158] showed that aging was associated with declining cognitive function across different domains. The areas include memory decline, poor working memory, reduced verbal span, and delayed speech recall [61, 159–161]. Language and communication are also affected by aging. Compared with the young people, the elderly exhibit a decline of language cognition, including vocabulary recognition [162], speech production [163], semantic extraction [164], text integration [165], and oral communication [166]. These factors not only restrict the cognitive activities of the elderly but also have adverse effects on their interpersonal communication, quality of life, and mental health.

Language learning engages an extensive network of the brain that overlaps with the decline caused by aging [167]. The new phonetic learning is related to the dorsal audiomotor interface covering posterior temporal and dorsal frontal lobe regions such as the ventral premotor cortex and posterior inferior frontal gyrus [168]. In the elderly, both potential declines in language ability and performance are associated with functional changes in the prefrontal-subcortical networks [169]. Wong et al. [170] showed that the acquisition of meaning involves the ventral stream, including medial, inferior, and anterior temporal regions, as well as regions in the inferior frontal gyrus for semantic retrieval and additional complex processing [171]. Training in a foreign language starting in adulthood causes structural changes in the brain. After 5 months of foreign language learning, learning-related structural changes in native English speakers revealed an association between Swiss German proficiency and grey matter density in the left inferior frontal gyrus and the left anterior temporal lobe [172]. Additionally, word learning often involves a rapid mapping process between sound and meaning, which is related to the medial temporal lobe [173]. In addition, the learning of grammar rules is related to the frontal-striatal system [174].

In studies involving older adults, a more advantageous approach may be to combine measures of categorical fluency into a language combination because of the high prevalence of age-related decline in language ability [175]. These semantic abilities often decline with age, and those abilities are among the earliest cognitive abilities that decrease during AD progression [176]. The language abilities of 74 older adults were quantified by examining biographies written in their youth. Language ability in early adulthood is negatively correlated with the incidence of AD in old age [177]. Antoniou et al. [178] suggest that foreign language learning may be particularly beneficial in promoting healthy cognitive function and preventing decline. Since the brains of older people are also malleable, language learning not only improves language-related functions but also improves cognitive functions of older people.

Multilingualism is a better predictor of cognitive ability than age, education, or sex, and the effects are observed throughout the life span development [179, 180]. Older adults who spoke two or more languages also performed better at a variety of cognitive tasks. This cognitive advantage depends on the experience of speaking two languages, which requires a slightly different set of attention and control procedures than those used in monolingual individuals [181]. As a result, older people who receive intensive foreign language training will benefit from the learning. Language learning is an ideal training activity for older learners because it has meaningful and related benefits, which may expand post-retirement activities [178].

3.5.2 Reasoning

Many aspects of cognition are adversely affected by aging, including processing speed, memory, and reasoning [145]. Among them, reasoning ability has always been considered as the core component of intelligence. Therefore, many researchers use reasoning as an index for measuring fluid intelligence [182, 183]. In recent decades, research on the mechanism of the development of reasoning ability has become a crucial topic in the field of cognitive development [184, 185].

The correlation between age and reasoning ability is often influenced by processing speed and working memory [186]. Fry and Hale confirmed the development cascade relationship between age and reasoning ability through horizontal comparative studies of 7–19-year-old individuals (see Fig. 3.3). According to this study, a series of processing stages within the effective-

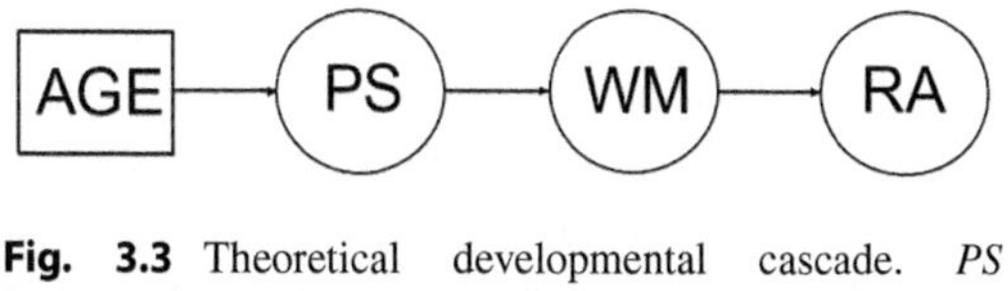

Fig. 3.3 Theoretical developmental cascade. *PS* Perceptual speed, *RA* reasoning ability, *WM* working memory [187]

ness of the processing at the first stage exerts a flow effect on the next stage, and the influences are transmitted. The study showed that improvements in information processing speed and working memory predicted inductive reasoning in children and adolescents [186].

The study by Kail [188] also showed that the cascading explanation for reasoning was valid longitudinally, as the results were replicated 1 year later. People with poorer fact memory and a reduced working memory capacity experience difficulty in reasoning based on the newly presented information [189].

A study of 240 children (ages 8–14) and 238 adults (ages 18–87) found that processing speeds increase as children age, affecting improvements in working memory, which subsequently affects reasoning [187]. Although children's cognitive development is substantially mediated by processing speed, the decline in reasoning ability in old age is affected by a slower processing speed. The reasoning ability of older people is also affected by age-related changes [189]. These changes affect working memory independently of processing speed [187].

3.5.3 Summary

In summary, language and reasoning change with age. Other factors contribute to the complex relationships between aging and language and between aging and reasoning. To some extent, maintaining good working memory can slow the process of declining language and reasoning skills.

3.6 Age-Related Decline of Spatial Navigation

Spatial navigation refers to the ability to select, determine, and execute navigation routes using different cues when moving from one to another location, including the involvement of multiple functions, for example, executive functions, planning, attention, spatial memory, and route learning [190]. As a complex cognitive activity, spatial navigation involves a large number of brain regions, including hippocampus, parahippocampal gyrus, striatum, entorhinal cortex, medial temporal lobe, prefrontal lobe, posterior parietal lobe, postcortical cortex, etc. [191–193].

3.6.1 Neural Basis of Spatial Navigation

Two fundamental representations of spatial navigation have been proposed: the allocentric representation and egocentric representation. The former decodes spatial information centered on its position to maintain the sense of direction when moving. Its neural basis mainly includes the inferior parietal lobe and striatum [192, 193]. The latter uses objects in the environment as a reference to navigate. It is mainly attributed to the functions of the hippocampus or the parahippocampus in the limbic system and the entorhinal cortex [193, 194].

Researchers suggest that brain regions such as the hippocampus and entorhinal cortex, which are involved in spatial navigation, are susceptible to aging [195, 196]. Therefore, during aging, changes occur in spatial navigation, and this deficit has different internal mechanisms.

3.6.2 Aging and Spatial Navigation

Healthy older adults often experience a deficit in spatial navigation. The explanation is the atrophy of the hippocampus and frontal cortex during aging, resulting in impaired functions [195, 197]. Individuals with AD exhibit more severe deficits in spatial navigation, both in allocentric and egocentric representations. Getting lost is one of the early manifestations of AD [198]. In patients with AD, the ability to learn the route and identify their own directions and landmarks significantly decreases [198–200]. The deterioration in the spatial navigation of patients with AD contributes to the loss of integrity in the hippocampus, which leads to failures in attempts to form and use cognitive maps [194]. Mild cognitive impairment (MCI) is the preclinical stage of AD;

some patients with MCI progress to AD, while others do not. In that case, normally aging patients and those with a high risk of AD conversion must be distinguished [199, 201]. In individuals with MCI, allocentric and egocentric navigation show greater impairments than in normally aging individuals and fewer deficits than in patients with AD [199]. Individuals with different subtypes of MCI show different levels of deterioration. Among them, individuals who have amnestic MCI in multiple domains suffer from the worst impairment in navigation, and their navigation performance is similar to the patients with AD [201], indicating that they are likely to convert to AD. Nevertheless, the navigation performance of individuals with non-amnestic MCI is similar to that of normal older adults [199]. As a result, spatial navigation has served as an indicator of AD in recent years.

3.6.3 Virtual Reality Technology and Space Navigation

With the development of virtual reality (VR) technology, the accuracy of screening has substantially improved [202]. At the same time, VR also helps us further understand the mechanisms and neural basis of spatial navigation in cognitive aging. VR technology has been combined with sophisticated design stimuli and some classic spatial navigation paradigms, such as the Morris water maze, eight-arm maze or ten-arm maze, and star-shaped maze. Different paradigms analyze different aspects; for example, the eight-arm maze paradigm focuses on the choice of spatial navigation strategies. Related research results show that elderly individuals are more inclined to choose self-centered navigation [203, 204]. This strategy involves less activation of the frontoparietal control network and uses fewer cognitive resources, which is consistent with the late-in, first-out theory of cortical degeneration to some extent [205]. In addition, the research results of the Morris water maze show that the elderly show a poor spatial navigation ability, including a slower spatial navigation speed, increased error rate, and mobile response time [206]. Combined with MRI technology, researchers first discovered grid cell-like features in the entorhinal cortex of the human brain. Moreover, in healthy adults who carry genes associated with a high risk of AD, grid cell-like characterization has decreased [207]. This decrease in the activation level of this signature was also observed in healthy elderly people, which may explain the mechanism underlying the impaired spatial navigation performance [208]. Additionally, aging, atrophy, and tau protein deposition in the entorhinal cortex area are also strongly correlated with AD, suggesting that related research on the role of aging in spatial navigation can provide references for an early predictive diagnosis and new ideas for interventional treatment for neurodegenerative diseases such as AD [196, 207].

3.6.4 Summary

Spatial navigation is an essential high-level cognitive function in daily life, which includes two representations, allocentric and egocentric navigation, with different neural basis. Specifically, the involved brain regions, such as the hippocampus and entorhinal cortex, are vulnerable to aging and exhibit structural atrophy and functional alterations. The ability to navigate differs from normal aging to AD. In addition, with technological advances, VR has been increasingly applied in age-related spatial navigation research, revealing the important roles of the medial temporal cortex and the hippocampus in age-related spatial navigation.

3.7 Conclusion

In conclusion, different domains of cognitive function show different trajectories during the aging process. Some typical age-related cognitive changes have been identified, such as a decline in episodic memory, processing speed, and spatial navigation ability. All these aspects provide us the opportunity to detect individuals experiencing abnormal aging. The measurement of cognitive performance and the identification of biomarkers related to cognitive impairments help us to understand the health status of elderly individuals. The identification of cognitive aging trajectories has benefited from longitudinal research carried out from a life span development perspective. Meanwhile, the analysis of development and aging provides new ideas for the interpretation of age-related research findings, and the internal mechanism of aging has also been explained by many neuroimaging studies. Additionally, distinct heterogeneity in cognitive aging, which includes pathological aging, normal aging, and successful aging processes, was discussed respectively. The purpose of exploring the age-related cognitive trajectory is to change the path of pathological aging and to improve the early prevention strategy. Therefore, the elderly individuals can experience less cognitive decline, remain in a healthy state, and avoid age-related diseases.

References

1. Fabiani M (2012) It was the best of times, it was the worst of times: a psychophysiologist's view of cognitive aging. Psychophysiology 49(3):283–304
2. Beard JR, Officer A, de Carvalho IA, Sadana R, Pot AM, Michel J-P et al (2016) The world report on ageing and health: a policy framework for healthy ageing. Lancet 387(10033):2145–2154
3. Kent P (2017) Fluid intelligence: a brief history. Appl Neuropsychol Child 6(3):193–203
4. Park DC, Reuter-Lorenz P (2009) The adaptive brain: aging and neurocognitive scaffolding. Annu Rev Psychol 60(1):173–196
5. Peng D (2001) General psychology fM1. Beijing Normal University Press, Beijing
6. Hirst RJ, Setti A, Kenny RA, Newell FN (2019) Age-related sensory decline mediates the sound-induced flash illusion: evidence for reliability weighting models of multisensory perception. Sci Rep 9(1):19347
7. Davis A, McMahon CM, Pichora-Fuller KM, Russ S, Lin F, Olusanya BO et al (2016) Aging and hearing health: the life-course approach. Gerontologist 56(Suppl 2):S256–S267
8. Albers MW, Gilmore GC, Kaye J, Murphy C, Wingfield A, Bennett DA et al (2015) At the interface of sensory and motor dysfunctions and Alzheimer's disease. Alzheimers Dement 11(1):70–98
9. Loh KY, Ogle J (2004) Age related visual impairment in the elderly. Med J Malaysia 59(4):562–568, quiz 569
10. MacDonald SWS, Keller CJC, Brewster PWH, Dixon RA (2018) Contrasting olfaction, vision, and audition as predictors of cognitive change and impairment in non-demented older adults. Neuropsychology 32(4):450–460
11. Beer AL, Plank T, Greenlee MW (2020) Aging and central vision loss: relationship between the cortical macro-structure and micro-structure. NeuroImage 212:116670
12. Zhuang J, Madden DJ, Duong-Fernandez X, Chen NK, Cousins SW, Potter GG et al (2018) Language processing in age-related macular degeneration associated with unique functional connectivity signatures in the right hemisphere. Neurobiol Aging 63:65–74
13. Lin FR, Albert M (2014) Hearing loss and dementia – who is listening? Aging Ment Health 18(6):671–673
14. Agrawal Y, Platz EA, Niparko JK (2009) Risk factors for hearing loss in US adults: data from the National Health and nutrition examination survey, 1999 to 2002. Otol Neurotol 30(2):139–145
15. Loughrey DG, Kelly ME, Kelley GA, Brennan S, Lawlor BA (2018) Association of age-related hearing loss with cognitive function, cognitive impairment, and dementia: a systematic review and meta-analysis. JAMA Otolaryngol Head Neck Surg 144(2):115–126
16. Lin FR, Ferrucci L, An Y, Goh JO, Doshi J, Metter EJ et al (2014) Association of hearing impairment with brain volume changes in older adults. NeuroImage 90:84–92
17. Luan Y, Wang C, Jiao Y, Tang T, Zhang J, Teng GJ (2019) Prefrontal-temporal pathway mediates the cross-modal and cognitive reorganization in sensorineural hearing loss with or without tinnitus: a multimodal MRI study. Front Neurosci 13:222
18. Attems J, Walker L, Jellinger KA (2015) Olfaction and aging: a mini-review. Gerontology 61(6):485–490
19. Dintica CS, Marseglia A, Rizzuto D, Wang R, Seubert J, Arfanakis K et al (2019) Impaired olfaction is associated with cognitive decline and neurodegeneration in the brain. Neurology 92(7):e700–e709
20. Doty RL, Kamath V (2014) The influences of age on olfaction: a review. Front Psychol 5:20
21. Larsson M, Farde L, Hummel T, Witt M, Lindroth NE, Backman L (2009) Age-related loss of olfactory sensitivity: association to dopamine transporter binding in putamen. Neuroscience 161(2):422–426
22. Meck WH (2005) Neuropsychology of timing and time perception. Brain Cogn 58(1):1–8
23. Burr D, Morrone C (2006) Time perception: space-time in the brain. Curr Biol 16(5):R171–R173
24. Pouthas V, Perbal S (2004) Time perception depends on accurate clock mechanisms as well as unimpaired attention and memory processes. Acta Neurobiol Exp 64(3):367–385
25. El Haj M, Kapogiannis D (2016) Time distortions in Alzheimer's disease: a systematic review and theoretical integration. NPJ Aging Mech Dis 2:16016
26. Connor CE, Knierim JJ (2017) Integration of objects and space in perception and memory. Nat Neurosci 20(11):1493–1503

27. Ariel R, Moffat SD (2018) Age-related similarities and differences in monitoring spatial cognition. Neuropsychol Dev Cogn B Aging Neuropsychol Cogn 25(3):351–377
28. Antonova E, Parslow D, Brammer M, Dawson GR, Jackson SH, Morris RG (2009) Age-related neural activity during allocentric spatial memory. Memory 17(2):125–143
29. Bruyer R, Scailquin J-C, Coibion P (1997) Dissociation between categorical and coordinate spatial computations: modulation by cerebral hemispheres, task properties, mode of response, and age. Brain Cogn 33(3):245–277
30. Gerhardstein P, Liu J, Rovee-Collier C (1998) Perceptual constraints on infant memory retrieval. J Exp Child Psychol 69(2):109–131
31. Nusbaum NJ (1999) Aging and sensory senescence. South Med J 92(3):267–275
32. Fischer ME, Cruickshanks KJ, Schubert CR, Pinto AA, Carlsson CM, Klein BEK et al (2016) Age-related sensory impairments and risk of cognitive impairment. J Am Geriatr Soc 64(10):1981–1987
33. Craik FI, Salthouse TA (2011) The handbook of aging and cognition. Psychology Press, London
34. Bäckman L, Small BJ, Fratiglioni L (2001) Stability of the preclinical episodic memory deficit in Alzheimer's disease. Brain 124(1):96–102
35. Nyberg L, Bäckman L (2011) Chapter 8–Memory changes and the aging brain: a multimodal imaging approach. In: Schaie KW, Willis SL (eds) Handbook of the psychology of aging. Elsevier Academic Press, pp 121–131
36. Bäckman L, Small BJ, Wahlin Å (2001) Aging and memory: cognitive and biological perspectives. In: Birren JE, Schaie KW (eds) Handbook of the psychology of aging. Academic Press, pp 349–377
37. Craik FI, McDowd JM (1987) Age differences in recall and recognition. J Exp Physcol 13(3):474–479
38. Nyberg L, Maitland SB, Ronnlund M, Backman L, Dixon RA, Wahlin A et al (2003) Selective adult age differences in an age-invariant multifactor model of declarative memory. Psychol Aging 18(1):149–160
39. Habib R, Nyberg L, Nilsson L-G (2007) Cognitive and non-cognitive factors contributing to the longitudinal identification of successful older adults in the Betula study. Aging Neuropsychol Cognit 14(3):257–273
40. Roediger IHL, Marsh EJ, Lee SC (2002) Kinds of memory. In: Steven's handbook of experimental psychology: memory and cognitive processes
41. Atkinson RC, Shiffrin RM (1971) The control of short-term memory. Sci Am 225(2):82–90
42. Cheng C-H, Lin Y-Y (2012) The effects of aging on lifetime of auditory sensory memory in humans. Biol Psychol 89(2):306–312
43. Poon LW, Fozard J, Cermak LS, Arenberg D, Thompson LW (2014) New directions in memory and aging (PLE: memory). In: proceedings of the George A. Talland Memorial Conference. Psychology Press
44. Abel ME (1973) The visual trace in relation to aging
45. Walsh DA, Thompson LW (1978) Age differences in visual sensory memory. J Gerontol 33(3):383–387
46. Kline DW, Schieber F (1981) What are the age differences in visual sensory memory? J Gerontol 36(1):86–89
47. Yang W (1999) Lifetime of human visual sensory memory: properties and neural substrate
48. Crick F, Koch C (1990) Some reflections on visual awareness. In: Cold Spring Harbor symposia on quantitative biology, vol 55. Cold Spring Harbor Laboratory Press, pp 953–962
49. Grossberg S, Mingolla E (1985) Neural dynamics of perceptual grouping: textures, boundaries, and emergent segmentations. Percept Psychophys 38(2):141–171
50. Pekkonen E, Jousmäki V, Partanen J, Karhu J (1993) Mismatch negativity area and age-related auditory memory. Electroencephalogr Clin Neurophysiol 87(5):321–325
51. Pekkonen E, Rinne T, Reinikainen K, Kujala T, Alho K, Naatanen R (1996) Aging effects on auditory processing: an event-related potential study. Exp Aging Res 22(2):171–184
52. Cooper RJ, Todd J, McGill K, Michie PT (2006) Auditory sensory memory and the aging brain: a mismatch negativity study. Neurobiol Aging 27(5):752–762
53. Alain C, Woods DL, Knight RT (1998) A distributed cortical network for auditory sensory memory in humans. Brain Res 812(1–2):23–37
54. Sperling G (1960) The information available in brief visual presentations. Psychol Monogr 74(11):1
55. Zhang SM (2015) A comparative study of visual memory and auditory memory. Examin Wkly 000(4):83–84
56. Wang S, Sun XW (2004) MRI study of visual and auditory memory function. J Med Imaging (5):5–7
57. Nilsson LG (2003) Memory function in normal aging. Acta Neurol Scand 107:7–13
58. Wahlin A, Backman L, Winblad B (1995) Free recall and recognition of slowly and rapidly presented words in very old age: a community-based study. Exp Aging Res 21(3):251–271
59. Corey-Bloom J, Wiederholt WC, Edelstein S, Salmon DP, Cahn D, Barrett-Connor E (1996) Cognitive and functional status of the oldest old. J Am Geriatr Soc 44(6):671–674
60. GrÉGoire J, Van Der Linden M (1997) Effect of age on forward and backward digit spans. Aging Neuropsychol Cogn 4(2):140–149
61. Park DC, Lautenschlager G, Hedden T, Davidson NS, Smith AD, Smith PK (2002) Models of visuospatial and verbal memory across the adult life span. Psychol Aging 17(2):299
62. Fandakova Y, Sander MC, Werkle-Bergner M, Shing YL (2014) Age differences in short-term memory binding are related to working memory performance across the lifespan. Psychol Aging 29(1):140
63. Naveh-Benjamin M, Guez J, Kilb A, Reedy S (2004) The associative memory deficit of older adults: fur-

ther support using face-name associations. Psychol Aging 19(3):541
64. Park DC, Welsh RC, Marshuetz C, Gutchess AH, Mikels J, Polk TA et al (2003) Working memory for complex scenes: age differences in frontal and hippocampal activations. J Cogn Neurosci 15(8):1122–1134
65. Reuter-Lorenz PA, Jonides J, Smith EE, Hartley A, Miller A, Marshuetz C et al (2000) Age differences in the frontal lateralization of verbal and spatial working memory revealed by PET. J Cogn Neurosci 12(1):174–187
66. Nordahl CW, Ranganath C, Yonelinas AP, DeCarli C, Fletcher E, Jagust WJ (2006) White matter changes compromise prefrontal cortex function in healthy elderly individuals. J Cogn Neurosci 18(3):418–429
67. Rypma B, D'Esposito M (2000) Isolating the neural mechanisms of age-related changes in human working memory. Nat Neurosci 3(5):509–515
68. Rypma B, Prabhakaran V, Desmond JE, Gabrieli JD (2001) Age differences in prefrontal cortical activity in working memory. Psychol Aging 16(3):371
69. Grady CL, McIntosh AR, Bookstein F, Horwitz B, Rapoport SI, Haxby JV (1998) Age-related changes in regional cerebral blood flow during working memory for faces. NeuroImage 8(4):409–425
70. Logan JM, Sanders AL, Snyder AZ, Morris JC, Buckner RL (2002) Under-recruitment and nonselective recruitment: dissociable neural mechanisms associated with aging. Neuron 33(5):827–840
71. Mitchell KJ, Johnson MK, Raye CL, D'Esposito M (2000) fMRI evidence of age-related hippocampal dysfunction in feature binding in working memory. Brain Res Cogn Brain Res 10(1–2):197–206
72. Ranganath C, Cohen MX, Brozinsky CJ (2005) Working memory maintenance contributes to long-term memory formation: neural and behavioral evidence. J Cogn Neurosci 17(7):994–1010
73. Schon K, Hasselmo ME, LoPresti ML, Tricarico MD, Stern CE (2004) Persistence of parahippocampal representation in the absence of stimulus input enhances long-term encoding: a functional magnetic resonance imaging study of subsequent memory after a delayed match-to-sample task. J Neurosci 24(49):11088–11097
74. Jeneson A, Squire LR (2012) Working memory, long-term memory, and medial temporal lobe function. Learn Mem 19(1):15–25
75. Craik FIM, Jennings JM (1992) Human memory. In: Craik FIM, Salthouse TA (eds) The handbook of aging and cognition. Lawrence Erlbaum Associates, Hillsdale, USA
76. Kramer JH, Yaffe K, Lengenfelder J, Delis DC (2003) Age and gender interactions on verbal memory performance. J Int Neuropsychol Soc 9(1):97–102
77. Charlton RA, Barrick T, Markus H, Morris R (2010) The relationship between episodic long-term memory and white matter integrity in normal aging. Neuropsychologia 48(1):114–122
78. Persson J, Nyberg L, Lind J, Larsson A, Nilsson L-G, Ingvar M et al (2006) Structure–function correlates of cognitive decline in aging. Cereb Cortex 16(7):907–915
79. Conway MA, Pleydell-Pearce CW, Whitecross SE, Sharpe H (2003) Neurophysiological correlates of memory for experienced and imagined events. Neuropsychologia 41(3):334–340
80. Morcom AM, Good CD, Frackowiak RS, Rugg MD (2003) Age effects on the neural correlates of successful memory encoding. Brain 126(1):213–229
81. Nyberg L, Sandblom J, Jones S, Neely AS, Petersson KM, Ingvar M et al (2003) Neural correlates of training-related memory improvement in adulthood and aging. Proc Natl Acad Sci 100(23):13728–13733
82. Anderson ND, Iidaka T, Cabeza R, Kapur S, McIntosh AR, Craik FI (2000) The effects of divided attention on encoding-and retrieval-related brain activity: a PET study of younger and older adults. J Cogn Neurosci 12(5):775–792
83. Grady CL, Bernstein LJ, Beig S, Siegenthaler AL (2002) The effects of encoding task on age-related differences in the functional neuroanatomy of face memory. Psychol Aging 17(1):7
84. Dennis NA, Daselaar S, Cabeza R (2007) Effects of aging on transient and sustained successful memory encoding activity. Neurobiol Aging 28(11):1749–1758
85. Gutchess AH, Welsh RC, Hedden T, Bangert A, Minear M, Liu LL et al (2005) Aging and the neural correlates of successful picture encoding: frontal activations compensate for decreased medial-temporal activity. J Cogn Neurosci 17(1):84–96
86. Grady CL, McIntosh AR, Horwitz B, Maisog JM, Ungerleider LG, Mentis MJ et al (1995) Age-related reductions in human recognition memory due to impaired encoding. Science 269(5221):218–221
87. Schiavetto A, Köhler S, Grady CL, Winocur G, Moscovitch M (2002) Neural correlates of memory for object identity and object location: effects of aging. Neuropsychologia 40(8):1428–1442
88. Iidaka T, Sadato N, Yamada H, Murata T, Omori M, Yonekura Y (2001) An fMRI study of the functional neuroanatomy of picture encoding in younger and older adults. Brain Res Cogn Brain Res 11(1):1–11
89. Cabeza R, Anderson ND, Houle S, Mangels JA, Nyberg L (2000) Age-related differences in neural activity during item and temporal-order memory retrieval: a positron emission tomography study. J Cogn Neurosci 12(1):197–206
90. Cabeza R, Anderson ND, Locantore JK, McIntosh AR (2002) Aging gracefully: compensatory brain activity in high-performing older adults. NeuroImage 17(3):1394–1402
91. Schacter DL, Savage CR, Alpert NM, Rauch SL, Albert MS (1996) The role of hippocampus and frontal cortex in age-related memory changes: a PET study. Neuroreport 7(6):1165–1169
92. Meulenbroek O, Petersson KM, Voermans N, Weber B, Fernández G (2004) Age differences in neural

correlates of route encoding and route recognition. NeuroImage 22(4):1503–1514
93. Maguire EA, Frith CD (2003) Aging affects the engagement of the hippocampus during autobiographical memory retrieval. Brain 126(7):1511–1523
94. Cabeza R, Daselaar SM, Dolcos F, Prince SE, Budde M, Nyberg L (2004) Task-independent and task-specific age effects on brain activity during working memory, visual attention and episodic retrieval. Cereb Cortex 14(4):364–375
95. Newcombe NS, Lloyd ME, Ratliff KR (2007) Development of episodic and autobiographical memory: a cognitive neuroscience perspective. Adv Child Dev Behav 35:37–85
96. Levine B, Svoboda E, Hay JF, Winocur G, Moscovitch M (2002) Aging and autobiographical memory: dissociating episodic from semantic retrieval. Psychol Aging 17(4):677
97. Hashtroudi S, Johnson MK, Chrosniak LD (1990) Aging and qualitative characteristics of memories for perceived and imagined complex events. Psychol Aging 5(1):119
98. Coxon P, Valentine T (1997) The effects of the age of eyewitnesses on the accuracy and suggestibility of their testimony. Appl Cogn Psychol 11(5):415–430
99. Holland CA, Rabbitt P (1990) Autobiographical and text recall in the elderly: an investigation of a processing resource deficit. Q J Exp Psychol 42(3):441–470
100. James LE, Burke DM, Austin A, Hulme E (1998) Production and perception of "verbosity" in younger and older adults. Psychol Aging 13(3):355
101. Burgess N, Maguire EA, Spiers HJ, O'Keefe J (2001) A temporoparietal and prefrontal network for retrieving the spatial context of lifelike events. NeuroImage 14(2):439–453
102. Donix M, Poettrich K, Weiss PH, Werner A, von Kummer R, Fink GR et al (2010) Age-dependent differences in the neural mechanisms supporting long-term declarative memories. Arch Clin Neuropsychol 25(5):383–395
103. Addis DR, Roberts RP, Schacter DL (2011) Age-related neural changes in autobiographical remembering and imagining. Neuropsychologia 49(13):3656–3669
104. St-Laurent M, Abdi H, Burianová H, Grady CL (2011) Influence of aging on the neural correlates of autobiographical, episodic, and semantic memory retrieval. J Cogn Neurosci 23(12):4150–4163
105. Jacques PLS, Rubin DC, Cabeza R (2012) Age-related effects on the neural correlates of autobiographical memory retrieval. Neurobiol Aging 33(7):1298–1310
106. Ge R, Fu Y, Wang D, Yao L, Long Z (2014) Age-related alterations of brain network underlying the retrieval of emotional autobiographical memories: an fMRI study using independent component analysis. Front Hum Neurosci 8:629
107. Nyberg L, Maitland SB, Rönnlund M, Bäckman L, Dixon RA, Wahlin Å et al (2003) Selective adult age differences in an age-invariant multifactor model of declarative memory. Psychol Aging 18(1):149
108. Allen PA, Sliwinski M, Bowie T, Madden DJ (2002) Differential age effects in semantic and episodic memory. J Gerontol Ser B 57(2):P173–P186
109. Spaniol J, Madden DJ, Voss A (2006) A diffusion model analysis of adult age differences in episodic and semantic long-term memory retrieval. J Exp Psychol Learn Memory Cogn 32(1):101
110. Rönnlund M, Nyberg L, Bäckman L, Nilsson L-G (2005) Stability, growth, and decline in adult life span development of declarative memory: cross-sectional and longitudinal data from a population-based study. Psychol Aging 20(1):3–18
111. Nyberg L, Lovden M, Riklund K, Lindenberger U, Backman L (2012) Memory aging and brain maintenance. Trends Cogn Sci 16(5):292–305
112. Smith CN, Squire LR (2009) Medial temporal lobe activity during retrieval of semantic memory is related to the age of the memory. J Neurosci 29(4):930–938
113. Nielson KA, Douville KL, Seidenberg M, Woodard JL, Miller SK, Franczak M et al (2006) Age-related functional recruitment for famous name recognition: an event-related fMRI study. Neurobiol Aging 27(10):1494–1504
114. Daselaar SM, Veltman DJ, Rombouts SA, Raaijmakers JG, Jonker C (2005) Aging affects both perceptual and lexical/semantic components of word stem priming: an event-related fMRI study. Neurobiol Learn Mem 83(3):251–262
115. Lustig C, Buckner RL (2004) Preserved neural correlates of priming in old age and dementia. Neuron 42(5):865–875
116. Aizenstein HJ, Butters MA, Clark KA, Figurski JL, Stenger VA, Nebes RD et al (2006) Prefrontal and striatal activation in elderly subjects during concurrent implicit and explicit sequence learning. Neurobiol Aging 27(5):741–751
117. Daselaar SM, Rombouts SA, Veltman DJ, Raaijmakers JG, Jonker C (2003) Similar network activated by young and old adults during the acquisition of a motor sequence. Neurobiol Aging 24(7):1013–1019
118. Fera F, Weickert TW, Goldberg TE, Tessitore A, Hariri A, Das S et al (2005) Neural mechanisms underlying probabilistic category learning in normal aging. J Neurosci 25(49):11340–11348
119. Buckner RL, Petersen SE, Ojemann JG, Miezin FM, Squire LR, Raichle M (1995) Functional anatomical studies of explicit and implicit memory retrieval tasks. J Neurosci 15(1):12–29
120. Winocur G, Moscovitch M, Stuss DT (1996) Explicit and implicit memory in the elderly: evidence for double dissociation involving medial temporal-and frontal-lobe functions. Neuropsychology 10(1):57
121. Bäckman L, Small BJ, Wahlin Å, Larsson M (2000) Cognitive functioning in very old age. In: Handbook of cognitive aging, vol 2

122. Churchill JD, Stanis JJ, Press C, Kushelev M, Greenough WT (2003) Is procedural memory relatively spared from age effects? Neurobiol Aging 24(6):883–892
123. Mitchell DB, Brown AS, Murphy DR (1990) Dissociations between procedural and episodic memory: effects of time and aging. Psychol Aging 5(2):264
124. Hirono N, Mori E, Ikejiri Y, Imamura T, Shimomura T, Ikeda M et al (1997) Procedural memory in patients with mild Alzheimer's disease. Dement Geriatr Cogn Disord 8(4):210–216
125. Collette F, Hogge M, Salmon E, Van der Linden M (2006) Exploration of the neural substrates of executive functioning by functional neuroimaging. Neuroscience 139(1):209–221
126. Collette F, Van der Linden M (2002) Brain imaging of the central executive component of working memory. Neurosci Biobehav Rev 26(2):105–125
127. Aron AR (2007) The neural basis of inhibition in cognitive control. Neuroscientist 13(3):214–228
128. Anguera JA, Gazzaley A (2012) Dissociation of motor and sensory inhibition processes in normal aging. Clin Neurophysiol 123(4):730–740
129. Hasher L, Zacks RT (1988) Working memory, comprehension, and aging: a review and a new view. Psychol Learn Motiv 22:193–225
130. West RL (1996) An application of prefrontal cortex function theory to cognitive aging. Psychol Bull 120(2):272
131. Baddeley AD, Baddeley H, Bucks R, Wilcock G (2001) Attentional control in Alzheimer's disease. Brain 124(8):1492–1508
132. Coubard OA, Ferrufino L, Boura M, Gripon A, Renaud M, Bherer L (2011) Attentional control in normal aging and Alzheimer's disease. Neuropsychology 25(3):353–367
133. Dulas MR, Duarte A (2014) Aging affects the interaction between attentional control and source memory: an fMRI study. J Cogn Neurosci 26(12):2653–2669
134. Milham MP, Erickson KI, Banich MT, Kramer AF, Webb A, Wszalek T et al (2002) Attentional control in the aging brain: insights from an fMRI study of the stroop task. Brain Cogn 49(3):277–296
135. Buitenweg JIV, van de Ven RM, Prinssen S, Murre JMJ, Ridderinkhof KR (2017) Cognitive flexibility training: a large-scale multimodal adaptive active-control intervention study in healthy older adults. Front Hum Neurosci 11:529
136. Wilson CG, Nusbaum AT, Whitney P, Hinson JM (2018) Age-differences in cognitive flexibility when overcoming a preexisting bias through feedback. J Clin Exp Neuropsychol 40(6):586–594
137. Giller F, Beste C (2019) Effects of aging on sequential cognitive flexibility are associated with fronto-parietal processing deficits. Brain Struct Funct 224(7):2343–2355
138. Hyafil A, Summerfield C, Koechlin E (2009) Two mechanisms for task switching in the prefrontal cortex. J Neurosci 29(16):5135–5142
139. Mizoguchi K, Shoji H, Tanaka Y, Maruyama W, Tabira T (2009) Age-related spatial working memory impairment is caused by prefrontal cortical dopaminergic dysfunction in rats. Neuroscience 162(4):1192–1201
140. Simon NW, LaSarge CL, Montgomery KS, Williams MT, Mendez IA, Setlow B et al (2010) Good things come to those who wait: attenuated discounting of delayed rewards in aged Fischer 344 rats. Neurobiol Aging 31(5):853–862
141. Karbach J, Verhaeghen P (2014) Making working memory work: a meta-analysis of executive-control and working memory training in older adults. Psychol Sci 25(11):2027–2037
142. Stopford CL, Thompson JC, Neary D, Richardson AM, Snowden JS (2012) Working memory, attention, and executive function in Alzheimer's disease and frontotemporal dementia. Cortex 48(4):429–446
143. Charlton RA, Barrick TR, Lawes IN, Markus HS, Morris RG (2010) White matter pathways associated with working memory in normal aging. Cortex 46(4):474–489
144. Berryhill ME, Jones KT (2012) tDCS selectively improves working memory in older adults with more education. Neurosci Lett 521(2):148–151
145. Salthouse TA (2016) Theoretical perspectives on cognitive aging. Psychology Press
146. McDowd JM, Shaw RJ (2000) Attention and aging: a functional perspective. In: Craik FIM, Salthouse TA (eds) The handbook of aging and cognition. Lawrence Erlbaum Associates Publishers, pp 221–292
147. Sarter M, Givens B, Bruno JP (2001) The cognitive neuroscience of sustained attention: where top-down meets bottom-up. Brain Res Rev 35(2):146–160
148. Geerligs L, Saliasi E, Maurits NM, Renken RJ, Lorist MM (2014) Brain mechanisms underlying the effects of aging on different aspects of selective attention. NeuroImage 91:52–62
149. Chou Y-H, Chen N-K, Madden DJ (2013) Functional brain connectivity and cognition: effects of adult age and task demands. Neurobiol Aging 34(8):1925–1934
150. Godefroy O, Rousseaux M (1996) Binary choice in patients with prefrontal or posterior brain damage. A relative judgement theory analysis. Neuropsychologia 34(10):1029–1038
151. Woods DL, Knight RT (1986) Electrophysiologic evidence of increased distractibility after dorsolateral prefrontal lesions. Neurology 36(2):212–212
152. Wilkins A, Shallice T, McCarthy R (1987) Frontal lesions and sustained attention. Neuropsychologia 25:359–365
153. Manes F, Sahakian B, Clark L, Rogers R, Antoun N, Aitken M et al (2002) Decision-making processes following damage to the prefrontal cortex. Brain 125(3):624–639
154. Mander BA, Rao V, Lu B, Saletin JM, Lindquist JR, Ancoli-Israel S et al (2013) Prefrontal atrophy, disrupted NREM slow waves and impaired

hippocampal-dependent memory in aging. Nat Neurosci 16(3):357
155. Raz N (2004) The aging brain: structural changes and their implications for cognitive aging. In: New frontiers in cognitive aging. pp 115–133
156. Drachman DA (2006) Aging of the brain, entropy, and Alzheimer disease. Neurology 67(8):1340–1352
157. Greenwood P (2007) Functional plasticity in cognitive aging: review and hypothesis. Neuropsychology 21(6):657
158. Cullum S, Huppert FA, McGee M, Dening T, Ahmed A, Paykel ES et al (2000) Decline across different domains of cognitive function in normal ageing: results of a longitudinal population-based study using CAMCOG. Int J Geriatr Psychiatry 15(9):853–862
159. Grady CL, Craik FI (2000) Changes in memory processing with age. Curr Opin Neurobiol 10(2):224–231
160. Bopp KL, Verhaeghen P (2005) Aging and verbal memory span: a meta-analysis. J Gerontol Ser B Psychol Sci Soc Sci 60(5):P223–P233
161. Chodosh J, Reuben DB, Albert MS, Seeman TE (2002) Predicting cognitive impairment in high-functioning community-dwelling older persons: MacArthur studies of successful aging. J Am Geriatr Soc 50(6):1051–1060
162. Gordon JK, Kurczek JC (2014) The ageing neighbourhood: phonological density in naming. Lang Cogn Neurosci 29(3):326–344
163. Burke DM, Shafto MA (2004) Aging and language production. Curr Dir Psychol Sci 13(1):21–24
164. Verhaegen C, Poncelet M (2013) Changes in naming and semantic abilities with aging from 50 to 90 years. J Int Neuropsychol Soc 19(2):119–126
165. Payne BR, Gao X, Noh SR, Anderson CJ, Stine-Morrow EA (2012) The effects of print exposure on sentence processing and memory in older adults: evidence for efficiency and reserve. Aging Neuropsychol Cognit 19(1–2, 122):-149
166. Wingfield A, Amichetti NM, Lash A (2015) Cognitive aging and hearing acuity: modeling spoken language comprehension. Front Psychol 6:684
167. Rodríguez-Fornells A, Cunillera T, Mestres-Missé A, de Diego-Balaguer R (2009) Neurophysiological mechanisms involved in language learning in adults. Philos Trans R Soc Lond B Biol Sci 364(1536):3711–3735
168. Wong PC, Perrachione TK, Parrish TB (2007) Neural characteristics of successful and less successful speech and word learning in adults. Hum Brain Mapp 28(10):995–1006
169. Wierenga CE, Benjamin M, Gopinath K, Perlstein WM, Leonard CM, Rothi LJG et al (2008) Age-related changes in word retrieval: role of bilateral frontal and subcortical networks. Neurobiol Aging 29(3):436–451
170. Wong PC, Ettlinger M, Sheppard JP, Gunasekera GM, Dhar S (2010) Neuroanatomical characteristics and speech perception in noise in older adults. Ear Hear 31(4):471
171. Thompson-Schill SL (2003) Neuroimaging studies of semantic memory: inferring "how" from "where". Neuropsychologia 41(3):280–292
172. Stein M, Federspiel A, Koenig T, Wirth M, Strik W, Wiest R et al (2012) Structural plasticity in the language system related to increased second language proficiency. Cortex 48(4):458–465
173. Davis MH, Gaskell MG (2009) A complementary systems account of word learning: neural and behavioural evidence. Philos Trans R Soc Lond B Biol Sci 364(1536):3773–3800
174. Ullman MT (2001) A neurocognitive perspective on language: the declarative/procedural model. Nat Rev Neurosci 2(10):717–726
175. Nation D, Wierenga C, Delano-Wood L, Jak A, Delis D, Salmon D et al (2010) Elevated pulse pressure is associated with age-related decline in language ability. J Int Neuropsychol Soc 16(5):933–938
176. Bondi MW, Jak AJ, Delano-Wood L, Jacobson MW, Delis DC, Salmon DP (2008) Neuropsychological contributions to the early identification of Alzheimer's disease. Neuropsychol Rev 18(1):73–90
177. Snowdon DA, Greiner LH, Markesbery WR (2000) Linguistic ability in early life and the neuropathology of Alzheimer's disease and cerebrovascular disease: findings from the nun study. Ann N Y Acad Sci 903(1):34–38
178. Antoniou M, Gunasekera GM, Wong PC (2013) Foreign language training as cognitive therapy for age-related cognitive decline: a hypothesis for future research. Neurosci Biobehav Rev 37(10):2689–2698
179. Kavé G, Eyal N, Shorek A, Cohen-Mansfield J (2008) Multilingualism and cognitive state in the oldest old. Psychol Aging 23(1):70
180. Zied KM, Phillipe A, Karine P, Valerie H-T, Ghislaine A, Arnaud R (2004) Bilingualism and adult differences in inhibitory mechanisms: evidence from a bilingual Stroop task. Brain Cogn 54(3):254–256
181. Green DW (1998) Mental control of the bilingual lexico-semantic system. Biling Lang Congn 1(2):67–81
182. Schweizer K, Moosbrugger H (2004) Attention and working memory as predictors of intelligence. Intelligence 32(4):329–347
183. Liu C (2004) The development of fluid intelligence: evidence for the role of working memory but not processing speed. Acta Psychol Sin 36(04):464–475
184. Burns NR, Nettelbeck T, McPherson J (2009) Attention and intelligence: a factor analytic study. J Individ Differ 30(1):44–57
185. Schweizer K, Moosbrugger H, Goldhammer F (2005) The structure of the relationship between attention and intelligence. Intelligence 33(6):589–611
186. Fry AF, Hale S (1996) Processing speed, working memory, and fluid intelligence: evidence for a developmental cascade. Psychol Sci 7(4):237–241
187. Nettelbeck T, Burns NR (2010) Processing speed, working memory and reasoning ability from childhood to old age. Personal Individ Differ 48(4):379–384

188. Kail RV (2007) Longitudinal evidence that increases in processing speed and working memory enhance children's reasoning. Psychol Sci 18(4):312
189. Light LL, Zelinski EM, Moore M (1982) Adult age differences in reasoning from new information. J Exp Psychol Learn Mem Cogn 8(5):435
190. Doeller CF, Barry C, Burgess N (2010) Evidence for grid cells in a human memory network. Nature 463(7281):657–661
191. Chersi F, Burgess N (2015) The cognitive architecture of spatial navigation: hippocampal and striatal contributions. Neuron 88(1):64–77
192. Lester AW, Moffat SD, Wiener JM, Barnes CA, Wolbers T (2017) The aging navigational system. Neuron 95(5):1019–1035
193. Colombo D, Serino S, Tuena C, Pedroli E, Dakanalis A, Cipresso P et al (2017) Egocentric and allocentric spatial reference frames in aging: a systematic review. Neurosci Biobehav Rev 80:605–621
194. Nedelska Z, Andel R, Laczo J, Vlcek K, Horinek D, Lisy J et al (2012) Spatial navigation impairment is proportional to right hippocampal volume. Proc Natl Acad Sci U S A 109(7):2590–2594
195. Lövdén M, Schaefer S, Noack H, Bodammer NC, Kühn S, Heinze H-J et al (2012) Spatial navigation training protects the hippocampus against age-related changes during early and late adulthood. Neurobiol Aging 33(3):620.e629–620.e622
196. Coutureau E, Di Scala G (2009) Entorhinal cortex and cognition. Prog Neuro-Psychopharmacol Biol Psychiatry 33(5):753–761
197. Head D, Isom M (2010) Age effects on wayfinding and route learning skills. Behav Brain Res 209(1):49–58
198. Cushman LA, Stein K, Duffy CJ (2008) Detecting navigational deficits in cognitive aging and Alzheimer disease using virtual reality. Neurology 71:888–895
199. delpolyi AR, Rankin KP, Mucke L, Miller BL, Gorno-Tempini ML (2007) Spatial cognition and the human navigation network in AD and MCI. Neurology 69:986–997
200. Monacelli AM, Cushman LA, Kavcic V, Duffy CJ (2003) Spatial disorientation in Alzheimer's disease. Neurology 61:1491–1497
201. Hort J, Laczó J, Vyhnálek M, Bojar M, Bures J, Vlček K (2007) Spatial navigation deficit in amnestic mild cognitive impairment. Proc Natl Acad Sci 104:4042–4047
202. Howett D, Castegnaro A, Krzywicka K, Hagman J, Marchment D, Henson R et al (2019) Differentiation of mild cognitive impairment using an entorhinal cortex-based test of virtual reality navigation. Brain 142(6):1751–1766
203. Bohbot VD, Lerch J, Thorndycraft B, Iaria G, Zijdenbos AP (2007) Gray matter differences correlate with spontaneous strategies in a human virtual navigation task. J Neurosci 27(38):10078–10083
204. Gazova I, Laczó J, Rubinova E, Mokrisova I, Hyncicova E, Andel R et al (2013) Spatial navigation in young versus older adults. Front Aging Neurosci 5:94
205. Fjell AM, McEvoy L, Holland D, Dale AM, Walhovd KB (2014) What is normal in normal aging? Effects of aging, amyloid and Alzheimer's disease on the cerebral cortex and the hippocampus. Prog Neurobiol 117:20–40
206. Driscoll I, Hamilton DA, Yeo RA, Brooks WM, Sutherland RJ (2005) Virtual navigation in humans: the impact of age, sex, and hormones on place learning. Horm Behav 47(3):326–335
207. Kunz L, Schröder TN, Lee H, Montag C, Lachmann B, Sariyska R et al (2015) Reduced grid-cell-like representations in adults at genetic risk for Alzheimer's disease. Science 350(6259):430–433
208. Stangl M, Achtzehn J, Huber K, Dietrich C, Tempelmann C, Wolbers T (2018) Compromised grid-cell-like representations in old age as a key mechanism to explain age-related navigational deficits. Curr Biol 28(7):1108–1115.e1106

The Overview of Cognitive Aging Models

4

Dandan Wang, Zhihao Tang, Jiawei Zhao, and Peng Lu

Abstract

To understand the cause of the age-related decline in cognitive function and its underlying mechanism, the cognitive aging model can provide us with important insights. In this section, we will introduce behavioral and neural models about age-related cognitive changes. Among behavioral models, several aging theories were discussed from the perspectives of educational, biological, and sociological factors, which could explain parts of the aging process. With the development of imaging technology, many studies have discussed the neural mechanism of aging and successively proposed neural models to explain the aging phenomenon. Behavioral models and neural mechanism models supplement each other, gradually unveiling the mystery of cognitive aging.

D. Wang · Z. Tang · J. Zhao · P. Lu (✉)
State Key Laboratory of Cognitive Neuroscience and Learning, Faculty of Psychology, Beijing Normal University, Beijing, China

Beijing Aging Brain Rejuvenation Initiative (BABRI) Centre, Beijing Normal University, Beijing, China
e-mail: 201931061030@mail.bnu.edu.cn; 201921061038@mail.bnu.edu.cn; PsyZhao@mail.bnu.edu.cn; peng.lu@bnu.edu.cn

Keywords

Cognitive function · Cognitive aging · Behavioral models · Neural models

4.1 Introduction

A major challenge for cognitive aging researchers is to understand the causes of the age-related decline in cognitive function and the underlying mechanism. In this section, cognitive aging models will be introduced from behavioral and neural perspectives of age-related cognitive changes. Behavioral models include psychological, educational, biological, and sociological theories, all of which were proposed in studies on aging from researchers in different disciplines. With the development of imaging technology, many studies have discussed the neural mechanism of aging and successively proposed neural models to explain the aging phenomenon. Behavioral models and neural mechanism models supplement each other, gradually unveiling the mystery of cognitive aging.

Z. Zhang (ed.), *Cognitive Aging and Brain Health*, Advances in Experimental Medicine and Biology 1419, https://doi.org/10.1007/978-981-99-1627-6_4

4.2 Behavioral Models of Cognitive Aging

4.2.1 Psychological Theories of Cognitive Aging

The psychology of aging is a very broad topic, and unsurprisingly, there is no single integrative psychological theory of aging. However, among the many psychological theories of aging, the commonality is that the related topics are within the psychological field (e.g., learning, memory, and language). Aging theories attempt to explain how multiple changes affect individuals and their behavior as people age [1].

Sensory function, processing speed, working memory, and inhibitory function are considered indicators of cognitive resources. Thus, when completing cognitive tasks, individuals must use the series of mental processing abilities mentioned above. Therefore, when we introduce *resource deficit theory*, we will integrate theories related to mental processing ability.

4.2.1.1 Sensory Deficit Theory

Sensory function appears to be even more fundamental in explaining age-related variance in performance on cognitive tasks. The key point of the theory is that age-related deficits in sensory processing play a major role in cognitive decline with age [2]. Some studies support this theory and suggest that older adults show considerable deficits in basic sensory function, including vision and auditory processing [3]. Meanwhile, visual and auditory acuity have strong explanatory power for cognitive aging [2, 4]. Impaired olfaction was also proposed as a predictor of faster cognitive decline [5].

Furthermore, the main evidence for this theory comprises findings of correlations between age-related declines in sensory and cognitive performance [2, 6, 7], such as the substantial increase in the relationships between vision, hearing, and cognitive functions from adulthood to old age. The intercorrelations of these three variables were higher in the old age group than in the younger group [4]. Another study showed that the perceptual burden is associated with hearing loss and affects cognitive processing, particularly in older adults with limited resources [8].

4.2.1.2 Processing Speed Theory

Processing speed theory is one of the most popular cognitive aging theories. It has been suggested as a general mechanism that underlies age-related differences virtually in all cognitive tasks [9]. The central hypothesis of this theory is that adults' processing speed for many available processing operations would decrease with age, potentially leading to impairments in cognitive function because relevant operations cannot be successfully executed in a limited period of time. In addition, since a cognitive task or complex cognitive activities require all processes to be performed simultaneously, the reduction in the processing of related operations will lead to a decrease in useful information, which is required at higher levels of simultaneous processing and served as the simultaneity mechanism. In summary, a reduction in speed is the major factor contributing to the negative age-related effects on fluid cognitive performance [10].

There are numerous studies supporting this theory, showing that after the speed control, the age-related variance was reduced in different cognitive areas, such as multitype memory [10–12], general intelligence [13–15], executive functions [16], and spatial abilities [17–19]. Furthermore, processing speed was associated with a large percentage of the age-related variance in memory, spatial rotation, matrix reasoning, etc [9, 20, 21]. Also, the result showed that even in very advanced age, cognitive abilities, including reasoning, memory, verbal fluency, and knowledge, were all mediated through processing speed [22].

4.2.1.3 Working Memory Theory

Numerous researchers have proposed age-related differences in a variety of different cognitive tasks, which might be interpretable in terms of age-related limitations in working memory [23]. Working memory theories of cognitive aging build on the definition that working memory is a memory type that processes and stores information [24–27]. In fact, the effects of a reduced working memory capacity on cognitive aging have been divided into three orientations.

The first orientation is the within-context assessment, and its evaluation is inferred from collected observations while research participants are performing other cognitive tasks. Researchers found that older adults showed the poorest performance on tasks with high processing demands, whereas tasks that demanded fewer resources showed smaller age-related differences [9]. Therefore, older adults appear to possess smaller effective working memory capacities than young adults [28].

The second orientation is the out-of-context assessment, which is also called as the measurement of individual and group differences. Numerous researchers have shown that performance on working memory tasks continuously decreases with age [29–31]. Researchers argue that age-related reductions in working memory might be responsible for the age-related differences in assorted comprehension tasks [32], and working memory is associated with 44% of the age-related variance in a speech comprehension task [33]. Furthermore, studies confirmed that age-related differences in language comprehension, verbal memory, and spatial memory are mediated through decreases in working memory [34, 35]. In addition, age-related differences in reasoning and problem solving have also been attributed to age-related reductions in working memory, and working memory mediates age-related differences in a variety of reasoning and other cognitive tasks [28, 36, 37].

The third orientation focuses on the contents of working memory rather than on its capacity [23], and age-related cognitive deficits may be due to a decrease in the inhibitory control of working memory contents [27]. Researchers regard this hypothesis as the inhibition deficit theory [38]. Moreover, we will elaborate on this theory in the next section.

4.2.1.4 Inhibition Deficit Theory

Another popular theory of cognitive aging is the inhibition deficit theory. Hasher and Zacks proposed that older adults are less able to inhibit unwanted or irrelevant information, which occupies the space of working memory, thereby reducing temporary storage and further processing abilities [38, 39]. In this case, aging weakens inhibitory processes that regulate attention and the contents of working memory, thereby affecting a broad range of cognitive performance [27, 38, 40]. For example, inhibition deficit theory might explain why older adults' performance suffers more from distracting stimuli during reading [41] or listening [40, 42] and why older adults' conversations are more likely to go off topic [43].

Numerous lines of evidence support this theory. According to one study, a decreased inhibitory response may explain the increased interference on Stroop tasks with age [44]. Age-related differences in inhibition at the level of selective attention have been observed, indicating that older adults may show impairments in the input and retrieval process due to inefficient inhibition that ultimately slows or impedes retrieval of target facts [45]. Older adults are also deficient in the deletion of information that is no longer relevant, providing evidence of greater proactive interference (PI) in older adults than younger adults [46]. Researchers also suggested that inhibitory dysfunction with aging is a source of general attentional dysregulation that might account for age-related deficits in other cognitive domains, such as task switching, response competition, and response suppression [47].

4.2.1.5 Executive Decline Hypothesis

The *executive decline hypothesis* is a theory that has emerged in recent years. This hypothesis suggests that executive function ability is more likely to decline with age, which is an important explanation for cognitive ability aging [48, 49]. Because the individual's behavioral performance on various cognitive tasks is often related to the effectiveness of controlling other cognitive operations, executive function may be an important factor contributing to individual differences in cognitive function, especially age-related individual differences. Age-related deficits in executive functions include reductions in the abilities to select relevant information and inhibit irrelevant stimuli inhibition, a decreased ability to perform task switching, and deficits in attentional processing resources [38, 40, 50, 51]. Differences in performance between younger and older adults are minimal on simple span tasks or item recognition that require only rote rehearsal, whereas older adults perform worse than younger adults when memory updating, reordering, or inhibition tasks are added. Thus, older people have worse executive function than younger adults [52].

In addition, some researchers have noted that the concepts of attention, inhibition, and working memory in traditional cognitive aging theory may only reflect one aspect of executive function, while an aging study examining executive control ability may provide a unified interpretation of the overall cognitive aging mechanism [53].

4.2.1.6 Summary

The aforementioned psychological theories of cognitive aging have evidence to support their interpretations of cognitive aging. Nevertheless, researchers also tended to explore their interactional relationship in cognitive aging. It was reported that there was a large attenuation of the age-related effects in working memory when the measures of inhibition were controlled. However, the attenuation was strong when the measures from the neutral condition and speed measures from separate tasks were controlled. Age differences in working memory may originate because of reduced inhibition abilities, thus statistical control of measures of inhibition should reduce the age-related effects on working memory. However, the attenuation was strong when the speed measures from separate tasks were controlled, which indicated that speed or even other factors appear to be more important than inhibition contributing to the age differences in working memory [54]. Other results suggested that after control, individual differences in perceptual speed, vision, and hearing, the age-related variance of the intellectual abilities was reduced. This result implies that the sensory function and perceptual speed may all account for age-related variance in the intellectual abilities [4]. Thus, the interaction between different mental processing abilities influences cognitive aging in subtle ways. Future research needs to integrate the four processing resources on cognitive aging and analyze the relationship between the four processing resources and executive decline.

4.2.2 Educational Theories of Cognitive Aging

Neural plasticity provides a neural basis for lifelong learning, as evidenced by phrases such as "Never too old to learn." Based on neural plasticity, research has proposed the scaffolding theory of aging and cognition, which is similar with the scaffolding theory in education.

4.2.2.1 The Scaffolding Theory of Aging and Cognition (STAC)

Decreases in brain structure, white matter integrity, dopaminergic receptors, and functional engagement in some brain areas, such as the hippocampus, have been observed with aging. However, compensatory increases have been detected in specific areas, such as frontal functional engagement, and correlate with better behavioral performance in older adults. Researchers proposed a homeostatic adaptive model called *"The Scaffolding Theory of Aging and Cognition (STAC)"* to explain these joint phenomena [47, 51, 52, 55].

As shown in this model [47], the process of aging potentially causes a deterioration of the brain structure or "neural challenges," including structural shrinkage, decreased white matter integrity, cortical thinning, and depletion of dopamine receptors [56–63]. In addition, aging could also cause "functional deterioration," including dedifferentiation [64], default network dysregulation [65], and decreased activation of the medial temporal lobe (MTL) structures [30, 66].

However, the brain might respond to these deteriorations by engaging in "compensatory scaffolding" through the recruitment of bilateral frontal or prefrontal or dorsolateral prefrontal cortex (DLPFC) and parietal areas [67–72], altering the form of structural changes (including neurogenesis) in the brain [73, 74], and engaging in more distributed processing when performing a task [67, 68, 71]. Furthermore, the aged brain is less efficient at generating scaffolding. Scaffolded networks are less efficient than highly developed cognitive networks, and even pathologies such as Alzheimer's disease may limit the ability to engage in scaffolding [75–77].

Scaffolding could be promoted or enhanced by training, improving learning ability, engaging in cognitive activities, sustaining abundant cognitive reserve, and other protective factors, such as higher education and increased levels of brain-derived neurotrophic factor and serotonin. Indeed, external experience and training serve to improve brain neuroplasticity. The effect is not limited to neural structural development in older adults, as it also increases neural resources and improves function in a more specific manner [51, 78–81]. The magnitude of neural deterioration combined with the effectiveness of new compensatory scaffolds predicts the overall level of cognitive function.

In summary, STAC suggests that despite neural challenges and functional deterioration, the behaviors of older adults are maintained at a relatively high level due to the continuous engagement of compensatory scaffolding.

4.2.2.2 A Life Course Model of the Scaffolding Theory of Aging and Cognition (STAC-R)

STAC-R is revisiting *The Scaffolding Theory of Aging and Cognition.* This conceptual model combines adverse and favorable effects on brain structure and function to determine cognitive status, which incorporates life-course factors that enhance, preserve, or compromise brain status, compensatory potential, and ultimately cognitive function over time rather than emphasizing adverse effects of aging together with beneficial effects of compensation. The model is compatible with concepts of reserve and hypothesizes that it is similar with compensatory scaffolding. These processes are affected positively or negatively by factors that enrich or deplete neural resources [55].

4.2.3 Biological Theories of Cognitive Aging

Biological theories examine aging from a biological perspective, which studies processes at many different levels and tends to be specific to the description of a particular biological process [1]. However, the studies' attempt to integrate biological theories of aging has been reported. We will introduce these theories from three general categories. The first two are stochastic theories and developmental–genetic theories, and the third category is evolutionary theories.

4.2.3.1 Stochastic Theories

Stochastic theories suggest that the loss of functions during aging may be due to the effects of environmental factors on living organisms that are random in nature and includes the subgroups known as the theory of somatic mutations, the theory of error catastrophe, the theory of DNA repair, the theory of the breaking of chemical bonds, the theory of advanced glycosylation, and the theory of oxidative stress [1, 82].

Somatic mutation theory is the best-known theory, which claims that normal radiation exposure over a person's lifetime increases the probability of acquiring diseases and diminishes life expectancy due to the destruction of biomolecules, mainly DNA [83]. The *error-catastrophe theory* suggests that the perpetuation of protein synthesis errors diminishes the reliability of its products and creates aberrant and/or lethal proteins that potentially affect DNA replication and increase the probability of somatic mutations [82, 84]. The *reparation of DNA theory* claims that the number of DNA replications determines the life span of a species [85]. The *theory of the breaking of biomolecular bonds* states that the modification of proteins generates the functional failure of cell metabolism [86] and leads to a decrease in physiological processes during aging [87].

According to the *theory of advanced glycosylation*, collagen glycosylation has been the theory studied most extensively and is associated with aging [88]. The glycosylation of proteins is defined as the cross-link between glucose and protein. These cross-links are caused by a high concentration of glucose in the blood and tissues, which results in functional deterioration [89]. The basis of the *theory of oxidative stress* is free radicals or reactive oxygen metabolites (ROMs), which lead to the destruction of biomolecules and then cause many degenerative alterations associated with aging [90, 91].

4.2.3.2 Non-Stochastic Theories/Development–Genetic Theories

According to non-stochastic theory, the environment and genetics function together during aging [92], including the subgroups known as the theory of programmed senescence, the theory of telomeres, the theory of intrinsic mutagenesis, the neuroendocrine theory, and the immunological theory [82]. These theories are developmental in nature and imply that developmental changes are genetically controlled.

The *theory of programmed senescence* suggests that aging is based on genetic programming that controls developmental changes [93]. The theory refers to the existence of an organic cell program that genetically determines the life span of each cell, and some genes are associated with diseases in the elderly, including some alleles of apolipoprotein E, which is also associated with AD [94]. The *theory of telomeres* suggests that the existence of complete telomeres depends on the activity of telomerase that decreases over time. In normal cells, shortening of telomeres and the disappearance of genes from the region may be associated with aging [95]. The *theory of intrinsic mutagenesis* claims that longevity depends on the reliable replication of genetic material [86]. Failure to replicate DNA might induce mutagenesis, indicating a loss of functions in the organism and thereby causing aging. The *neuroendocrine theory* claims that decreased levels of many hormones of the hypothalamic–pituitary–adrenal axis cause problems with metabolism [96], subsequently resulting in the aging phenomenon. *Immunologic theory* focuses on the decreasing function of the immune system as people age [97]. Specifically, aging results in reduced resistance to diseases and an increased incidence of autoimmune diseases [1].

4.2.3.3 Evolutionary Theories

Evolutionary theories indicate how the process of natural selection may affect the aging process. Evolutionary theories of aging align closely with the aforementioned theories, suggesting that aging occurs due to the decreasing effect of natural selection on the aging process [1].

The three specific evolutionary theories are described as the following [98]:

The *mutation accumulation theory* might explain why aging exists. Children with progeria do not reach adulthood; thus, the mutation would not be passed to the next generation. However, if the mutations are exclusively expressed during old age, they will not be selected because carriers have already transmitted their genes to the next generation. Mutations in certain genes expressed only in old age will have no real effect because most of them are already inactivated at this age. If the whole process is repeated multiple times with mutations and many generations, deleterious mutations will accumulate in old age, which leads to many health problems if the individuals reach old age [98].

4.2.3.4 The Antagonistic Pleiotropy Theory

The *antagonistic pleiotropy theory* suggests that no actual negative effect would be observed on creatures in the wild because death would occur before negative effects become observable. However, if animals could live beyond the mean age in the wild, the negative effects of aging would occur because of the antagonistic effects of some alleles [98]. For example, in early childhood, calcium in bones is beneficial for protecting organs and maintaining physical stability; after giving birth, the same genes that calcify bones could lead to coronary artery disease and heart attacks.

4.2.3.5 The Disposable Soma Theory

The disposable soma theory is supported by Kirkwood, who argued that the germline is immortal but that the soma is disposable [99]. This theory complements the mutation accumulation theory from a microscopic perspective because it suggests that organisms should balance reproduction and maintenance of the somatoplasm. Because lifetimes are short in the wild, an investment of more energy in body maintenance mechanisms is useless compared to the expected life span in the wild. Thus, the soma is disposable. If extrinsic mortality is high, more resources should be invested in reproduction, and subjects who live beyond the mean life span in the wild will undergo aging by chance. If extrinsic mortality decreases in successive generations, a greater investment in maintenance of the somatoplasm would be useful [98].

4.2.4 Sociological Theory of Cognitive Aging

Early sociological theories included disengagement theory [100] and subculture theory [101]. Interestingly, most of these theories are no longer viewed as viable by sociological-aging researchers and theorists. Here, we only introduce a very broad and multidisciplinary theory, the life course theory. *Life course theory* postulates that to understand the situation of elderly people, one must not only consider individuals but also the life experiences, including the social, cultural, and structural contexts of a person's lifelong development. Additionally, the four key principles of historical time and place, timing of events in lives, linked lives, and human agency to make decisions are considered [1, 102]. Although a person's life experiences might provide information that helps us to understand their current health status, adverse childhood experiences (ACEs) increase the risk of negative health outcomes across cohorts, despite social changes. The life course perspective recognizes that the timing of these experiences affects health and aging trajectories and may continue to influence health even in older adults [102].

4.2.5 Summary

Altogether, several aging theories were discussed from the perspectives of psychological, social, and biological factors. Some parts of the aging process are explained by these theories, but the features of aging cannot be fully explained. Therefore, a more helpful approach is to develop an integrative model that considers the genetic and environmental factors across the life span to understand the causes of age-related deficits.

4.3 Neural Mechanism Models of Cognitive Aging

4.3.1 Cognitive and Brain Reserve Theory

From a life span perspective, researchers found that individuals with higher childhood IQs and better educational attainment are less likely to suffer from severe cognitive decline and less likely to be diagnosed with dementia [103], resulting in the proposal of cognitive reserve theory. *Cognitive reserve theory* indicates that individuals with a high cognitive reserve (higher childhood IQ, greater educational and occupational attainment in early life) have a higher cognitive peak. Cognitive decline presents later in individuals with a high cognitive reserve because pathology is tolerated longer than people with a low cognitive reserve. However, the decline rate of individuals with high reserves is more rapid than in those with low reserves after the turning point because the accumulated pathology of the high reserve group is much more severe than that of the low reserve group.

The concept of the reserve has been advanced to *Brain Reserve Theory*, which refers to the difference of age-related brain changes and pathology in individuals with different levels of cognitive reserve [104]. Individuals with high brain reserve may have a larger brain size, especially in some critical regions such as the hippocampus, which is related closely to memory function [105]. And the highly reserved brain may be the underlying mechanism of the pathological toleration of high cognitive reserve individuals.

Scholars have debated the effect of cognitive reserve on cognitive performance. One proposal is differential preservation, which refers to the tendency of individuals with different levels of risk factors to preserve their cognitive function differentially with adult aging. Namely, cognitive reserve (including factors such as high IQ, high level of education and better occupation) influences the rate of cognitive decline. The other hypothesis is preserved differentiation. Here, educational attainment is not a protective factor against aging-related cognitive decline, but is associated with the level of cognitive development during childhood and adolescence. For example, cognitive reserve is associated with higher cognitive function throughout adulthood (i.e., the cognitive peak), but not the rate of cognitive decline. Moreover, accumulating evidence supports this hypothesis, indicating that a higher cognitive reserve is not related to the rate of cognitive decline [106].

4.3.2 "Last in, first out" Theory

The changing trajectory of the human brain throughout the life span exhibits an inverted U shape with three stages: the development stage with increasing brain growth and reserve, the stable stage with a peak during several years in early or middle age, and the decline stage with decreasing brain integrity. This trajectory is not exactly the same in different regions in the brain. According to multiple studies, cortical regions with complex architectures, such as frontal and parietal regions, are among the slowest regions to develop and the fastest to decline. The findings have been named the *"last in, first out" theory* [107, 108]. These regions usually support higher-order cognitive processes, including executive function and memory.

Cerebral cortical thickness is a biomarker for brain gray matter, driven by the density of cells in a cortical column. However, the changing trajectory of cerebral cortical thickness in the human life span is nonlinear. The peak cerebral cortical thickness usually appears at approximately 2 years age, followed by a decrease of approximately 1% per year in childhood and adolescence. The explanations for these changes are synaptic pruning and optimization of neuronal structures. At midlife, the cerebral cortical thickness maintains a stable rate of decline between 0.1 and 0.5% per year, and in old age, the rate increases to the same or a higher level than in early age [109]. Although most of the cerebral regions follow the trajectory described above, exceptions still exist. The entorhinal cortex, a region crucial for memory, peaks at approximately 40 years age [110]. Studies have revealed that the entorhinal cortex is a central region for age-related changes and pathological accumulation of age-related neurodegeneration [111, 112]. The development and decline of white matter tracts in the human brain also follow the "last in, first out" rule in some regions. Tracts connecting frontal regions show delayed maturation, and this evidence supports the hypothesis that basic sensory processing is the quickest to mature. In old age, commissural fibers connecting association cortices are the most vulnerable during aging [113].

4.3.3 Posterior–Anterior Shift in Aging (PASA)

The brain reserve theory and the "last in, first out" theory were proposed based on studies of changes in brain structure during the human life span. In addition to the structural morphology of the brain, the functional performance in the activation of brain regions during completing cognitive tasks in functional magnetic resonance (fMRI) revealed several rules of aging. Over the last several decades, neuroimaging research has identified age-related neural changes that occur during cognitive tasks, one of which is the "*posterior-anterior shift in aging (PASA).*" This shift is often described as compensatory recruitment of prefrontal regions due to age-related sensory-processing deficits in posterior regions [114].

In 1994, Grady et al. [115] first examined facial perception and location under two task conditions. The elderly presented weaker activation in the occipital temporal area compared with the younger participants, while the activation of anterior brain areas, including the prefrontal cortex, was enhanced. The researchers proposed that the elderly compensated for defects in visual processing (decreased occipital lobe activation) by recruiting more advanced cognitive processes (enhanced prefrontal lobe activation). In this study, the task accuracy rates of the elderly and young people were similar, but the elderly had longer response times than young adults. Therefore, the researchers suggested that the additional recruitment of the prefrontal lobe enabled the elderly to maintain accuracy at a higher level at the expense of a slower response time. Some researchers explain the PASA pattern observed in cognitive processing using the *dedifferentiation hypothesis of aging*. During the aging process, the specificity of the anterior brain executive control areas and the posterior brain visual perception areas are weakened, and their covariation is enhanced, which reflects the dedifferentiation phenomenon of aging. Age was recently shown to be associated with spatial and temporal shifts in recruitment, in which younger and older adults recruit the same neural region at different points in a task trial [116]. This work reflects that older adults are not unable to engage in sensory processing, but rather have a temporal change when they recruit related brain regions.

4.3.4 Hemispheric Asymmetry Reduction in Older Adults (HAROLD)

Similar to PASA, the *hemispheric asymmetry reduction in older adults (HAROLD)* phenomenon is observed when older adults complete cognitive tasks. A study with a memory task conducted using fMRI by Cabeza found that young people recruited the left prefrontal lobe more in the encoding stage and the right prefrontal region in the extraction stage, indicating asymmetric brain activation characteristics. While older people tended to recruit both sides of the frontal lobe in the extraction stage of memory, a significant decrease in functional hemispheric asymmetry was observed [117]. This symmetry activation pattern in elderly individuals is observed in simple action tasks such as key press tasks [118] or complex tasks such as word working memory and memory extraction tasks [31].

Numerous studies support the application of the compensatory hypothesis and the differentiation hypothesis, which explain the reduction in brain asymmetry during the aging process. Research supporting the compensation hypothesis revealed that elderly individuals with poor memory performance had a similar brain activation pattern to young people. Older people with better performance tend to recruit more contralateral brain regions to compensate for the decline in cognitive function caused by aging. The relationship between the degree of contralateral activation and cognitive performance supported this compensation hypothesis [119]. Studies have proposed that the weakening of asymmetry in aging is only a dedifferentiation phenomenon, and the activation of the contralateral brain regions is less specific in the elderly [120]. The term of compensation has been normalized by a recent review proposed by Cabeza and his colleagues. The authors mentioned that compensation should be used to refer to the cognition-enhancing recruitment of neural resources in response to relatively high cognitive demand. Two basic criteria must be fulfilled to attribute any more intense activity or connectivity observed in older adults to compensation: the aspect being compensated for should be clear, and evidence should be available for increased activation in older adults, which is a beneficial effect on cognitive performance [121]. The relationship between the extra recruitment of contralateral regions and cognitive performance is critical to identify whether the HAROLD model is explained by the compensation hypothesis.

4.3.5 Compensation-Related Utilization of Neural Circuits Hypothesis (CRUNCH)

Some researchers have suggested the "compensation-related utilization of neural circuits hypothesis (CRUNCH)" to propose a further explanation of the relationship between increased brain activity and cognitive aging. According to the CRUNCH hypothesis, insufficient cognitive processing efficiency causes the aging brain to recruit more neural resources to achieve the same computing output as the brain of younger people. This compensatory brain activity is only efficient at the easier task levels, but when the task requirements increase and reach the maximum limit of neural resources, a decrease in cognitive processing efficiency and cognitive performance on difficult tasks will be observed accordingly [31]. Compensation is an increase in cognitive resources in the elderly. Older people start to increase their neural resources at a lower task difficulty, and the platform period is lower and starts to fall earlier than younger people [121, 122].

4.3.6 Summary

Along with the development of structural and functional imaging technology, neural mechanism models of cognitive aging have been investigated and proposed, respectively. In the early years of the human life span, factors such as higher IQ and high educational attainment benefit cognitive and brain reserves, which prepare a high cognitive peak to cope with age-related cognitive decline. Inspired by cognitive and brain reserve theory and during the exploration of neuroplasticity in old age, researchers found that cognitive training increases the reserve of elderly individuals and provide a new possibility for the intervention of age-related neurodegeneration [123, 124]. Comparing the developmental and aging periods from the life span perspective, researchers established rules to explain their correspondence, such as the "last in, first out" theory. This theory is closely related to evolutionary theories, indicating the importance of large-scale longitudinal studies in revealing life span changes in cognition and the brain. Functional MRI provides researchers with an opportunity to study brain activation pattern during cognitive processing in elderly individuals. PASA, HAROLD, CRUNCH, and many other hypotheses intended to explain the brain processing pattern of the elderly while completing cognitive tasks, and comparative studies with more elaborate experimental designs are needed to improve our understanding of brain function and plasticity in aging.

References

1. Baltes BB, Rudolph CW, Bal AC (2012) A review of aging theories and modern work perspectives. In: Borman WC, Hedge JW (eds) The Oxford handbook of work and aging. Oxford Handbooks Online, London, pp 1–34
2. Lindenberger U, Baltes PB (1994) Sensory functioning and intelligence in old age: a strong connection. Psychol Aging 9(3):339
3. Schneider BA, Pichora-Fuller MK (2000) Implications of perceptual deterioration for cognitive aging research. In: Craik FIM, Salthouse TA (eds) Handbook of cognitive aging II. Lawrence Erlbaum Associates, Mahwah
4. Baltes PB, Lindenberger U (1997) Emergence of a powerful connection between sensory and cognitive functions across the adult life span: a new window to the study of cognitive aging? Psychol Aging 12(1):12
5. Dintica CS, Marseglia A, Rizzuto D, Wang R, Seubert J, Arfanakis K et al (2019) Impaired olfaction is associated with cognitive decline and neurodegeneration in the brain. Neurology 92(7):e700–e709
6. Elyashiv SM, Shabtai EL, Belkin M (2014) Correlation between visual acuity and cognitive functions. Br J Ophthalmol 98(1):129–132
7. Heyl V, Wahl H-W (2012) Managing daily life with age-related sensory loss: cognitive resources gain in importance. Psychol Aging 27(2):510
8. Tun PA, Williams VA, Small BJ, Hafter ER (2012) The effects of aging on auditory processing and cognition. Am J Audiol 21(2):344–350
9. Park DC, Smith AD, Lautenschlager G, Earles JL, Frieske D, Zwahr M et al (1996) Mediators of long-term memory performance across the life span. Psychol Aging 11(4):621–637
10. Salthouse TA (1996) The processing-speed theory of adult age differences in cognition. Psychol Rev 103(3):403
11. Bryan J, Luszcz MA (1996) Speed of information processing as a mediator between age and free-recall performance. Psychol Aging 11(1):3
12. Clarys D, Isingrini M, Gana K (2002) Mediators of age-related differences in recollective experience in recognition memory. Acta Psychol 109(3):315–329
13. Hertzog C, Bleckley MK (2001) Age differences in the structure of intelligence: influences of information processing speed. Intelligence 29(3):191–217
14. Zimprich D, Martin M (2002) Can longitudinal changes in processing speed explain longitudinal age changes in fluid intelligence? Psychol Aging 17(4):690
15. Ritchie SJ, Tucker-Drob EM, Deary IJ (2014) A strong link between speed of visual discrimination and cognitive ageing. Curr Biol 24(15):R681–R683
16. Albinet CT, Boucard G, Bouquet CA, Audiffren M (2012) Processing speed and executive functions

in cognitive aging: how to disentangle their mutual relationship? Brain Cogn 79(1):1–11
17. Finkel D, Reynolds CA, McArdle JJ, Pedersen NL (2007) Age changes in processing speed as a leading indicator of cognitive aging. Psychol Aging 22(3):558
18. Verhaeghen P, Salthouse TA (1997) Meta-analyses of age–cognition relations in adulthood: estimates of linear and nonlinear age effects and structural models. Psychol Bull 122(3):231
19. Robitaille A, Muniz G, Piccinin AM, Johansson B, Hofer SM (2012) Multivariate longitudinal modeling of cognitive aging: associations among change and variation in processing speed and visuospatial ability. Geropsych 25(1):15–24
20. Salthouse TA (1993) Speed mediation of adult age differences in cognition. Dev Psychol 29(4):722–738
21. Salthouse TA (1994) The nature of the influence of speed on adult age differences in cognition. Dev Psychol 30(2):240–259
22. Lindenberger U, Mayr U, Kliegl R (1993) Speed and intelligence in old age. Psychol Aging 8:207–220
23. Salthouse TA (1990) Working memory as a processing resource in cognitive aging. Dev Rev 10(1):101–124
24. Baddeley AD (1986) Working memory. Oxford University Press, London
25. Engle RW, Tuholski SW, Laughlin JE, Conway AR (1999) Working memory, short-term memory, and general fluid intelligence: a latent-variable approach. J Exp Psychol Gen 128(3):309–331
26. Just MA, Carpenter PAJPR (1992) A capacity theory of comprehension: individual differences in working memory. Psychol Rev 98(1):122–149
27. Burke DM, Shafto MA (2011) Language and aging. In: Craik FI, Salthouse TA (eds) The handbook of aging and cognition. Psychology Press, London
28. Salthouse TA, Eric S (1992) Within-context assessment of age differences in working memory. J Gerontol Psychol Sci 47:110–120
29. Park DC, Lautenschlager G, Hedden T, Davidson NS, Smith AD, Smith PK (2002) Models of visuospatial and verbal memory across the adult life span. Psychol Aging 17(2):299–320
30. Park DC, Welsh RC, Marshuetz C, Gutchess AH, Mikels J, Polk TA et al (2003) Working memory for complex scenes: age differences in frontal and hippocampal activations. J Cogn Neurosci 15(8):1122–1134
31. Reuter-Lorenz PA, Jonides J, Smith EE, Hartley A, Miller A, Marshuetz C et al (2000) Age differences in the frontal lateralization of verbal and spatial working memory revealed by PET. J Cogn Neurosci 12(1):174–187
32. Light LL, Capps JL (1986) Comprehension of pronouns in young and older adults. Dev Psychol 22:580–585
33. Stine EL, Wingfield A (1987) Process and strategy in memory for speech among younger and older adults. Psychol Aging 2(3):272–279
34. Van Der Linden M, Hupet M, Feyereisen P, Schelstraete M-A, Bestgen Y, Bruyer R et al (1999) Cognitive mediators of age-related differences in language comprehension and verbal memory performance. Aging Neuropsychol Cognit 6(1):32–55
35. Cherry KE, Park DC (1993) Individual difference and contextual variables influence spatial memory in younger and older adults. Psychol Aging 8(4):517–526
36. Salthouse TA (1993) Influence of working memory on adult age differences in matrix reasoning. Br J Psychol 84:171–199
37. Salthouse TA (1992) Working-memory mediation of adult age differences in integrative reasoning. Mem Cogn 20(4):413–423
38. Hasher L, Zacks RT (1988) Working memory, comprehension, and aging: a review and a new view. Psychol Learn Motiv 22:193–225
39. Anderson ND, Craik FI (2017) 50 years of cognitive aging theory. J Gerontol 72(1):1–6
40. Hasher L, Lustig C, Zacks R (2007) Inhibitory mechanisms and the control of attention. In: Conway A, Jarrold C, Kane M, Miyake A, Towse J (eds) Variation in working memory. Oxford University Press, New York
41. Connelly SL, Hasher L, Zacks RT (1991) Age and reading: the impact of distraction. Psychol Aging 6(4):533–541
42. Tun PA, O'Kane G, Wingfield A (2002) Distraction by competing speech in young and older adult listeners. J Psychol Aging 17(3):453–467
43. Arbuckle TY, Nohara-Leclair M, Pushkar D (2000) Effect of off-target verbosity on communication efficiency in a referential communication task. Psychol Aging 15(1):65–77
44. Troyer AK, Leach L, Strauss E (2006) Aging and response inhibition: normative data for the Victoria Stroop test. Neuropsychol Dev Cogn B Aging Neuropsychol Cogn 13(1):20–35
45. Hasher L, Stoltzfus ER, Zacks RT, Rypma B (1991) Age and inhibition. J Exp Psychol Learn Mem Cogn 17(1):163–169
46. Lustig C, May CP, Hasher L (2001) Working memory span and the role of proactive interference. J Exp Psychol Gen 130:199–207
47. Park DC, Reuter-Lorenz P (2009) The adaptive brain: aging and neurocognitive scaffolding. Annu Rev Psychol 60(1):173–196
48. Baddeley A (1996) Exploring the central executive. Q J Exp Psychol Sect A 49(1):5–28
49. West R, L. (1996) An application of prefrontal cortex function theory to cognitive aging. Psychol Bull 120(2):272–292
50. Craik FIM, Byrd M (1982) Aging and cognitive deficits the role of attentional resources. In: Craik FIM, Trehub S (eds) Aging and cognitive processes. Plenum Press, New York, pp 191–211
51. Goh JO, Park DC (2009) Neuroplasticity and cognitive aging: the scaffolding theory of aging and cognition. Restor Neurol Neurosci 27(5):391–403

52. Reuter-Lorenz PA, Park DC (2010) Human neuroscience and the aging mind: a new look at old problems. J Gerontol Physchol Sci 4:4
53. Miyake A, Friedman NP, Emerson MJ, Witzki AH, Howerter A et al (2000) The unity and diversity of executive functions and their contributions to complex "frontal lobe" tasks: a latent variable analysis. Cogn Psychol 41(1):49–100
54. Salthouse TA, Meinz EJ (1995) Aging, inhibition, working memory, and speed. J Gerontol Ser B Psychol Sci Soc Sci 50(6):297–306
55. Reuter-Lorenz PA, Park DC (2014) How does it STAC up? Revisiting the scaffolding theory of aging and cognition. Neuropsychol Rev 24(3):355–370
56. Kaasinen V, Vilkman H, Hietala J, Någren K, Helenius H, Olsson H et al (2000) Age-related dopamine D2/D3 loss in extrastriatal regions of the human brain. Neurobiol Aging 21(5):683–688
57. Davatzikos C, Resnick SM (2002) Degenerative age changes in white matter connectivity visualized in vivo using magnetic resonance imaging. Cereb Cortex 12(7):767–771
58. Head D, Buckner RL, Shimony JS, Williams LE, Akbudak E, Conturo TE et al (2004) Differential vulnerability of anterior white matter in nondemented aging with minimal acceleration in dementia of the Alzheimer type: evidence from diffusion tensor imaging. Cereb Cortex 14(4):410–423
59. Moseley M (2002) Diffusion tensor imaging and aging – a review. NMR Biomed 15(7-8):553–560
60. Raz N, Lindenberger U, Rodrigue KM, Kennedy KM, Head D, Williamson A et al (2005) Regional brain changes in aging healthy adults: general trends, individual differences and modifiers. Cereb Cortex 15(11):1676–1689
61. Resnick SM, Pham DL, Kraut MA, Zonderman AB, Davatzikos C (2003) Longitudinal magnetic resonance imaging studies of older adults: a shrinking brain. J Neurosci Off J Soc Neurosci 23(8):3295–3301
62. Salat DH, Buckner RL, Snyder AZ, Greve DN, Desikan SR, Evelina B et al (2004) Thinning of the cerebral cortex in aging. Cereb Cortex 7:7
63. Wen W, Sachdev P (2004) The topography of white matter hyperintensities on brain MRI in healthy 60- to 64-year-old individuals. NeuroImage 22(1):144–154
64. Park DC, Polk TA, Park R, Minear M, Savage A, Smith MR (2004) Aging reduces neural specialization in ventral visual cortex. Proc Natl Acad Sci U S A 101(35):13091–13095
65. Persson J, Lustig C, Nelson JK, Reuter-Lorenz PA (2007) Age differences in deactivation: a link to cognitive control? J Cogn Neurosci 19(6):1021–1032
66. Gutchess AH, Welsh RC, Hedden T, Bangert A, Minear M, Liu LL et al (2005) Aging and the neural correlates of successful picture encoding: frontal activations compensate for decreased medial-temporal activity. J Cogn Neurosci 17(1):84–96
67. Cabeza R, Grady CL, Nyberg L, McIntosh AR, Tulving E, Kapur S et al (1997) Age-related differences in neural activity during memory encoding and retrieval: a positron emission tomography study. J Neurosci 17(1):391–400
68. Grady CL, McIntosh AR, Rajah MN, Beig S, Craik FIM (1999) The effects of age on the neural correlates of episodic encoding. Cereb Cortex 9(8):805–814
69. Fera F, Weickert TW, Goldberg TE, Tessitore A, Hariri AR, Das S et al (2005) Neural mechanisms underlying probabilistic category learning in Normal aging. J Neurosci 25(49):11340–11348
70. Morcom AM, Good CD, Frackowiak RSJ, Rugg MD (2003) Age effects on the neural correlates of successful memory encoding. Brain 126(1):213–229
71. Reuterlorenz PA, Jonides J, Smith EE, Hartley AA, Miller A, Marshuetz C et al (2000) Age differences in the frontal lateralization of verbal and spatial working memory revealed by PET. J Cogn Neurosci 12(1):174–187
72. Rossi S, Miniussi C, Pasqualetti P, Babiloni C, Rossini PM, Cappa SF (2004) Age-related functional changes of prefrontal cortex in long-term memory: a repetitive transcranial magnetic stimulation study. J Neurosci 24(36):7939–7944
73. Kempermann G, Gast D, Gage FH (2002) Neuroplasticity in old age: sustained fivefold induction of hippocampal neurogenesis by long-term environmental enrichment. Ann Neurol 52(2):135–143
74. Maguire EA, Gadian DG, Johnsrude IS, Good CD, Ashburner J, Frackowiak RSJ et al (2000) Navigation-related structural change in the hippocampi of taxi drivers. Proc Natl Acad Sci U S A 97(8):4398–4403
75. Buckner RL, Snyder AZ, Shannon BJ, Larossa GN, Sachs R, Fotenos AF et al (2005) Molecular, structural, and functional characterization of Alzheimer's disease: evidence for a relationship between default activity, amyloid, and memory. J Neurosci 25(34):7709–7717
76. Lustig C, Snyder AZ, Bhakta MN, Obrien KC, Mcavoy MP, Raichle ME et al (2003) Functional deactivations: change with age and dementia of the Alzheimer type. Proc Natl Acad Sci U S A 100(24):14504–14509
77. Tsai S (2003) Brain-derived neurotrophic factor: a bridge between major depression and Alzheimer's disease? Med Hypotheses 61(1):110–113
78. Colcombe SJ, Kramer AF (2003) Fitness effects on the cognitive function of older adults: a meta-analytic study. Psychol Sci 14(2):125–130
79. Colcombe SJ, Kramer AF, Erickson KI, Scalf PE, Mcauley E, Cohen NJ et al (2004) Cardiovascular fitness, cortical plasticity, and aging. Proc Natl Acad Sci U S A 101(9):3316–3321
80. Colcombe SJ, Erickson KI, Scalf PE, Kim JS, Prakash RS, Mcauley E et al (2006) Aerobic exercise training increases brain volume in aging humans. J Gerontol Ser A Biol Sci Med Sci 61(11):1166–1170

81. Boyke J, Driemeyer J, Gaser C, Buchel C, May A (2008) Training-induced brain structure changes in the elderly. J Neurosci 28(28):7031–7035
82. Trevisan K, Cristina-Pereira R, Silva-Amaral D, Aversi-Ferreira TA (2019) Theories of aging and the prevalence of Alzheimer's disease. Biomed Res Int 2019:1–9
83. Curtis HJ (1963) Biological mechanisms underlying the aging process. Science 141(3582):686–694
84. Orgel LE (1963) The maintenance of the accuracy of protein synthesis and its relevance to ageing. Proc Natl Acad Sci U S A 49:517–521
85. Hart RW, Setlow RB (1974) Correlation between deoxyribonucleic acid excision-repair and life-span in a number of mammalian species. Proc Natl Acad Sci USA 71(6):2169–2173
86. Cristofalo VJ, Gerhard GS, Pignolo RJ (1994) Molecular biology of aging. Surg Clin North Am 74(1):1–21
87. Lee A, Cerami A (1990) Modifications of proteins and nucleic acids by reducing sugars: possible role in aging. In: Schneider EL, Rowe JW (eds) Handbook of the biology of aging. Academic Press, New York, pp 116–130
88. Cerami A (1985) Hypothesis. Glucose as a mediator of aging. J Am Geriatr Soc 33(9):626–634
89. Hayoz D, Ziegler T, Brunner HR, Ruiz J (1999) Diabetes mellitus and vascular lesions. Metab Clin Exp 47(1):16–19
90. Krause KH (2007) Aging: a revisited theory based on free radicals generated by NOX family NADPH oxidases. Exp Gerontol 42(4):256–262
91. Harman D (1991) The aging process: major risk factor for disease and death. Proc Natl Acad Sci USA 88(12):5360–5363
92. Teixeira INDAO, Guariento ME (2010) Biologia do envelhecimento: teorias, mecanismos e perspectivas. Ciência Saúde Coletiva 15(6):2845–2857
93. Hayflick L (1968) Human cells and aging. Sci Am 218(3):32–37
94. Curtsinger JW, Fukui HH, Khazaeli A, AW, K., & Tatar, M. (1995) Genetic variation and aging. Annu Rev Genet 29:553–575
95. Kurenova EV, Mason JM (1997) Telomere functions. A review. Biochemistry (Mosc) 62:1242–1253
96. Levin ER (2001) Cell localization, physiology, and nongenomic actions of estrogen receptors. J Appl Physiol 91(4):1860–1867
97. Finch CE, Rose MR (1995) Hormones and the physiological architecture of life history evolution. Q Rev Biol 70:1–52
98. Bourg EL (2014) Evolutionary theories of aging can explain why we age. Interdiscip Top Gerontol 39:8–23
99. Kirkwood TB (2008) Understanding ageing from an evolutionary perspective. J Intern Med 263:117–127
100. Cumming E, Henry WE (1962) The process of disengagement. J Nerv Ment Dis 134(4):377
101. Rose AM (1964) A current theoretical issue in social gerontology. Geron 4:46–50
102. Hasworth SB, Cannon ML (2015) Social theories of aging: a review. Dis Mon 61(11):475–479
103. Baumgart M, Snyder HM, Carrillo MC, Fazio S, Kim H, Johns H (2015) Summary of the evidence on modifiable risk factors for cognitive decline and dementia: a population-based perspective. Alzheimers Dement 11(6):718–726
104. Stern Y (2012) Cognitive reserve in ageing and Alzheimer's disease. Lancet Neurol 11(11):1006–1012
105. Katzman R, Terry R, DeTeresa R, Brown T, Davies P, Fuld P et al (1988) Clinical, pathological, and neurochemical changes in dementia: a subgroup with preserved mental status and numerous neocortical plaques. Ann Neurol 23(2):138–144
106. Tucker-Drob E (2019) Cognitive aging and dementia: a life-span perspective. Ann Rev Dev Psychol 1:177–196
107. Douaud G, Groves AR, Tamnes CK, Westlye LT, Duff EP, Engvig A et al (2014) A common brain network links development, aging, and vulnerability to disease. Proc Natl Acad Sci 111(49):17648
108. Fjell AM, McEvoy L, Holland D, Dale AM, Walhovd KB, Alzheimer's Disease Neuroimaging Initiative (2014) What is normal in normal aging? Effects of aging, amyloid and Alzheimer's disease on the cerebral cortex and the hippocampus. Prog Neurobiol 117:20–40
109. Fjell AM, Westlye LT, Grydeland H, Amlien I, Espeseth T, Reinvang I et al (2013) Critical ages in the life course of the adult brain: nonlinear subcortical aging. Neurobiol Aging 34(10):2239–2247
110. Hasan KM, Mwangi B, Cao B, Keser Z, Tustison NJ, Kochunov P et al (2016) Entorhinal cortex thickness across the human lifespan. J Neuroimaging 26(3):278–282
111. Fyhn M, Molden S, Witter MP, Moser EI, Moser MB (2004) Spatial representation in the entorhinal cortex. Science 305(5688):1258–1264
112. Fjell AM, Westlye LT, Grydeland H, Amlien I, Espeseth T, Reinvang I et al (2014) Accelerating cortical thinning: unique to dementia or universal in aging? Cereb Cortex 24(4):919–934
113. Lebel C, Beaulieu C (2011) Longitudinal development of human brain wiring continues from childhood into adulthood. J Neurosci 31(30):10937
114. Zhang H, Lee A, Qiu A (2017) A posterior-to-anterior shift of brain functional dynamics in aging. Brain Struct Funct 222(8):3665–3676
115. Grady CL, Maisog JM, Horwitz B, Ungerleider LG, Mentis MJ, Salerno JA et al (1994) Age-related changes in cortical blood flow activation during visual processing of faces and location. J Neurosci 14(3):1450
116. Ford JH, Kensinger EA (2017) Age-related reversals in neural recruitment across memory retrieval phases. J Neurosci 37(20):5172
117. Cabeza R (2002) Hemispheric asymmetry reduction in older adults: the HAROLD model. Psychol Aging 17(1):85–100

118. Mattay VS, Fera F, Tessitore A, Hariri AR, Das S, Callicott JH et al (2002) Neurophysiological correlates of age-related changes in human motor function. Neurology 58(4):630
119. Cabeza R, Anderson ND, Locantore JK, McIntosh AR (2002) Aging gracefully: compensatory brain activity in high-performing older adults. NeuroImage 17(3):1394–1402
120. Hyodo K, Dan I, Kyutoku Y, Suwabe K, Byun K, Ochi G et al (2016) The association between aerobic fitness and cognitive function in older men mediated by frontal lateralization. NeuroImage 125:291–300
121. Cabeza R, Albert M, Belleville S, Craik FIM, Duarte A, Grady CL et al (2018) Maintenance, reserve and compensation: the cognitive neuroscience of healthy ageing. Nat Rev Neurosci 19(11):701–710
122. Cappell KA, Gmeindl L, Reuter-Lorenz PA (2010) Age differences in prefrontal recruitment during verbal working memory maintenance depend on memory load. Cortex 46(4):462–473
123. Tang Y, Xing Y, Zhu Z, He Y, Li F, Yang J et al (2019) The effects of 7-week cognitive training in patients with vascular cognitive impairment, no dementia (the cog-VACCINE study): a randomized controlled trial. Alzheimers Dement 15(5):605–614
124. Hill NT, Mowszowski L, Naismith SL, Chadwick VL, Valenzuela M, Lampit A (2017) Computerized cognitive training in older adults with mild cognitive impairment or dementia: a systematic review and meta-analysis. Am J Psychiatry 174(4):329–340

Part III

Sociological Studies of Cognitive Aging

Emotional and Affective Disorders in Cognitive Aging

5

Jialing Fan, Chao Du, and Xin Li

Abstract

The increased aging population who have aging-related diseases poses a serious challenge to health services, including mental health services. Due to the changes of body, brain, living environment, and lifestyle, the elderly will have different psychological changes from other age stages, some of which would develop into mental disorders, and affect the cognition of the elderly in turn. This elderly mental health condition has drawn wide attention from scientists. This chapter introduces the two most common emotional and affective disorders, late-life depression and anxiety, and focuses on their epidemiology and impact on the elderly. Furthermore, this chapter also reviews the effects of these two disorders on cognitive function and cognitive impairment in the elderly, and tries to explain the underlying mechanism of this effect from the perspective of related disease, cerebral circuit, and molecular biology.

J. Fan · C. Du · X. Li (✉)
State Key Laboratory of Cognitive Neuroscience and Learning, Faculty of Psychology, Beijing Normal University, Beijing, China

Beijing Aging Brain Rejuvenation Initiative (BABRI) Centre, Beijing Normal University, Beijing, China
e-mail: jialingfan@mail.bnu.edu.cn; 201931061028@mail.bnu.edu.cn; lixin99@bnu.edu.cn

Keywords

Late-life depression · Anxiety · Cognition · Underlying mechanism

With the development of science and technology, the average life expectancy has increased, inducing worldwide population aging. The increased aging population, who have aging-related diseases, poses a serious challenge to health services, including mental health services. This elderly mental health condition has drawn wide attention from scientists. Late-life depression and anxiety are the most common emotional and affective disorders and influence cognitive function, physical health, and quality of life. In this chapter, we introduce emotional function in elderly individuals and focus on emotional and affective disorders and their effects on cognitive aging.

5.1 Emotional and Affective Disorders in the Elderly

5.1.1 Emotional Function in the Elderly

Despite the decline in physical and cognitive functions, emotional function seems to be stable or even enhanced in old age. Compared to young and midlife adults, depression and anxiety rates

Z. Zhang (ed.), *Cognitive Aging and Brain Health*, Advances in Experimental Medicine and Biology 1419, https://doi.org/10.1007/978-981-99-1627-6_5

are lower in older people [1]. In general, aging is associated with more positive overall emotional well-being, better emotional stability, and more emotional complexity [2]. Negative emotions, including anger, sadness, and fear, generally decrease with age.

Elderly individuals may tend to be more sensitive and remember positive relative information better than negative information. Charles and colleagues presented participants with positive, negative, and neutral images and asked them to recall as many images as they could. The older adults recalled fewer images in general than young adults, but they recalled a greater number of positive images than negative images. There were no significant differences between the positive and negative images that younger adults recalled. Thus, with aging, compared to the recalled neutral and positive images, the recalled number of negative images decreased [3]. In addition, these age-related shifts in the ratio of recalled or attended positive to negative materials (e.g., images from the experiment mentioned above) can potentially improve the longer-term well-being and momentary mood of elderly individuals [4].

In addition to the information itself, the way of recognizing and reacting to the information is also very important in emotional function. Compared to younger adults, the elderly may also be good at emotional reactivity and regulation. For example, elderly individuals' daily anger experiences decreased with age [5]. Additionally, when elderly adults were required to regulate their emotions in the mood induction response, they performed equally well or even better than younger adults who followed the instructions [6, 7]. Furthermore, older adults appear superior to younger adults in strategies used in emotional regulation. From a social perspective, older adults reported less distress in interpersonal conflicts than younger adults. Meanwhile, elderly adults use passive strategies more often to avoid conflict involvement and select more efficient strategies to solve interpersonal problems than younger adults [8–10]. This result indicates that using different strategies can also help explain why daily stress levels generally decrease with age [11] and why affective well-being improves in elderly individuals.

5.1.2 Common Emotional and Affective Disorders in the Elderly

Although, in general, there is no emotional function decline in normal aging, severe abnormal emotional function and affective disorders can still result in accelerated cognitive decline, social isolation, frailty, and increased illness severity. In the following section, we introduce common affective disorders in elderly individuals: late-life depression and anxiety.

5.1.2.1 Late-Life Depression

Depression includes symptoms of persistent sadness and loss of interests, which impacts the lives of individuals from the level of emotions, cognition, and behavior. It may also lead to a variety of physical problems. Depression diagnosis procedures often follow the Diagnostic and Statistical Manual of Mental Disorders, fifth edition (DSM-5). If an individual experiences a depressed mood or was unable to enjoy or take an interest in activities, along with the other four related symptoms for most of the day for 2 weeks, they could be diagnosed with a major depressive episode. The other symptoms of the major depressive episode include (a) change in appetite or weight, (b) insomnia or hypersomnia, (c) psychomotor agitation or retardation, (d) fatigue or loss of energy, (e) feelings of worthlessness or excessive or inappropriate guilt, (f) poor concentration, and (g) recurrent thoughts of death and recurrent suicidal ideation. When one or more major depressive episodes occur that are not attributable to medical or substance-induced conditions and cannot be better explained by a psychotic disorder without mania or hypomania episodes, the individual can be diagnosed with major depression disorder (MDD) [12].

Depression is not a part of normal aging. As mentioned above, aging is associated with a reduced prevalence of depression [1]; however, clinical depression can have serious consequences. Two forms of depression are included in late-life depression. One is late-onset depression, which is defined as depression that occurs in late life for the first time (in patients approximately 60 years or older). The other is depressive patients diagnosed before reaching elderly age.

Approximately half or more geriatric patients with depression have late-onset depression. The prevalence of MDD varies among different studies and populations, ranging from 7.2 to 38% in the elderly [13, 14]. Rates of depression are modulated by gender, with a higher prevalence found in female elderly individuals than in males. Furthermore, the depression prevalence rate of older adults is substantially higher in those who live in medical institutions and long-term care facilities, while prevalence in the community is lower. Risk factors leading to the development of late-life depression are likely to comprise complex interactions with certain vulnerabilities, including genetic factors, cognitive diathesis, age-associated neurobiological changes, physical illness and disability, and stressful events, such as bereavement and caregiving [15, 16].

There are age differences in the presentation of depression symptoms, i.e., older adults may present depressive symptoms differently than younger adults. Compared to younger or middle-aged adults who are diagnosed with depression, symptoms such as sleep disturbance, fatigue, psychomotor retardation, loss of interest in living, and hopelessness about the future tend to occur more frequently in late-life depression [17]. Similar results have also been reported by Variend and colleagues when reviewing clinical differences between early- and late-onset symptoms of depression. The results indicate that older individuals present the following symptoms more frequently: anhedonia (an inability to experience pleasure from activities normally found enjoyable), cognitive deficits, psychomotor retardation, and weight loss [18]. Cognitive impairment is very common in late-life depression. For instance, poor memory and concentration, slower cognitive processing speed exhibition, and executive dysfunction were frequently reported in depressed elderly individuals [19]. (For details regarding the effects of depression on cognitive aging, see the second section of Chap. 5.)

Depression is also associated with many medical conditions, including heart disease, arthritis, asthma, back pain, chronic pulmonary disease, migraine, cardiovascular and cerebrovascular diseases, dementia, and cancer. The comorbidity rate of anxiety and depression is also high. As measured by the Canadian Community Health Study, the lifetime prevalence of comorbidity rate is 39.2% [20]. The co-occurrences of depression and other diseases can cause heavy burdens to individual health and their caregivers. Additionally, depression is an affective disorder associated with suicide. It has been estimated that approximately 15% of patients with MDD commit suicide. Additionally, those who experience late-onset depression are at higher risk [21], and older adults over the age of 65 have the highest suicide rate compared to other age groups [14]. Comorbidity with other disorders, such as anxiety and agitation, and rapid changes in depressive status also increase the risk of suicide. Thankfully, studies have proven that identifying and treating depression can effectively prevent suicide among elderly individuals. Considering the aforementioned findings, late-life depression should be given more attention.

5.1.2.2 Anxiety Disorder

Anxiety disorders display as uncontrolled feelings of apprehension and persistent worries. Normally, experiencing anxiety, worry, and stress does not mean that one has anxiety disorder. These symptoms evolve into a mental disorder only when the symptoms become excessive and persistent, such as anxiety responses or panic attacks, eliciting avoidance behavior that impairs our daily life functioning. Generalized anxiety disorder, panic disorder, obsessive-compulsive disorder, post-traumatic stress disorder, social anxiety disorder, and specific phobias are all included in the anxiety disorder category [12]. The physical symptoms of anxiety include palpitations, shortness of breath, dry mouth, trembling, sweating, gastrointestinal discomfort, diarrhea, muscle tension, and blushing. In some types of anxiety disorder, dizziness, fainting, sleep disturbance, and concentration difficulties are also very common [22].

As mentioned above, depression and anxiety often coexist, and longitudinal studies have shown that anxiety more often precedes depression if comorbid anxiety and depression occur [23]. Worry, poor concentration and cognitive function, sleep disturbance, feelings of guilt, hopelessness, and a sense of worthlessness are negative impacts frequently observed in both anxiety and depression [22].

In general, gender modulates the prevalence of anxiety disorders, with females being more susceptible to anxiety disorders. However, this gender difference tends to diminish with increasing age. Additionally, the prevalence of anxiety disorders generally decreases with age [24]. The prevalence of anxiety disorders in elderly individuals varies among different studies and populations. For example, Andreas and colleagues investigated the prevalence of anxiety disorder in a European sample of 3142 people aged 65 to 84 years old, and the prevalence of lifetime anxiety disorder was 25.6%, 12-month anxiety was 17.2%, and current anxiety disorder was 11.4% [25]. However, in the United States, a study in a sample of 2575 individuals aged 55 years and older reported a 12-month prevalence of 11.6% [26]. Furthermore, in a study with an Australian sample aged 65 years and older, the 12-month prevalence decreased to 4.4% [27].

Anxiety disorder can be caused by a complex combination of factors, including genetic factors, childhood and personality development, and social and environmental factors. Abuse and trauma are also factors that cause anxiety disorder, while major losses are considered the foremost causes. In the elderly, the symptoms of anxiety disorder are similar to younger patterns, but with dominant somatic symptoms and complaints. Some topics seem to need more attention, which are fear of death and disease and displacement phobias related to the fear of falling or being attacked [28].

Apart from depression, anxiety is also associated with other medical conditions and disorders, including substance abuse disorders, psychosis, cognitive impairment, and neurodegenerative disease (for details regarding to the effects of anxiety in cognitive aging, see the second section of Chap. 5.). There is also an association between anxiety disorders and suicide, similar to depression. For instance, when examining the relationship between anxiety disorders and suicide attempts, Nepon and colleagues found that among individuals who reported a lifetime history of suicide attempts, > 70% had a history of anxiety disorder. Additionally, suicide risk varies in different types of anxiety disorders [29]. Similar results were also found by Sareen et al., in which individuals with post-traumatic stress disorder were significantly associated with suicidal ideation and attempts [30]. For older adults, anxiety disorders occurred in one of six patients who died due to suicide in the United Kingdom. Compared to younger patients who died from suicide, physical decline and social isolation played a more prominent role in elderly individuals who committed suicide [31].

5.2 Effects of Emotional and Affective Disorders on Cognitive Aging

According to aforementioned information, we realize that emotional and affective disorders negatively impact the daily life and body health of elderly individuals (e.g., sleep disturbance, fatigue, and weight loss). The following section will focus on the impacts of affective disorders (i.e., late-life depression and anxiety) on cognitive aging, including physiological and pathological impacts and the possible underlying mechanism of these effects.

People have noted the relationships between emotion and cognition, especially in the elderly. Research findings from the 1990s reported that both anxiety and late-life depression occur as a reaction to the onset of cognitive decline rather than as a cause of cognitive decline [32]. Late-life depression has been repeatedly found to be a predictor of preclinical cognitive decline [32, 33]. Moreover, an increasing number of recent studies have found that late-life depression and anxiety in the elderly are likely to cause cognitive decline, even MCI (mild cognitive impairment) and dementia. Some theoretical and empirical evidence may explain the mechanism of this phenomenon [34].

5.2.1 Effects of Common Emotional and Affective Disorders on Cognitive Aging in the Elderly

5.2.1.1 Late-Life Depression

Considering the relatively high prevalence of late-life depression in older adults, many depressed older adults often complain about cognitive symptoms. Therefore, studies have begun to probe the complex relationship between late-life depression and cognitive impairment. After decades of research, studies have found that older adults with depressive symptoms or clinical depression have an increased risk of cognitive decline, MCI, and dementia [35–37].

A negative correlation between late-life depression and cognitive function has been previously reported in clinical studies in both younger and elderly individuals [38–41]. Similarly, cross-sectional studies with community samples also reported a negative correlation between late-life depression and cognitive function in old age [42–44] and indicate that late-life depression is associated with more precipitous cognitive decline over time [36, 45]. This impacts various cognitive functions in the elderly, such as memory, language, attention, and executive function [36].

Late-life depression not only affects the cognitive function of healthy elderly individuals but also seems to affect pathological cognitive aging, such as MCI and dementia. The prevalence of late-life depression in the range of MCI is very broad. The population-based prevalence ranges from 3 to 83%, with a median prevalence of 44.3% in hospital-based settings and 15.7% in population-based settings. Conversely, the prevalence of MCI in late-life depression ranges from approximately 30% to 50%, and many studies have found that depression predicts the occurrence of MCI [46, 47]. According to a review from Byers and Yaffe, several studies found that late-life depression is strongly associated with dementia [34]. Meanwhile, late-life depression could be a risk factor for predicting the progression of dementia [34, 37]. In fact, depressive symptoms are common in dementia patients, with a prevalence of approximately 30% [48]. More importantly, several researchers found that depression at the baseline level was associated with an increased risk of incident dementia, including the increased precipitation of certain biomarkers of dementia [49].

5.2.1.2 Anxiety Disorder

Several studies suggest not only a negative association between anxiety and cognitive performance in elderly individuals [42, 44, 50] but also that they covary over time [51]. Additionally, anxiety is negatively associated with cognitive performance. Akin to late-life depression, almost all aspects of cognitive function are negatively affected by anxiety.

Similar to late-life depression affecting the elderly, anxiety also has a negative effect on elderly individuals' cognitive function [44]. However, some comparative studies have found that anxiety and late-life depression may affect different elderly people in different ways. Depression may be characteristic of neuropathological aging, whereas anxiety may be more common in normal aging [52].

A recent review found that the prevalence rate of anxiety in MCI was 31.2% in hospital-based settings and 14.3% in population-based settings [53], which is slightly lower than the prevalence rate of depression mentioned above. Researchers have found that late-life depression and anxiety often occur as comorbidities in abnormal cognitive aging, which means the two disorders are frequently studied together. These trends have led some studies to compare the common effects and differences between late-life depression and anxiety on pathological cognitive aging. A previous study showed that the inverse association between anxiety and reduced cognitive performance was explained by adjustment for comorbid depression [54]. Additionally, several studies found that elderly individuals diagnosed with Alzheimer's Disease (AD) reported less anxiety, and anxiety increased the probability of dementia in the future [55]. Therefore, the effects of anxiety on pathological cognitive aging may not be as severe as depression. Additionally, part of the impact comes from the comorbidity of depression.

5.2.2 Possible Underlying Mechanism of the Effects of Emotional Disorders on Cognitive Aging

Although the underlying mechanism of how emotional disorders affect cognitive aging has been studied for decades, an explanation has still not been revealed. Research studies have proposed a series of hypotheses and theories attempting to explain this phenomenon, but most of them are still controversial and need to be further verified.

According to a review published in Nature Reviews Neurology [34], vascular disease, glucocorticoid steroid levels, hippocampal atrophy, increased deposition of amyloid-β plaques, inflammatory changes, and deficits of nerve growth factors or neurotrophins are likely to underlie depression and dementia. Dementia is a late stage of pathological development of cognitive aging, and the effects of emotional disorders on cognitive aging include normal, MCI, and dementia stages. Moreover, anxiety is often associated with depression and has a similar effect on cognition. Therefore, we will focus on the possible underlying mechanism of the effect of emotional disorders on cognitive aging primarily from the perspective of neurological review.

5.2.2.1 Vascular Disease

Late-life depression is related to subsequent vascular disease through multiple proposed mechanisms, primarily including hypothalamic–pituitary–adrenal (HPA)-axis dysregulation and elevated cortisol levels, which are associated with metabolic syndrome, disruption of normal endothelial function and development of hypertension, and proinflammatory cytokines [56, 57]. As vascular disease has been shown to contribute to the clinical manifestation of dementia symptoms, ischemic damage, largely in the frontostriatal brain regions, may lead to substantial cognitive deficits [57].

5.2.2.2 Cortisol–Hippocampal Pathway

In this pathway, late-life depression and depressive symptoms activate the HPA axis and increase glucocorticoid production, which may in turn damage the hippocampus. Some studies had found that loss of hippocampal volume increases the risk of cognitive decline [58], even dementia [59] in adults with late-life depression. However, the relationship between late-life depression and hippocampal atrophy is confirmed, and evidence showed that this link is mediated by increased cortisol levels is inconsistent. Thus, further research on elucidating these pathways is necessary.

5.2.2.3 Amyloid-β Plaque Formation

One hypothesized pathway links depression and Aβ together. This entails increased Aβ production initiated by a stress response associated with depression and glucocorticoid levels [60]. In addition, depression might influence Aβ accumulation through direct interaction with amyloidogenic processing early in AD [61].

5.2.2.4 Inflammatory Changes

Two potential pathways involved in inflammatory changes within the central nervous system (CNS) connect depression and dementia. The first pathway is increased level of cytokines. For example, increased interleukin (IL)-6 and tumor necrosis factor (TNF) in depression patients may lead to a decrease in anti-inflammatory and immunosuppressant regulation, an increase in proinflammatory changes within the CNS and, ultimately, cognitive deficits and dementia [62]. Second, proinflammatory cytokines, such as IL-6 and TNF, interfere with 5-hydroxytryptamine metabolism and reduce both synaptic plasticity and hippocampal neurogenesis [60].

5.2.2.5 Nerve Growth Factors

Impairment in brain-derived neurotrophic factor (BDNF) signaling has been detected in stress-induced animal models of depression and in individuals with depression, as well as in patients with AD [63]. In fact, studies of both depressed and AD patients have demonstrated decreased messenger RNA levels of BDNF in the hippocampus [64].

Taken together, we find that emotional disorders may affect the cognition of the elderly in several ways. First, the nervous system response to emotional distress may activate the HPA axis and decrease messenger RNA levels of BDNF, resulting in increased plasma levels of glucocorticoids, increased incidence of vascular diseases, proinflammatory changes, and increased levels of amyloid plaques. These changes at the molecular level of the brain eventually lead to the atrophy and death of neurons, which is reflected in the reduction of brain region volumes, such as the hippocampus and frontal lobe. Finally, the physiological and pathological effects of emotional disorders on the cognition of the elderly are reflected by the decline of cognitive behavior.

In summary, this chapter discusses the definition, features, measurement methods, and influencing factors of some common emotional diseases (primarily depression and anxiety) in the process of cognitive aging, as well as their effects and possible mechanisms in the physiological and pathological process of cognitive aging. However, many issues regarding the effect of emotional disorders on cognitive aging have not yet to be answered, especially in the interpretation of the underlying mechanism. Therefore, research should further explore these issues in the future.

References

1. Scott KM, Von Korff M, Alonso J, Angermeyer M, Bromet EJ, Bruffaerts R et al (2008) Age patterns in the prevalence of DSM-IV depressive/anxiety disorders with and without physical co-morbidity. Psychol Med 38(11):1659–1669
2. Carstensen LL, Turan B, Scheibe S, Ram N, Ersner-Hershfield H, Samanez-Larkin GR et al (2011) Emotional experience improves with age: evidence based on over 10 years of experience sampling. Psychol Aging 26(1):21–33
3. Charles ST, Mather M, Carstensen LL (2003) Aging and emotional memory: the forgettable nature of negative images for older adults. J Exp Psychol Gen 132(2):310–324
4. Scheibe S, Carstensen LL (2010) Emotional aging: recent findings and future trends. J Gerontol B Psychol Sci Soc Sci 65B(2):135–144
5. Phillips LH, Henry JD, Hosie JA, Milne AB (2006) Age, anger regulation and well-being. Aging Ment Health 10(3):250–256
6. Kunzmann U, Kupperbusch CS, Levenson RW (2005) Behavioral inhibition and amplification during emotional arousal: a comparison of two age groups. Psychol Aging 20(1):144–158
7. Phillips LH, Henry JD, Hosie JA, Milne AB (2008) Effective regulation of the experience and expression of negative affect in old age. J Gerontol B Psychol Sci Soc Sci 63(3):P138–P145
8. Charles ST, Carstensen LL (2008) Unpleasant situations elicit different emotional responses in younger and older adults. Psychol Aging 23(3):495–504
9. Charles ST, Piazza JR, Luong G, Almeida DM (2009) Now you see it, now you don't: age differences in affective reactivity to social tensions. Psychol Aging 24(3):645–653
10. Birditt KS, Fingerman KL, Almeida DM (2005) Age differences in exposure and reactions to interpersonal tensions: a daily diary study. Psychol Aging 20(2):330–340
11. Stawski RS, Sliwinski MJ, Almeida DM, Smyth JM (2008) Reported exposure and emotional reactivity to daily stressors: the roles of adult age and global perceived stress. Psychol Aging 23(1):52–61
12. American Psychiatric Association (2013) Diagnostic and statistical manual of mental disorders (DSM-5®). American Psychiatric Association Publishing
13. Valiengo Lda C, Stella F, Forlenza OV (2016) Mood disorders in the elderly: prevalence, functional impact, and management challenges. Neuropsychiatr Dis Treat 12:2105–2114
14. Dines P, Hu W, Sajatovic M (2014) Depression in later-life: an overview of assessment and management. Psychiatr Danub 26(Suppl 1):78–84
15. Fiske A, Wetherell JL, Gatz M (2009) Depression in older adults. Annu Rev Clin Psychol 5:363–389
16. Cole MG, Dendukuri N (2003) Risk factors for depression among elderly community subjects: a systematic review and meta-analysis. Am J Psychiatr 160(6):1147–1156
17. Christensen H, Jorm AF, Mackinnon AJ, Korten AE, Jacomb PA, Henderson AS et al (1999) Age differences in depression and anxiety symptoms: a structural equation modelling analysis of data from a general population sample. Psychol Med 29(2):325–339

18. Variend H, Gopal YV (2008) Late-onset depression: issues affecting clinical care. Adv Psychiatr Treat 14(2):152–158
19. Butters MA, Whyte EM, Nebes RD, Begley AE, Dew MA, Mulsant BH et al (2004) The nature and determinants of neuropsychological functioning in late-life depression. Arch Gen Psychiatry 61(6):587–595
20. Patten SB, Williams JV, Lavorato DH, Wang JL, McDonald K, Bulloch AG (2015) Descriptive epidemiology of major depressive disorder in Canada in 2012. Can J Psychiatry 60(1):23–30
21. Angst J, Angst F, Stassen HH (1999) Suicide risk in patients with major depressive disorder. J Clin Psychiatry 60(Suppl 2):57–62; discussion 75–56, 113–116
22. O'Donnell RM, Kaszniak AW (2011) Charting late-life affective disorders. Generations 35(2):46–57
23. Byrne GJ (2017) Anxiety in late life. Focus (Am Psychiatr Publ) 15(2):157–161
24. Krasucki C, Howard R, Mann A (1998) The relationship between anxiety disorders and age. Int J Geriatr Psychiatry 13(2):79–99
25. Andreas S, Schulz H, Volkert J, Dehoust M, Sehner S, Suling A et al (2017) Prevalence of mental disorders in elderly people: the European MentDis_ICF65+ study. Br J Psychiatry 210(2):125–131
26. Byers AL, Yaffe K, Covinsky KE, Friedman MB, Bruce ML (2010) High occurrence of mood and anxiety disorders among older adults: the National Comorbidity Survey Replication. Arch Gen Psychiatry 67(5):489–496
27. Trollor JN, Anderson TM, Sachdev PS, Brodaty H, Andrews G (2007) Prevalence of mental disorders in the elderly: the Australian National Mental Health and Well-being survey. Am J Geriatr Psychiatry 15(6):455–466
28. Rangaraj J, Pelissolo A (2006) Anxiety disorders in the elderly: clinical and therapeutic aspects. Psychol Neuropsychiatr Vieil 4(3):179–187
29. Nepon J, Belik SL, Bolton J, Sareen J (2010) The relationship between anxiety disorders and suicide attempts: findings from the National Epidemiologic Survey on alcohol and related conditions. Depress Anxiety 27(9):791–798
30. Sareen J, Houlahan T, Cox BJ, Asmundson GJ (2005) Anxiety disorders associated with suicidal ideation and suicide attempts in the National Comorbidity Survey. J Nerv Ment Dis 193(7):450–454
31. Voshaar RC, van der Veen DC, Kapur N, Hunt I, Williams A, Pachana NA (2015) Suicide in patients suffering from late-life anxiety disorders: a comparison with younger patients. Int Psychogeriatr 27(7):1197–1205
32. Devanand DP, Sano M, Tang MX, Taylor S, Gurland BJ, Wilder D et al (1996) Depressed mood and the incidence of Alzheimer's disease in the elderly living in the community. Arch Gen Psychiatry 53(2):175–182
33. Alexopoulos GS, Meyers BS, Young RC, Mattis S, Kakuma T (1993) The course of geriatric depression with "reversible dementia": a controlled study. Am J Psychiatry 150(11):1693–1699
34. Byers AL, Yaffe K (2011) Depression and risk of developing dementia. Nat Rev Neurol 7(6):323–331
35. Verdelho A, Madureira S, Moleiro C, Ferro JM, O'Brien JT, Poggesi A et al (2013) Depressive symptoms predict cognitive decline and dementia in older people independently of cerebral white matter changes: the LADIS study. J Neurol Neurosurg Psychiatry 84(11):1250–1254
36. Bielak AA, Gerstorf D, Kiely KM, Anstey KJ, Luszcz M (2011) Depressive symptoms predict decline in perceptual speed in older adulthood. Psychol Aging 26(3):576–583
37. Steenland K, Karnes C, Seals R, Carnevale C, Hermida A, Levey A (2012) Late-life depression as a risk factor for mild cognitive impairment or Alzheimer's disease in 30 US Alzheimer's disease centers. J Alzheimers Dis 31(2):265–275
38. Lyness SA, Eaton EM, Schneider LS (1994) Cognitive performance in older and middle-aged depressed outpatients and controls. J Gerontol 49(3):P129–P136
39. Beats BC, Sahakian BJ, Levy R (1996) Cognitive performance in tests sensitive to frontal lobe dysfunction in the elderly depressed. Psychol Med 26(3):591–603
40. Elliott R, Sahakian BJ, McKay AP, Herrod JJ, Robbins TW, Paykel ES (1996) Neuropsychological impairments in unipolar depression: the influence of perceived failure on subsequent performance. Psychol Med 26(5):975–989
41. Kindermann SS, Brown GG (1997) Depression and memory in the elderly: a meta-analysis. J Clin Exp Neuropsychol 19(5):625–642
42. Bunce D, Handley R, Gaines SO Jr (2008) Depression, anxiety, and within-person variability in adults aged 18 to 85 years. Psychol Aging 23(4):848–858
43. Elderkin-Thompson V, Mintz J, Haroon E, Lavretsky H, Kumar A (2006) Executive dysfunction and memory in older patients with major and minor depression. Arch Clin Neuropsychol 21(7):669–676
44. Reppermund S, Brodaty H, Crawford JD, Kochan NA, Slavin MJ, Trollor JN et al (2011) The relationship of current depressive symptoms and past depression with cognitive impairment and instrumental activities of daily living in an elderly population: the Sydney memory and ageing study. J Psychiatr Res 45(12):1600–1607
45. Kohler S, van Boxtel MP, van Os J, Thomas AJ, O'Brien JT, Jolles J et al (2010) Depressive symptoms and cognitive decline in community-dwelling older adults. J Am Geriatr Soc 58(5):873–879
46. Bhalla RK, Butters MA, Becker JT, Houck PR, Snitz BE, Lopez OL et al (2009) Patterns of mild cognitive impairment after treatment of depression in the elderly. Am J Geriatr Psychiatry 17(4):308–316
47. Reinlieb M, Ercoli LM, Siddarth P, St Cyr N, Lavretsky H (2014) The patterns of cognitive and functional impairment in amnestic and non-amnestic mild cognitive impairment in geriatric depression. Am J Geriatr Psychiatry 22(12):1487–1495

48. Lyketsos CG, Lopez O, Jones B, Fitzpatrick AL, Breitner J, DeKosky S (2002) Prevalence of neuropsychiatric symptoms in dementia and mild cognitive impairment: results from the cardiovascular health study. JAMA 288(12):1475–1483
49. Richard E, Reitz C, Honig LH, Schupf N, Tang MX, Manly JJ et al (2013) Late-life depression, mild cognitive impairment, and dementia. JAMA Neurol 70(3):374–382
50. Stawski RS, Sliwinski MJ, Smyth JM (2006) Stress-related cognitive interference predicts cognitive function in old age. Psychol Aging 21(3):535–544
51. Sliwinski MJ, Smyth JM, Hofer SM, Stawski RS (2006) Intraindividual coupling of daily stress and cognition. Psychol Aging 21(3):545–557
52. Bunce D, Batterham PJ, Mackinnon AJ, Christensen H (2012) Depression, anxiety and cognition in community-dwelling adults aged 70 years and over. J Psychiatr Res 46(12):1662–1666
53. Chen C, Hu Z, Jiang Z, Zhou F (2018) Prevalence of anxiety in patients with mild cognitive impairment: a systematic review and meta-analysis. J Affect Disord 236:211–221
54. Biringer, E., Mykletun, A., Dahl, A. A., Smith, A. D., Engedal, K., Nygaard, H. A., et al. (2005). The association between depression, anxiety, and cognitive function in the elderly general population--the Hordaland Health Study. Int J Geriatr Psychiatry, 20(10), 989–997
55. Bierman EJ, Comijs HC, Jonker C, Beekman AT (2007) Symptoms of anxiety and depression in the course of cognitive decline. Dement Geriatr Cogn Disord 24(3):213–219
56. Butters MA, Young JB, Lopez O, Aizenstein HJ, Mulsant BH, Reynolds CF 3rd et al (2008) Pathways linking late-life depression to persistent cognitive impairment and dementia. Dialogues Clin Neurosci 10(3):345–357
57. Alexopoulos GS (2006) The vascular depression hypothesis: 10 years later. Biol Psychiatry 60(12):1304–1305
58. Hickie I, Naismith S, Ward PB, Turner K, Scott E, Mitchell P et al (2005) Reduced hippocampal volumes and memory loss in patients with early- and late-onset depression. Br J Psychiatry 186:197–202
59. Steffens DC, Payne ME, Greenberg DL, Byrum CE, Welsh-Bohmer KA, Wagner HR et al (2002) Hippocampal volume and incident dementia in geriatric depression. Am J Geriatr Psychiatry 10(1):62–71
60. Caraci F, Copani A, Nicoletti F, Drago F (2010) Depression and Alzheimer's disease: neurobiological links and common pharmacological targets. Eur J Pharmacol 626(1):64–71
61. Sun X, Steffens DC, Au R, Folstein M, Summergrad P, Yee J et al (2008) Amyloid-associated depression: a prodromal depression of Alzheimer disease? Arch Gen Psychiatry 65(5):542–550
62. Sorrells SF, Sapolsky RM (2007) An inflammatory review of glucocorticoid actions in the CNS. Brain Behav Immun 21(3):259–272
63. Murer MG, Yan Q, Raisman-Vozari R (2001) Brain-derived neurotrophic factor in the control human brain, and in Alzheimer's disease and Parkinson's disease. Prog Neurobiol 63(1):71–124
64. Karege F, Vaudan G, Schwald M, Perroud N, La Harpe R (2005) Neurotrophin levels in postmortem brains of suicide victims and the effects of antemortem diagnosis and psychotropic drugs. Brain Res Mol Brain Res 136(1–2):29

The Mechanism of Socioeconomic Status Effects on Cognition

6

Chen Liu and Xin Li

Abstract

Socioeconomic status (SES) is a measurement of the sociological and economic statuses of individuals compared to others within the social and economic hierarchies. The common indicators of SES are income, education, and occupation statuses. Recently, researchers have used mixed measurements of SES, such as the MacArthur Scale. Numerous researches have proven the influence of SES on human development. Individuals who are less educated, have lower job status, and earn less or no income are at greater risk of poor health than their higher SES counterparts. SES has also been proven to influence life satisfaction, academic achievement, emotion regulation, cognitive function, and decision-making tendencies. SES has life span influence, which correlates with the level of cognition, rate of cognitive decline, and incidence of Alzheimer's disease among elderly individuals. Besides the individual level of SES, neighborhood SES can also affect cognitive function as an environmental factor. Low-SES individuals exhibit hypoactivation of the executive network and hyperactivation of the reward network, indicating low-SES individuals tend to focus more on monetary issues, while neglecting other non-monetary issues, which is consistent with the scarcity hypothesis.

Keywords

Socioeconomic status (SES) · Development · Cognition · Environmental factors · Scarcity hypothesis

6.1 History and Definition

6.1.1 The Origin and Development of Socioeconomic Status

Social scientists and psychologists have shown great interest in socioeconomic status (SES) and its influence on life span development. However, there has never been a complete consensus on precise representations and containment [1]. Brown-Iannuzzi et al. [2] agreed that socioeconomic status is a measurement of the sociological and economic statuses of individuals compared to others within the social and economic hierarchies. Some researchers believe that SES represents a social class (or economic position), and others believe that SES represents social status

C. Liu · X. Li (✉)
State Key Laboratory of Cognitive Neuroscience and Learning, Faculty of Psychology, Beijing Normal University, Beijing, China

Beijing Aging Brain Rejuvenation Initiative (BABRI) Centre, Beijing Normal University, Beijing, China
e-mail: c_liu@mail.bnu.edu.cn; lixin99@bnu.edu.cn

Z. Zhang (ed.), *Cognitive Aging and Brain Health*, Advances in Experimental Medicine and Biology 1419, https://doi.org/10.1007/978-981-99-1627-6_6

(or prestige). Perhaps the current meaning that social scientists and psychologists hold for SES is the idea of capital [3]. Since individuals' well-being is directly affected by access to material capital, nonmaterial capital, and social capital, capital (resources, assets) has become a preferred indicator for SES [4]. Capital is not only linked to material status but also brings into focus nonmaterial experiences, such as education and social network resources [5].

There have been numerous researches on the influence of SES on human development. Worldwide economic and political changes during the past four decades (e.g., increasing income inequality) have demonstrated how social position and economic resources affect the development of children and their families. According to the social researches on health disparities, or the general trend that is studied by the developmental scholars, the adults and children who are in the disadvantaged economic position are at increased risk of physical, emotional, and behavioral problems [4, 6]. Concerning the influence of SES on children and adolescents, there is evidence for a solid relationship between poverty and mental health [7], between SES and cognitive development [8, 9], and between social class position and physical well-being [10].

In the past few centuries, multiple disciplines, such as medicine, epidemiology, sociology, demography, and economics, have participated in a strong and consistent relationship between SES and health. One of the research directions is comparing diseased and healthy groups' living conditions. Across diverse cases, individuals who are less educated, have lower job status, and earn less or no income are at greater risk of poor health than their higher SES counterparts [11]. Moreover, these associations have a monotonic or stepwise functional relationship with health, and these associations are not only due to income.

In the domain of developmental psychology, scientists have been aware of the influence of academic achievement on SES since 1990. Recently, a meta-analysis research reviewed the studies on SES and academic achievement. In the sample of 101,157 students, there is a medium to enhance SES-achievement relation, which is moderated by the range of SES variable, the measurement of SES, and the geographical location [12].

While SES has been shown to affect health conditions, life satisfaction [13], and academic achievement [4, 12], psychological studies have also revealed the influences of SES on an individual's thought processes [14]. For instance, well-being, income, and SES have been shown to affect emotion regulation [15], cognitive function [16], and decision-making tendencies [17].

6.1.2 Indicators and Measurements of Socioeconomic Status

SES is a multidimensional structure that has been measured in multiple ways. One structure captures various dimensions of social position, including prestige, power, and economic well-being [6, 18]. Researches [4] claimed that income, education, and occupational status are the three indicators that provide a more comprehensive coverage of the areas of interest. Although these three quantitative indicators are positively associated with themselves [19], there is also a general agreement that they should not be combined into simple composite scores. For example, Duncan and Magnuson [20] suggested that these three indicators separately presented different levels of stability across time, and they differentially predicted childhood academics, suggesting three indicators: achievement, life satisfaction, and life span health condition. Therefore, income, education, and occupational status are not interchangeable indicators of SES. Investigating and analyzing these variables separately provide the insights to understanding the uniqueness and combined contributions to human development.

Education is one of the foundational elements of SES that prioritizes later indicators, such as occupation, income, and wealth. Education is typically completed during the early stage of human development (except in childhood); therefore, the impact of reverse causation is less vulnerable than other markers. For example, poor health may lead to lower income, in which poor health is the cause of low SES, while education is always considered to be a cause of health conditions because it was accomplished before the onset of major health problems. In addition, low refusal rates are the typical and accurate answer to education questions. However, education studies have some limitations, and the implications for earnings and occupation potential may vary across demographic groups (e.g., age, gender, ethnicity/race, or national origins). In addition, education is a relatively crude indicator with few categories and does not include information about educational quality [11].

Income is another widely used SES indicator. Some studies use absolute income as an indicator of SES, while others consider family size and family construction and use adjusted income as a measurement of SES. Household income estimates adjusted for the supported numbers provided more complete information regarding the standard of living. In addition to absolute and adjusted income levels, income dynamics (changes in income across the life span) provide important changing information that may relate to life span variables. The limitation of income information is that it is subject to high nonresponse rates, and biased reporting can be less relevant in individuals who are not in the labor force. Likewise, income does not provide comprehensive information about consumption level since this may vary across community contexts and demographic characteristics.

Occupations are not a quantitative measure of SES, they are categorized hierarchically according to characteristics such as earned income, required education, associated skills, or level of influence. For example, International Standard Classification of Occupations (ISCO) has categorized the occupations into ten levels according to the required skills. The limitations of occupational measurements include the high subjectiveness of occupational status ranking and the measurements cannot be used for individuals who are not in the labor market. Besides, considerable variation in prestige, earnings, and other factors were frequently contained in the occupation categories [21].

Recently, researchers have tended to use mixed measurement of SES due to its multidimensional structure and comprehensive characteristics. There are some measurements of SES that are widely accepted. One is Hollinghead's four-factor index of social status, which is a survey designed based on four domains of SES: marital status, retired/employed status, educational attainment, and occupational prestige [22]. According to the description of the scale, the education code is rated on a 7-point scale, in which 7 = graduate/professional training; 6 = standard college or university graduation; 5 = partial college, at least 1 year of specialized training; 4 = high school graduate; 3 = partial high school, tenth or 11th grade; 2 = junior high school, including ninth grade; and 1 = less than seventh grade, and 0 = not applicable or unknown, respectively. The occupational code is rated on a 9-point scale, in which 9 = higher executive, proprietor of large businesses, major professional; 8 = administrators, lesser professionals, proprietor of a medium-sized business; 7 = smaller business owners, farm owners, managers, minor professionals; 6 = technicians, semi-professionals, small business owners (business valued at $50,000–70,000), 5 = clerical and sales workers, small farm and business owners (business valued at $25,000–50,000); 4 = smaller business owners (<$25,000), skilled manual laborers, craftsmen, tenant farmers; 3 = machine operators, and semi-skilled workers; 2 = unskilled workers; and 1 = farm laborers, menial service workers, students, housewives (dependent on welfare, no regular occupation); and 0 = not applicable or unknown. An SES score is then calculated for a total SES score [22].

MacArthur Scale of Subjective Social Status (MacArthur SSS Scale) is another comprehensive scale to measure SES. The MacArthur SSS Scale was created by health psychologist [23]. It is a single-item measurement that assesses a person's perceived rank relative to others in their group. During the process of measurement, the drawing of a ladder with ten rungs will be presented to the participants. The representation of the ladder where people stand can be determined in the way of hearing or reading. Individuals further read or hear: "At the top of the ladder are the people who are the best off, those who have the most money, most education, and best jobs. At the bottom are the people who are the worst off, those who have the least money, least education, worst jobs, or no job. Please place an 'X' on the rung that best represents where you think you stand on the ladder." At some point, the MacArthur SSS Scale is better than objective measures of SES because, during reporting, many individuals were not clear about their absolute income or education level, and may hide the real data. Moreover, compared to the objective SES measures, the MacArthur SSS Scale sometimes predicts health and well-being better [24].

6.2 The Effects of Socioeconomic Status on Cognition in the Elderly

Socioeconomic status seems to have a life span influence. Various studies have shown that SES affects cognitive ability, school performance, and mental health among children and adolescents. Recently, researchers have found that SES also has similar effects on the elderly. The correlation between early life socioeconomic status and the level of cognition, rate of cognitive decline, and incidence of Alzheimer's disease among elderly individuals was studied by Wilson and colleagues [25]. The baseline was set by 859 older Catholic clergy members without dementia, which is a part of the Religious Orders Study. The results indicated that the higher the individual and community socioeconomic levels in early life were, the higher the level of cognition in late life. However, neither of these two variables were related to the risk of AD or the rate of cognitive decline. The results suggest that early-life socioeconomic level and late-life cognition level are correlated. Zhang and colleagues [26] studied whether the risk of cognitive impairment is affected by SES. During the study, the influences of early-life SES on the risk of cognitive impairment in the elderly individuals in China were examined, and the mediating effect of adult status was considered. The relationship between living in urban areas in early life and lower incidence of cognitive impairment at baseline was found to be significant by analyzing data from the Chinese Longitudinal Healthy Longevity Survey. The same effect was found among education and incidence of cognitive impairment. For the longitude analysis, Zhang and colleagues [26] found that advantaged childhood SES has a modest protective effect on the incidence of cognitive impairment onset during the 2-year follow-up, especially in women. These results suggest that the risk of cognitive impairment and incidence of Alzheimer's disease (AD) in elderly individuals can be impacted by the socioeconomic environment in the overall life span development, especially during early life.

Brayne and Calloway [27] investigated the distribution of scores in the Mini Mental State Examination (MMSE); all the participants are women aged 70–79 years from a rural area of Cambridgeshire. This research investigates its relationship with SES. The results showed that lower socioeconomic and less education groups were all found to be associated with lower scores on MMSE, indicating the SES effect on cognition for the elderly. As for multidomain of cognitive ability, Angela and coworkers found that after adjusting for demographic information such as age, sex, and race, education had the strongest association with global cognition, episodic memory, semantic memory, and visuospatial ability [28]. Brain mechanism is behind the SES effect on cognition for adults and the elderly. Fotenos and coworkers [29] investigated the relationship between SES and structural brain changes in nondemented older adults which is a potential indicator of preclinical AD. For the sample of 362 nondemented adults aged 18–93 years, after controlling the effects of age and sex in older cognitive normal adults (>65 years), higher SES was associated with smaller whole-brain volume and more severe volume loss. Follow-up data analysis revealed that compared with those who remained nondemented, those who later developed mild dementia has a greater association between brain volume reduction and SES. However, nondemented participants with high SES were found to have reduced brain volume in the cross-sectional analysis and accelerated volume loss in the longitudinal analysis. It is in line with the cognitive reserve theory, in which more privileged individuals have the capacity to cope longer with brain pathology before manifesting dementia may contribute to this association relationship.

Besides the individual level of SES, neighborhood SES can also affect cognitive function as an environmental factor. In a group of community-dwelling elderly individuals, researchers assessed cognitive function in a low-SES Asian neighborhood (owner-occupied housing in Singapore) and compared them to a high-SES neighborhood (rental flats in Singapore). They found that neighborhood SES has an influence on cognitive function in the elderly, and living in a low-SES community was independently associated with poorer cognitive function and cognitive impairment [30].

6.3 The Factors between Socioeconomic Status and Cognition

6.3.1 Functional and Structural Brain

SES represents overall life span and societal position, which strongly affect one's individual experiences over the life span development. Understanding the mechanisms of how early childhood to life span environmental experiences affect the structure and function of the brain has become a popular topic for researchers. Many studies have focused on early childhood and specifically investigated the link between SES and academic performance. Hackman and colleagues [31] claimed that the underlying mechanisms for the link between SES and academic performance are difficult to identify since the function of multiple cognitive and socioemotional systems were reflected by various outcome variables: IQ, school achievement, and diagnostic classifications. Thus, identifying the differences in underlying cognitive and affective neural systems at the beginning as SES-related is a promising approach for studying the mechanisms of how SES affects these outcome variables.

Studies have shown an effect of SES on language development. The degree of hemispheric specialization in the left inferior frontal gyrus (IFG), a region which includes Broca's area, was positively correlated with SES, and this brain area is activated during the language task in young children [32]. This result indicated that children with low SES have decreased specialization in language function in the left hemisphere. Additionally, among low-SES children, phonological awareness was positively associated with left fusiform activity during reading [33].

There are also SES-related differences in executive function, the degree to which specific neural systems are recruited during tasks are different, even when there are no task performance differences between SES groups. For example, event-related potentials (ERPs) revealed that low-SES children have difficulty suppressing distraction early during the processing stream and thus exhibit larger responses to unattended stimuli [34, 35]. In addition, low-SES children do not recruit prefrontal attention circuits in response to novel distracter stimuli compared to high-SES children [36]. In a task that requires the subject to apply familiar stimulus-response rules and to learn new rules, low-SES children preferentially recruit the right middle frontal gyrus when learning novel rules according to functional MRI arousal.

As for the neural processing of emotion, there is also evidence indicating SES-related differences. Low-SES adolescents exhibit lower left-sided brain activity at rest, which is a pattern that is typically observed in patients with depression [37]. Increased amygdala response to angry faces is associated with lower subjective social status in college students [38]. In adults, lower subjective social status is related to the regulation of emotional states and the high risk of having affective disorders, revealing by a smaller volume of the perigenual anterior cingulate cortex [39], a region that is functionally connected with the amygdala.

A meta-regression analysis conducted recently proposed a general perspective. In summary, low-SES individuals exhibit hypoactivation of the executive network and hyperactivation of the reward network [40]. These results of the meta-analysis demonstrated that hypoactivity of the frontoparietal/cinguloopercular executive networks and hyperactivity of the right caudate nucleus both occur across age and as a function of age. Structural differences in individuals with low SES were complementary to the functional brain analyses. Most notably, the anterior cingulate cortex was reduced in gray matter volume in individuals with low SES as a function of age, which has been reported in many empirical studies investigating the neural structures of individuals residing in impoverished conditions [41]. The findings are consistent with the scarcity hypothesis, which states that individuals with low SES tend to focus more on monetary issues, while neglecting other nonmonetary issues [42].

6.3.2 Environmental Factors

A growing number of researchers have studied one or more candidate-mediating factors in the field of SES neuroscience. These studies measured and tested whether those factors statically account for the relationship between SES and brain structure or activity. Luby and colleagues [43] reported that mother-reported life stress and quality of parenting behavior both fully mediated the relationship between SES and hippocampal volume in children. Glen Elder [44] proposed the family stress model to explain the influence of low-SES children. According to this model, daily expenditures were forced to be reduced in families with low income. At the same time, those families struggle to pay bills and purchase necessary goods and services. Meanwhile, other stressful events, such as reduced social support and low self-esteem, are coupled with low economic status. Among poor parents, these factors normally cause high levels of psychological distress, such as depressive and hostile feelings [45].

As evidence from the psychoneuroimmunology field, experiencing chronic and elevated physiological stress responses may impact the stress response system in children to healthy development, including the brain regions responsible for self-regulation. It has been documented in an animal model that such stress has harmful effects on brain development. Stress hormones, such as cortisol, increase during exposure to stressful events and negatively influence the cognitive function of animals. For instance, brain structures, such as the hippocampus, which is crucial for memory, are impaired due to increased stress hormones [46]. Decreased immunological function is correlated with high levels of psychological stress, which influences many late life inflammatory diseases [47]. For instance, recent evidence indicates that brain regions that support cognitive skills, such as executive functioning and self-regulation, are related to the body stress response system [48].

Numerous studies have focused on the process of SES influencing the brain. These studies indicated health-relevant (or lifestyle) behaviors as mediators between SES and the brain. More tobacco and alcohol were used, but less diet and exercise were performed among individuals of low SES. Wills and the colleagues [49] indicated that the likelihood of smoking was related to adolescent academic and behavioral competence, and smoking was positively correlated with low parental education level, which increases the possibility of social interactions with those involved in drug uses and adolescent drug abuse.

6.3.3 Decision-Making Process: A New Perspective

According to the scarcity hypothesis, the "scarcity of resources affects how people allocate attention; individuals living in poverty tend to focus more on problems associated with money while neglecting non-monetary problems." Predisposition to reward sensitivity could also develop among individuals raised in a low-SES environment [50]. Additionally, risky decision-making behavior, such as overborrowing, increase [51]. Further, different neural responses and gray matter volume densities were observed in the area responsible for the choice evaluation and reward processing called the orbital frontal cortex in adults exposed to urbanicity [52]. This suggests that research investigating and interpreting the relationships of poverty, cognition, and the brain is in need of a life span perspective.

There are two competing neurobehavioral decision systems models that influence decision-making: impulsive decision systems and executive decision systems. An impulsive decision system encodes immediate reinforcers, and an executive decision system inhibits impulses and prefers long-term outcomes [53]. Some studies claim that low SES might be connected with a less effective decision system, and the balance between the two systems collapses [40]. The functional and structural brain changes in low SES support the idea of two competing decision-making systems. A meta-analysis revealed hypoactivity of the frontoparietal/cinguloopercular executive networks both across age and as a function of age [40]. Hypoactivity of these executive networks reflects the lack of efficient executive abilities in individuals with low SES.

The meta-analysis also found the hyperactivity of the right caudate nucleus both across age and as a function of age, which is an area within the anterior striatum associated with stimulus–action–reward-associative learning [40]. Other regions that showed increased gray matter volume in low SES individuals were the orbital frontal cortex, dorsal lateral prefrontal cortex, and hippocampus, key regions of the reward circuit. Left orbital frontal cortex is associated with the ability to assign value and pleasure to stimuli, facilitating decision-making process [54]. The left dorsal prefrontal cortex is associated with volitional risky decision-making activity. When facing uncertain decisions, the hippocampus is sensitive to approach–avoidance conflict processing [55].

In summary, low SES tends to decrease executive function and increase reward-related function. Due to the scarcity of resources, the balance between the impulsive decision system and the executive decision system is damaged accordingly. Low SES makes individuals pay more attention to immediate rewards, affecting the decision-making process. It is worthwhile to explore how reward-related biases (e.g., delay discounting bias, framing bias) and the corresponding reward-related brain regions are affected by SES. In addition, additional mediators, including early life stress, fear of learning, neglect, urbanicity, environmental toxins, nutritional deficiencies, and exposure to other hostile environmental factors, may also contribute to the decision-making process in low SES.

References

1. McLoyd VC (1997) The impact of poverty and low socioeconomic status on the socioemotional functioning of African-American children and adolescents: mediating effects. In: Taylor RD, Wang MC (eds) Social and emotional adjustment and family relations in ethnic minority families, pp 7–34
2. Brown-Iannuzzi JL, Lundberg KB, McKee S (2017) The politics of socioeconomic status: how socioeconomic status may influence political attitudes and engagement. Curr Opin Psychol 18:11–14
3. Guo G, Harris KM (2000) The mechanisms mediating the effects of poverty on children's intellectual development. Demography 37(4):431–447
4. Bradley RH, Corwyn RF (2002) Socioeconomic status and child development. Annu Rev Psychol 53(1):371–399
5. Krieger N, Williams DR, Moss NE (1997) Measuring social class in US public health research: concepts, methodologies, and guidelines. Annu Rev Public Health 18(1):341–378
6. Oakes JM, Rossi PH (2003) The measurement of SES in health research: current practice and steps toward a new approach. Soc Sci Med 56(4):769–784
7. Ackerman BP, Brown ED, Izard CE (2004) The relations between persistent poverty and contextual risk and children's behavior in elementary school. Dev Psychol 40(3):367
8. Hoff E (2003) The specificity of environmental influence: socioeconomic status affects early vocabulary development via maternal speech. Child Dev 74(5):1368–1378
9. Mezzacappa E (2004) Alerting, orienting, and executive attention: developmental properties and sociodemographic correlates in an epidemiological sample of young, urban children. Child Dev 75(5):1373–1386
10. Evans GW, English K (2002) The environment of poverty: multiple stressor exposure, psychophysiological stress, and socioemotional adjustment. Child Dev 73(4):1238–1248
11. Matthews KA, Gallo LC (2011) Psychological perspectives on pathways linking socioeconomic status and physical health. Annu Rev Psychol 62:501–530
12. Sirin SR (2005) Socioeconomic status and academic achievement: a meta-analytic review of research. Rev Educ Res 75(3):417–453

13. Gerdtham U-G, Johannesson M (2001) The relationship between happiness, health, and socio-economic factors: results based on Swedish microdata. J Socio-Econ 30(6):553–557
14. Lipina SJ, Posner MI (2012) The impact of poverty on the development of brain networks. Front Hum Neurosci 6:238
15. Kim P, Evans GW, Angstadt M, Ho SS, Sripada CS, Swain JE et al (2013) Effects of childhood poverty and chronic stress on emotion regulatory brain function in adulthood. Proc Natl Acad Sci 110(46):18442–18447
16. Mani A, Mullainathan S, Shafir E, Zhao J (2013) Poverty impedes cognitive function. Science 341(6149):976–980
17. Spears D (2011) Economic decision-making in poverty depletes behavioral control. BE J Econ Anal Policy 11(1):2973
18. Liu WM, Ali SR, Soleck G, Hopps J, Pickett T Jr (2004) Using social class in counseling psychology research. J Couns Psychol 51(1):3
19. Ensminger ME, Fothergill KE, Bornstein M, Bradley R (2003) A decade of measuring SES: what it tells us and where to go from here. In: Bornstein MH, Bradley RH (eds) Socioeconomic status, parenting, and child development. Lawrence Erlbaum Associates Publishers, pp 13–27
20. Duncan GJ, Magnuson KA (2003) Off with Hollingshead: socioeconomic resources, parenting, and child development. In: Bornstein MH, Bradley RH (eds) Socioeconomic status, parenting, and child development. Lawrence Erlbaum Associates Publishers, pp 83–106
21. Braveman PA, Cubbin C, Egerter S, Chideya S, Marchi KS, Metzler M et al (2005) Socioeconomic status in health research: one size does not fit all. JAMA 294(22):2879–2888
22. Hollingshead AB (1975) Four factor index of social status. Unpublished manuscript, Yale University, New Haven, CT.
23. Adler NE, Epel ES, Castellazzo G, Ickovics JR (2000) Relationship of subjective and objective social status with psychological and physiological functioning: preliminary data in healthy, white women. Health Psychol 19(6):586
24. Singh-Manoux A, Marmot MG, Adler NE (2005) Does subjective social status predict health and change in health status better than objective status? Psychosom Med 67(6):855–861
25. Wilson RS, Scherr PA, Hoganson G, Bienias JL, Evans DA, Bennett DA (2005) Early life socioeconomic status and late life risk of Alzheimer's disease. Neuroepidemiology 25(1):8–14
26. Zhang Z, Gu D, Hayward MD (2008) Early life influences on cognitive impairment among oldest old Chinese. J Gerontol Ser B Psychol Sci Soc Sci 63(1):S25–S33
27. Brayne C, Calloway P (1990) The association of education and socioeconomic status with the mini mental state examination and the clinical diagnosis of dementia in elderly people. Age Ageing 19(2):91–96
28. Jefferson AL, Gibbons LE, Rentz DM, Carvalho JO, Manly J, Bennett DA et al (2011) A life course model of cognitive activities, socioeconomic status, education, reading ability, and cognition. J Am Geriatr Soc 59(8):1403–1411
29. Fotenos AF, Mintun MA, Snyder AZ, Morris JC, Buckner RL (2008) Brain volume decline in aging: evidence for a relation between socioeconomic status, preclinical Alzheimer disease, and reserve. Arch Neurol 65(1):113–120
30. Wee LE, Yeo WX, Yang GR, Hannan N, Lim K, Chua C et al (2012) Individual and area level socioeconomic status and its association with cognitive function and cognitive impairment (low MMSE) among community-dwelling elderly in Singapore. Dement Geriatr Cogn Dis Extra 2(1):529–542
31. Hackman DA, Farah MJ, Meaney MJ (2010) Socioeconomic status and the brain: mechanistic insights from human and animal research. Nat Rev Neurosci 11(9):651–659
32. Raizada RD, Richards TL, Meltzoff A, Kuhl PK (2008) Socioeconomic status predicts hemispheric specialisation of the left inferior frontal gyrus in young children. NeuroImage 40(3):1392–1401
33. Noble KG, Wolmetz ME, Ochs LG, Farah MJ, McCandliss BD (2006) Brain–behavior relationships in reading acquisition are modulated by socioeconomic factors. Dev Sci 9(6):642–654
34. D'Angiulli A, Herdman A, Stapells D, Hertzman C (2008) Children's event-related potentials of auditory selective attention vary with their socioeconomic status. Neuropsychology 22(3):293
35. Stevens C, Lauinger B, Neville H (2009) Differences in the neural mechanisms of selective attention in children from different socioeconomic backgrounds: an event-related brain potential study. Dev Sci 12(4):634–646
36. Kishiyama MM, Boyce WT, Jimenez AM, Perry LM, Knight RT (2009) Socioeconomic disparities affect prefrontal function in children. J Cogn Neurosci 21(6):1106–1115
37. Tomarken AJ, Dichter GS, Garber J, Simien C (2004) Resting frontal brain activity: linkages to maternal depression and socio-economic status among adolescents. Biol Psychol 67(1–2):77–102
38. Gianaros PJ, Horenstein JA, Hariri AR, Sheu LK, Manuck SB, Matthews KA et al (2008) Potential neural embedding of parental social standing. Soc Cogn Affect Neurosci 3(2):91–96
39. Gianaros PJ, Horenstein JA, Cohen S, Matthews KA, Brown SM, Flory JD et al (2007) Perigenual anterior cingulate morphology covaries with perceived social standing. Soc Cogn Affect Neurosci 2(3):161–173
40. Yaple ZA, Yu R (2020) Functional and structural brain correlates of socioeconomic status. Cereb Cortex 30(1):181–196
41. Yang J, Liu H, Wei D, Liu W, Meng J, Wang K et al (2016) Regional gray matter volume mediates the relationship between family socioeconomic status and

depression-related trait in a young healthy sample. Cogn Affect Behav Neurosci 16(1):51–62
42. Mullainathan S, Shafir E (2013) Scarcity: why having too little means so. Macmillan, much
43. Luby J, Belden A, Botteron K, Marrus N, Harms MP, Babb C et al (2013) The effects of poverty on childhood brain development: the mediating effect of caregiving and stressful life events. JAMA Pediatr 167(12):1135–1142
44. Elder GH (2018) Children of the great depression. Routledge
45. Duncan GJ, Magnuson K, Votruba-Drzal E (2017) Moving beyond correlations in assessing the consequences of poverty. Annu Rev Psychol 68:413–434
46. McEwen BS (2000) The neurobiology of stress: from serendipity to clinical relevance. Brain Res 886(1–2):172–189
47. Miller GE, Chen E, Parker KJ (2011) Psychological stress in childhood and susceptibility to the chronic diseases of aging: moving toward a model of behavioral and biological mechanisms. Psychol Bull 137(6):959
48. Blair C, Granger DA, Willoughby M, Mills-Koonce R, Cox M, Greenberg MT et al (2011) Salivary cortisol mediates effects of poverty and parenting on executive functions in early childhood. Child Dev 82(6):1970–1984
49. Wills TA, McNamara G, Vaccaro D (1995) Parental education related to adolescent stress-coping and substance use: development of a mediational model. Health Psychol 14(5):464
50. Gonzalez MZ, Allen JP, Coan JA (2016) Lower neighborhood quality in adolescence predicts higher mesolimbic sensitivity to reward anticipation in adulthood. Dev Cogn Neurosci 22:48–57
51. Shah AK, Mullainathan S, Shafir E (2012) Some consequences of having too little. Science 338(6107):682–685
52. Krämer B, Diekhof EK, Gruber O (2017) Effects of city living on the mesolimbic reward system—an fMRI study. Hum Brain Mapp 38(7):3444–3453
53. Bickel WK, Jarmolowicz DP, Mueller ET, Gatchalian KM, McClure SM (2012) Are executive function and impulsivity antipodes? A conceptual reconstruction with special reference to addiction. Psychopharmacology 221(3):361–387
54. Grabenhorst F, Rolls ET (2011) Value, pleasure and choice in the ventral prefrontal cortex. Trends Cogn Sci 15(2):56–67
55. O'Neil EB, Newsome RN, Li IH, Thavabalasingam S, Ito R, Lee AC (2015) Examining the role of the human hippocampus in approach–avoidance decision making using a novel conflict paradigm and multivariate functional magnetic resonance imaging. J Neurosci 35(45):15039–15049

Part IV

Neuroimaging Studies of Brain Aging

The Aging Patterns of Brain Structure, Function, and Energy Metabolism

7

Mingxi Dang, Feng Sang, Shijie Long, and Yaojing Chen

Abstract

The normal aging process brings changes in brain structure, function, and energy metabolism, which are presumed to contribute to the age-related decline in brain function and cognitive ability. This chapter aims to summarize the aging patterns of brain structure, function, and energy metabolism to distinguish them from the pathological changes associated with neurodegenerative diseases and explore protective factors in aging. We first described the normal atrophy pattern of cortical gray matter with age, which is negatively affected by some neurodegenerative diseases and is protected by a healthy lifestyle, such as physical exercise. Next, we summarized the main types of age-related white matter lesions, including white matter atrophy and hyperintensity. Age-related white matter changes mainly occurred in the frontal lobe, and white matter lesions in posterior regions may be an early sign of Alzheimer's disease. In addition, the relationship between brain activity and various cognitive functions during aging was discussed based on electroencephalography, magnetoencephalogram, and functional magnetic resonance imaging. An age-related reduction in occipital activity is coupled with increased frontal activity, which supports the posterior–anterior shift in aging (PASA) theory. Finally, we discussed the relationship between amyloid-β deposition and tau accumulation in the brain, as pathological manifestations of neurodegenerative disease and aging.

Keywords

Brain aging · Structural and functional change · Posterior–anterior shift

7.1 Cortical Gray Matter Atrophy in Aging

Gradual gray matter atrophy with aging can be potentially caused by various reasons, for example, the loss of neurons and the decrease of synaptic spines and synapses. Many factors, such as age and neurodegenerative diseases, affect gray matter atrophy, whereas some other factors are protective and slow down the brain atrophy process.

M. Dang · F. Sang · S. Long · Y. Chen (✉)
State Key Laboratory of Cognitive Neuroscience and Learning, Faculty of Psychology, Beijing Normal University, Beijing, China

Beijing Aging Brain Rejuvenation Initiative (BABRI) Centre, Beijing Normal University, Beijing, China
e-mail: chenyaojing@bnu.edu.cn

Z. Zhang (ed.), *Cognitive Aging and Brain Health*, Advances in Experimental Medicine and Biology 1419, https://doi.org/10.1007/978-981-99-1627-6_7

7.1.1 Normal Atrophy of Cortical Gray Matter with Aging

As verified in cross-sectional and longitudinal studies, the whole-brain volume decreases significantly with age, with an annual shrinkage rate of the whole brain volume of less than 1%, as verified in cross-sectional and longitudinal studies [1–4]. The specific shrinkage rate varies among different studies. A study reported an annual whole brain shrinkage rate of 0.32% [2], while another study reported a rate of 0.56% [1]. The pattern of brain atrophy also differs among studies. In a cross-sectional study, brain volume decreased linearly with age [3], while a longitudinal study reported a significant increase in the atrophy rate after the age of 70 [2]. The rate of atrophy also varies in different brain regions. A previous study [1] shows that the temporal lobe exhibits the highest annual atrophy rate, followed by the occipital and parietal lobes, and the frontal lobe shows the lowest atrophy rate. In particular, a significant relationship between the decrease in temporal lobe volume and the decline in cognitive ability has been reported [5].

7.1.2 The Pathological Processes Affect the Atrophy of Gray Matter during Aging

Neurodegenerative diseases, characterized by a gradual loss of neuronal structure and function, including Parkinson's disease, Alzheimer's disease (AD), and multiple lateral sclerosis, will accelerate the atrophy of gray matter. Chen et al. [6] analyzed the density of gray matter (similar to the volume of gray matter, indicating the possibility of gray matter at a certain voxel) by measuring the density of gray matter in 37 patients with Parkinson's disease and 21 healthy controls. The gray matter density in the frontal, temporal, parietal, bilateral cerebral areas, and other areas of cognitively normal patients with Parkinson's disease was significantly lower than that in the control group; patients with Parkinson's disease presenting a mild cognitive impairment displayed more severe atrophy in areas such as the left hippocampus an insula. In addition, Alzheimer's disease [7], Lewy body dementia [8], and multiple lateral sclerosis [9] also cause abnormal atrophy of gray matter.

Atrophy of the cerebral cortex has been reported in abnormal cognitively impaired subjects. Driscoll et al. [10] studied a pattern of abnormal volume changes in mild cognitive impairment based on a longitudinal aging cohort. They found that the subjects with mild cognitive impairment had faster volume changes in the whole brain, cerebrospinal fluid, temporal lobe, and orbitofrontal-temporal cortex (including hippocampus).

7.1.3 Protective Factors for Gray Matter Atrophy

A good lifestyle slows gray matter atrophy. Meditation is one of the most popular factors that has been studied. A systematic review in 2017 confirmed the role of meditation in slowing gray matter atrophy [11]. In terms of specific studies, researchers found larger volumes of the left hippocampus [12] and the orbitofrontal cortex in the meditation group [13]. In addition, some researchers have found that physical exercise slows down the rate of change in gray matter volume. Erickson et al. [14] reviewed related studies and indicated that performing more physical activities is associated with larger gray matter volumes, including the prefrontal cortex and hippocampus.

Gray matter atrophy is a normal phenomenon in the process of human aging. Some neurodegenerative diseases affect this process and shift this pattern from the normal aging process. However, lifestyle changes and physical exercise can slow the process of gray matter atrophy.

7.2 Age-Related Changes in White Matter

White matter, which is generally composed of bundled myelinated or unmyelinated axons and myelin-producing glial cells and other cell types, occupies approximately 50% of the total cerebral volume in humans [15]. White matter plays important roles, such as information transmission and network construction. Studies using diffusion tensor imaging (DTI) showed that white matter microstructural alterations in neurodegeneration are related to tissue degeneration (i.e., axonal damage and demyelination) and excessive extracellular fluid [16–19]. Age-related changes in white matter are very common. Next, we intend to summarize the main types of age-related white matter damage, differences in age-related white matter damage between brain regions, and the relationship between white matter changes and cognitive decline.

7.2.1 Main Types of Age-Related White Matter Lesions

White matter lesions (WMLs), caused by chronic cerebral hypoperfusion, are frequently observed in older adults and are closely related to cognitive decline. WMLs mainly include axonal loss and demyelination. Chronic diseases, such as hypertension [20], diabetes, and hyperlipidemia, have been shown to increase the WMLs. Common types of WMLs include white matter atrophy and hyperintensity.

7.2.1.1 White Matter Atrophy

Previous studies have linked the functional decline in aging brains with a significant progressive neuronal loss in gray matter. However, subsequent studies documented only a 10% loss in the total number of neurons in the cortex of both sexes from 20 to 90 years of age [21], and the association with age-related cognitive decline was not strong. In contrast, normal aging led to a 28% reduction in the white matter volume, and age-related white matter changes were associated with functional impairments, such as sensorimotor cognitive impairments and mental disorders. Therefore, the changes in white matter are associated with more severe neurobehavior and cognitive impairments during aging.

In normal aging, white matter atrophy may begin later and progress more rapidly than gray matter atrophy, and the volume change of white matter generally follows an inverted u-shaped curve. Remarkably, white matter volume gradually increased before 40 years of age, reaching the peak around 40–50 years of age, and rapidly declined after 60 years of age [22]. One study indicated that white matter volume increased before the age of 40 years and thereafter declined at an annual rate of 0.5%, which was accelerated after 70 years of age [23]. Further, another study found that white matter volume in a person 70 years old was only reduced by approximately 6% compared to persons of 30 years of age, but declined by as much as 25% in persons 80 years of age. Thus, white matter volume loss could be considered as a feature of more advanced age.

7.2.1.2 White Matter Hyperintensity

White matter hyperintensities (WMHs) are common in old people, even those without any apparent disease [24], and the association between WMHs and vascular risk factors in nondemented older adults is well established. In the general population, the prevalence of white matter hyperintensities ranges from 11% to 21% in adults aged around 64 years to 94% at the age of 82 years [25]. White matter hyperintensities are more common and more extensive in patients with cardiovascular risk factors and symptomatic cerebrovascular disease. Hypertension and a decreased normalized peak expiratory flow rate are the principal predictors of deep white matter hyperintensities in non-diabetic subjects [26]. A longitudinal study [27] reported that childhood intelligence and brain white matter hyperintensities could predict fluid intelligence at the age of 78–81 years, which suggests that the WMH level or burden would affect cognitive decline.

7.2.2 Differences in Brain Regions of White Matter Age-Related Damage

Age-related changes in white matter volumes differ by brain region and exhibit a roughly anterior-to-posterior gradient pattern: the age-related changes in frontal volume were more significant than those in posterior regions. A study with a large sample size of nondemented elderly persons investigated age-related changes in tract-specific diffusion characteristics in 25 tracts using probabilistic tractography and reported the loss of microstructural organization in the association, commissural, and limbic tracts [28]. Some studies found that the bilateral frontal lobe is the most vulnerable region in the brain, while the occipital lobe shows a slower decline in white matter volume [22], suggesting that the myelination occurs last in the frontal lobe, while it starts earlier in the occipital lobe. In the frontal lobe, the myelin sheath is thinner, the repair mechanism is less efficient, and oligodendrocytes are prone to metabolic changes.

Fractional Anisotropy (FA), a measurement from diffusion tensors, indicates white matter integrity. Compared with young people, the frontal lobes of the elderly showed a broad decrease in FA, including the orbitofrontal cortex, the genu of the corpus callosum, forceps major, and the anterior corona radiata of the prefrontal cortex, as well as the middle and superior temporal gyri and posterior parietal cortex [29]. Moreover, the pattern of fractional anisotropy changes in frontal lobes was consistent with the reduced frontal white matter volume. It increased white matter hyperintensities of older adults reported in other studies [30, 31].

The observed interaction effect between aging and brain regions on changes in the white matter showed that the effect of aging on the white matter differs among regions. The temporal and occipital lobes showed significant age-related changes in white matter, but the effect was much smaller than that observed in the frontal lobes [32]. Selected striatal regions also showed age-related decreases in the white matter volume [33, 34].

Notably, patients with early dementia show little or no change in white matter in the anterior region and greater deterioration in the posterior region [22]. The dissociation between the regional effects of age and dementia status suggests a relation between AD and age-associated cognitive decline.

7.2.3 The Relationship between Changes in White Matter and Cognitive Function

White matter lesions in different areas of the brain may lead to deficits in different cognitive functions. For example, lesions in the frontal lobe are responsible for several cognitive impairments in the speed of information processing, visual-motor function, verbal fluency, classification, and mental sequences.

The decline of several frontal functions is related to white matter lesions in the frontal lobe [35]. Ziegler et al. found episodic memory was related to the integrity of white matter in the temporal and parietal lobe [29]. In addition, subcortical white matter lesions are mainly correlated with depression in the elderly [36], whereas periventricular white matter lesions are mainly related to cognitive decline. Thus, the age-related impairments in specific cognitive capacities may arise from degenerative processes that affect the underlying connections of their respective neural networks.

In summary, various macrostructural changes in the white matter have been identified during normal aging, and these changes are presumed to contribute to the age-related decline in brain function. Even during healthy aging, elderly individuals also develop white matter lesions, such as white matter hyperintensities, and white matter lesions in posterior regions may be an early sign of AD.

7.3 Electroencephalogram and Magnetoencephalogram Recordings during Aging

Over the last decade, electroencephalography (EEG) and magnetoencephalogram (MEG) have emerged as noninvasive alternative techniques for the study of AD and aging that compete with more expensive neuroimaging tools, such as MRI and positron emission tomography (PET). At the same time, EEG and MEG have excellent temporal resolution and good spatial resolution, and thus they have been widely used in research on aging mechanisms and treatments.

7.3.1 Electroencephalogram Recordings

EEG is a method of recording brain activity using electrophysiological indicators. During brain activity, many neurons synchronously generate the sum of postsynaptic potentials. EEG records changes in electrical waves during brain activity and reflects the overall electrophysiological activity of brain nerve cells on the surface of the cerebral cortex or scalp. The complexity of EEG signals reflects the irregularity of waveform and dynamics and is affected by healthy aging, as well as brain dysfunction such as AD.

Oscillations of the alpha frequencies are probably mainly affected by the aging process. In the process of physiological aging, posterior cortical alpha rhythms decrease with both linear and nonlinear trends [37], and rhythm changes are associated with global cognitive levels [38]. The median frequency, peak frequency, and alpha-to-theta ratio of resting-state EEG measured in prefrontal regions also decrease with the reduction in the global cognitive score of elderly individuals [39]. Compared with healthy older adults, the peak alpha frequency is significantly decreased in the posterior region of the brain in individuals with mild cognitive impairment and positively correlates with hippocampal volume (a common marker for early structural AD-related damage) [40]. Alpha source abnormalities are also helpful for the early detection of cognitive impairment and discrimination of healthy elderly people from patients with cognitive impairment and dementia [41, 42].

Some studies have suggested that EEG theta activity might indicate early abnormal degenerative changes in the brain. In cognitively healthy elderly subjects, cerebral spinal fluid (CSF) T-tau, P-tau, and the combined P-tau/Aβ42 ratio are correlated with relative EEG theta power, especially in the posterior quadrants, and which is correlated with slowing of cognitive speed [43]. Interestingly, in addition to some early studies showing that the sleep rhythm changes of patients with early AD were related to episodic memory disorders [44], a recent study showed that frontotemporal synchronous theta tuning stimulation could improve the performance of working memory in the elderly [45]. These studies on the physiological basis of age-related cognitive impairment provide insights for future non-drug interventions aiming at cognitive decline.

7.3.2 Magnetoencephalogram Recordings

In recent years, MEG has been widely used to study healthy aging and pathological aging. Due to its excellent temporal resolution, MEG provides valuable information about the temporal dynamics of age-related brain function in cognitive neuroscience models of aging that other neuroimaging techniques are unable to provide.

Some research studies using MEG have focused on explaining age-related memory deficits—a major feature of aging. Interference from new items onto old memories (retroactive interference) affects accuracy in older adults to a greater extent than in young adults when performing an interference-based working memory recognition task. Additionally, MEG results revealed decreased neural activity in the posterior frontal, temporo-occipital, and superior parietal lobes of the elderly. Based on this result, age-related deficits in inhibitory mechanisms that reduce the ability to suppress irrelevant information are associated with under-recruitment in a high-interference task [46, 47]. Another study showed that elderly adults generate enhanced prefrontal activity during the encoding process of an episodic memory task compared to younger adults [48]. Thus, the activation of the brain compensation mechanism occurs in the coding process of the elderly, indicating that the elderly require more executive control resources in the coding process of working memory to obtain sufficient performance on the task.

Other studies using MEG revealed deviations in brain oscillations in patients with AD and cognitive impairment. Patients with AD showed an increased MEG dipole density in delta and theta bands, and increased low-wave activity in temporoparietal regions is associated with cognitive and daily life functional declines [49]. Furthermore, the left hippocampal volume is related to the left temporal and parietal delta dipole density and the left temporal theta dipole density. Additionally, combining the left hippocampal volume and the left temporal theta dipole density accurately distinguishes patients with AD from healthy controls [50]. Another study compared the oscillatory brain activity of patients with the amnestic mild cognitive impairment (MCI) subtype with the relative power values of MEG data. Increases in relative power in lower frequency bands (delta and theta frequency ranges) and decreases in power values in higher frequency bands (alpha and beta frequency ranges) were observed in the MCI groups compared to the control group. In addition, different spectral distributions were noted between groups, as individuals with MCI showed smaller frequency peaks and broader frequency bands.

7.4 Functional MRI Studies of Brain Activation in Aging

7.4.1 Changes in Task-Related Brain Activation Change during Aging

Aging is associated with functional activation alterations in the human brain as functional magnetic resonance imaging (fMRI) has been used to clarify neural activity. Compared with younger adults, older adults perform more poorly on a variety of motor, perceptual, and cognitive tasks, and the decline of cognitive abilities is related to brain activation.

7.4.1.1 Memory Task fMRI Studies

A number of studies have investigated the effects of age on memory functions [51] and reported altered activation in several brain regions of older subjects compared with younger subjects, although performance levels were comparable in both groups [52–54]. Many recent neuroimaging studies have assessed verbal memory. Young adults showed increased activity in the left and right prefrontal cortex during coding and retrieval, respectively [55], and older adults showed reduced prefrontal activation during both coding and retrieval [56]. Increased prefrontal activity may help older adults perform tasks adequately. In the aging brain, these increased task-related activations reflect compensatory complementation of additional pathways that have a positive impact on task performance [57–59]. For example, Cabeza and his colleagues [60] found an increased prefrontal activation only in those older subjects that performed as well as young adults during recall and source memory of recently studied words. There is increasing evidence that age-related differences in brain activation during various cognitive tasks may reflect compensatory recruitment of cortical units required to produce the same performance.

Little is known about the brain function of older people, despite the rapid growth of the elderly population in developed countries and its vital public health importance. One interesting study [61] investigated brain response to a specific memory task in healthy participants over the age of 90 years compared with healthy younger elderly participants. The task led to activation in the posterior temporal, posterior frontal, and parietal cortices in older adults. However, it was more intense in younger participants, particularly in the right hippocampus, temporal, and parietal cortices. Despite significant low activation and brain atrophy, the phenomenon of relatively maintained performance can be explained by the cognitive reserve theory, which may be specific to successful aging in older adults.

7.4.1.2 Motor Task fMRI Studies

Normal aging is generally associated with decreased motor performance. Studies have reported more extended activation patterns in the elderly despite the identical performance of hand movements compared with young controls [62, 63]. In fact, only tasks of increasing complexity provide insights into the functional significance of age-related differences in movement representation [64, 65]. Another study [66] documents that during the execution of a simple paced motor task, the older the subject, the less lateralized the primary motor cortex (M1) activation balance. As a result, activation increases bilaterally, suggesting that motor performance requires bilateral M1 recruitment in the aging brains. These findings indicate a tendency for the aging brain to reduce its functional lateralization, perhaps from less efficient transcallosal connections.

The attractive compensatory hypothesis may meet its limitation when explaining other brain functions, such as the motor system. One study found that shorter response times on visually cued motor tasks (i.e., better performance) were positively correlated with the degree of activation of the motor system in older adults [67]. Other studies suggested that although young and older people perform the same performance tasks, older people have a broader pattern of activation during hand movements [62]. Another result [68] indicates that age-related brain activation patterns within the motor system are independent of functional needs and do not fully reflect the neurobiological changes of reorganization to compensate for aging.

7.4.1.3 Other Task-Based fMRI Studies

Age-related hyperactivation of frontal and parietal regions and hypoactivation of occipital regions have been observed in a variety of tasks, such as attention, visual perception, language processing, and emotion processing [69], as well as the memory tasks and motor tasks described above. A meta-analysis of brain activation across multiple cognitive domains [70] reported greater activation in the prefrontal regions in older adults and greater activation in occipital regions in younger adults.

7.4.2 The Posterior–Anterior Shift in Aging (PASA) Theory

Before the era of functional imaging, behavioral methods and interpretive logic of clinical neuropsychology-guided brain-based theories of cognitive aging assumed that minimal age differences in behavioral performance infer minimal alterations in underlying cognitive mechanisms. However, with the development of PET and fMRI, which enabled researchers to measure neural activity patterns during cognitive processing, some authors [71, 72] discovered that the assumption that cognitive aging is a process of progressive mental loss that is linear throughout life is incomplete or even erroneous.

PASA was first proposed by Grady et al. [71] in a PET study that explored the relationship between position and face perception. In both cases, older adults showed less activity in the occipitotemporal regions than younger adults, but more activity in anterior regions such as the prefrontal cortex (PFC). This work suggested that older adults used anterior regions to compensate for sensory information processing deficits in occipitotemporal regions [71]. Over the last two decades, numerous fMRI studies have detected the PASA pattern in cognitive function in several cognitive domains, including visual perception [73, 74], attention [58, 75], visuospatial processing [76, 77], working memory [54, 78], episodic memory encoding [79, 80], and episodic memory retrieval [58, 81]. A consistent finding from functional neuroimaging studies of cognitive aging is that an age-related reduction in occipital activity is coupled with increased frontal activity. Although not all studies have detected age-related increases in frontal activity [55, 74, 82], the PASA pattern is nonetheless a well-established aging phenomenon in the functional neuroimaging literature.

The PASA shift is often described as compensatory recruitment of prefrontal regions due to age-related sensory-processing deficits in posterior regions. However, age is associated with both spatial and temporal shifts in recruitment, as younger and older adults recruit the same neural region at different points in a task trial. The results reported by Ford [83] suggested that the oft-reported posterior–anterior shift may not reflect the inability of older adults to engage in sensory processing, but rather a change when they recruit regions responsible for this processing. Evidence from the region of interest (ROI) and functional connectivity analyses suggests that these posterior regions function in comparable ways during search and elaboration processes, contributing to overall memory vividness ratings and interacting with similar neural networks. Critically, these findings suggest that the posterior–anterior shift does not reflect an overall failure of older adults to engage posterior regions during cognitive tasks, but rather a delay in engagement that has not been captured by prior designs. Dew et al. [84] proposed that this effect, called the "early-to-late shift in aging" (ELSA), might not be specific to specific regions or cognitive tasks, but rather a universal feature of aging. Specifically, ELSA can be detected in different neural regions based on the specific demands of cognitive tasks, which helps explain several age-related differences reported in the cognitive neuroscience literature.

7.5 Brain Amyloid and Tau Accumulation in Normal Aging

7.5.1 Amyloid-β in the Aging Brain

The National Institute on Aging–Alzheimer's Association (NIA-AA) Research Framework has proposed the amyloid protein as a crucial vivo biomarker of AD [85]. Amyloid-β (Aβ) deposition is the earliest pathological change among a series of neurodegenerative events that eventually lead to clinical symptoms and dementia [86, 87]. Ten to thirty percent of cognitively normal elderly individuals have Aβ deposition in their brains, and the prognosis of these individuals is worse than that of those with no Aβ deposition [88]. Therefore, studies are investigating whether Aβ deposition in the brain is a part of normal aging or an early response to the neurodegenerative disease. Further clarifying the relationship between Aβ and normal aging is necessary.

Aβ is distributed differently in the brains of normal elderly individuals and patients with AD. An Aβ-atrophy correlation was observed both in cognitively normal older adults and patients with MCI/AD, but was detected in different regions. In cognitively healthy elderly individuals, an Aβ-atrophy correlation was almost exclusively observed in the entorhinal cortex [89–92]; in contrast, in patients with MCI/AD, an Aβ-atrophy correlation mainly existed in the medial and lateral temporal lobes [93, 94]. The selective lack of an Aβ-atrophy correlation in the medial temporal lobe during normal aging may be the key mechanism to understanding AD progression.

Weak relationships between Aβ deposition and cognition have been shown in aging. Although some elderly people show a significantly increased Aβ burden, but still maintain normal cognitive function [95, 96], others show a slight cognitive decline over time or even undetectable changes until approximately 4 years later [97, 98]. The inconsistency of the effect of Aβ on cognition may be due to the region or sequence of Aβ deposition. Aβ deposition was initially detected in the medial frontal and parietal cortices, followed by the lateral temporal lobe [99]. The elderly in the early stages of Aβ accumulation exhibit intact cognition may be contributed to the fact that lesions targeting the temporal lobe tend to produce intractable memory problems, increasing the likelihood of cognitive impairment, whereas lesions in other parts of the cerebral cortex are less directly related to memory problems and more easily remedied.

Aβ is associated with AD risk factors in cognitively normal individuals. Rowe and colleagues [100] found that 18% of cognitively normal adults aged 60–69 years were Aβ-positive, but this value increased to 65% in individuals aged over 80 years. Even in healthy individuals who were Aβ-negative, Aβ deposition was linearly correlated with age when the age range was wide from 23 to 80 years [101]. Aβ deposition appears inevitable, even in healthy older adults [102]. The ApoE4 status, another genetic AD risk factor, plays a crucial role in generating a higher Aβ burden. In normal adults, the Aβ burden is associated with the ApoE status, with significantly higher Aβ deposition observed in ApoE4 allele carriers than in noncarriers [103, 104], especially in the temporoparietal regions [105]. Individuals with a family history of AD are more likely to have a higher Aβ burden [106]. In addition, environmental/acquired factors affect Aβ deposition or regulate the relationship between brain Aβ and cognition, such that cognitive reserve mitigates the adverse effects of Aβ on neuropsychological performance [107], and physical activity and cognitive involvement decrease brain amyloid deposition in cognitively normal APOE4-positive individuals [108].

7.5.2 Tau in the Aging Brain

Tau pathology, another biomarker of AD, also occurs in normal cognitive aging. Normal aging is related to the pathological accumulation of tau in the medial temporal lobe and its assembly into neurofibrillary tangles (NFTs), particularly in the perirhinal and entorhinal cortex, even in the absence of Aβ [109]. This phenomenon is called "primary age-related tauopathy" (PART), a common pathology associated with human aging, and individuals with PART usually show symptoms ranging from normal to amnestic cognitive changes, with only a few showing severe impairments [110].

Tau accumulation is the most commonly observed phenomenon in the entorhinal cortex and is typical of Braak stages I/II and spread stepwise to the basal and lateral temporal, inferior parietal, posterior cingulate, and other association cortices consistent with Braak stages III/IV, and then ultimately to the primary cortical regions, consistent with Braak stages V/VI [111, 112]. NFTs are universal in the brains of older adults; approximately 47% of cognitively normal elderly individuals have moderate or frequent neuritic plaque scores in all Braak stages [113], and a significant number of individuals (approximately 20%) do not develop AD, although they may have tau pathology early in life [109].

However, other studies suggest that compared with Aβ, tau, a downstream product of AD pathology, is not as common in normal older adults. Additionally, tau production is generally accompanied by cognitive decline or impairments in the brain structure and function. In cognitively normal adults, tau pathology in the medial temporal lobe (MTL) is associated with episodic memory performance and MTL atrophy [114]. Tau pathology in the entorhinal cortex is associated with frontal hypometabolism [115]. Tau deposits are also associated with functional hypoconnectivity in the aging brain [116]. Therefore, when high-density tau accumulation is associated with some cognitive deficits and brain lesions, the disorder should attract our attention, which may be tangle-predominant senile dementia (TPSD) [117] or "tangle-only dementia" [118].

References

1. Leong RLF, Lo JC, Sim SKY, Zheng H, Tandi J, Zhou J et al (2017) Longitudinal brain structure and cognitive changes over eight years in an east Asian cohort. NeuroImage 147:852–860
2. Scahill RI, Frost C, Jenkins R, Whitwell JL, Rossor MN, Fox NC (2003) A longitudinal study of brain volume changes in normal aging using serial registered magnetic resonance imaging. Arch Neurol 60(7):989–994
3. Good CD, Johnsrude IS, Ashburner J, Henson RN, Friston KJ, Frackowiak RS (2001) A voxel-based morphometric study of ageing in 465 normal adult human brains. NeuroImage 14(1 Pt 1):21–36
4. Kim RE, Yun CH, Thomas RJ, Oh JH, Johnson HJ, Kim S et al (2018) Lifestyle-dependent brain change: a longitudinal cohort MRI study. Neurobiol Aging 69:48–57
5. Fletcher E, Gavett B, Harvey D, Farias ST, Olichney J, Beckett L et al (2018) Brain volume change and cognitive trajectories in aging. Neuropsychology 32(4):436–449
6. Chen FX, Kang DZ, Chen FY, Liu Y, Wu G, Li X et al (2016) Gray matter atrophy associated with mild cognitive impairment in Parkinson's disease. Neurosci Lett 617:160–165
7. Phillips JS, Da Re F, Dratch L, Xie SX, Irwin DJ, McMillan CT et al (2018) Neocortical origin and progression of gray matter atrophy in nonamnestic Alzheimer's disease. Neurobiol Aging 63:75–87
8. van der Zande JJ, Steenwijk MD, Ten Kate M, Wattjes MP, Scheltens P, Lemstra AW (2018) Gray matter atrophy in dementia with Lewy bodies with and without concomitant Alzheimer's disease pathology. Neurobiol Aging 71:171–178
9. Eshaghi A, Prados F, Brownlee WJ, Altmann DR, Tur C, Cardoso MJ et al (2018) Deep gray matter volume loss drives disability worsening in multiple sclerosis. Ann Neurol 83(2):210–222
10. Driscoll I, Davatzikos C, An Y, Wu X, Shen D, Kraut M et al (2009) Longitudinal pattern of regional brain volume change differentiates normal aging from MCI. Neurology 72(22):1906–1913

11. Last N, Tufts E, Auger LE (2017) The effects of meditation on gray matter atrophy and neurodegeneration: a systematic review. J Alzheimers Dis 56(1):275–286
12. Luders E, Kurth F, Toga AW, Narr KL, Gaser C (2013) Meditation effects within the hippocampal complex revealed by voxel-based morphometry and cytoarchitectonic probabilistic mapping. Front Psychol 4:398
13. Luders E, Toga AW, Lepore N, Gaser C (2009) The underlying anatomical correlates of long-term meditation: larger hippocampal and frontal volumes of gray matter. NeuroImage 45(3):672–678
14. Erickson KI, Leckie RL, Weinstein AM (2014) Physical activity, fitness, and gray matter volume. Neurobiol Aging 35(Suppl 2):S20–S28
15. Zhang K, Sejnowski TJ (2000) A universal scaling law between gray matter and white matter of cerebral cortex. PNAS 97(10):5621–5626
16. Acosta-Cabronero J, Williams GB, Pengas G, Nestor PJ (2010) Absolute diffusivities define the landscape of white matter degeneration in Alzheimer's disease. Brain 133(2):529–539
17. Bosch B, Arenaza-Urquijo EM, Rami L, Sala-Llonch R, Junqué C, Solé-Padullés C et al (2012) Multiple DTI index analysis in normal aging, amnestic MCI and AD. Relationship with neuropsychological performance. Neurobiol Aging 33(1):61–74
18. Huang J, Friedland RP, Auchus AP (2007) Diffusion tensor imaging of Normal-appearing white matter in mild cognitive impairment and early Alzheimer disease: preliminary evidence of axonal degeneration in the temporal lobe. Am J Neuroradiol 28(10):1943–1948
19. O'Dwyer L, Lamberton F, Bokde ALW, Ewers M, Faluyi YO, Tanner C et al (2011) Multiple indices of diffusion identifies white matter damage in mild cognitive impairment and Alzheimer's disease. PLoS One 6(6):e21745
20. SPRINT MIND Investigators for the SPRINT Research Group, Nasrallah IM, Pajewski NM, Auchus AP, Chelune G, Cheung AK, Cleveland ML et al (2019) Association of Intensive vs standard blood pressure control with cerebral white matter lesions. JAMA 322(6):524–534
21. Pakkenberg B, Gundersen HJG (1998) Neocortical neuron number in humans: effect of sex and age. J Comp Neurol 384(2):312–320
22. Geng Z, Liu H, Wang L, Zhu Q, Lv H (2016) A voxel-based morphometric study of age- and sex-related changes in white matter volume in the normal aging brain. Neuropsych Dis Treat 12(Issue 1):453
23. Shankar SK (2010) Biology of aging brain. Indian J Pathol Microbiol 53(4):595–604
24. Boone KB, Miller BL, Lesser IM, Mehringer CM, Hill-Gutierrez E, Goldberg MA et al (1992) Neuropsychological correlates of white-matter lesions in healthy elderly subjects. A threshold effect. Arch Neurol 49(5):549–554
25. Garde E, Mortensen EL, Krabbe K, Rostrup E, Larsson HB (2000) Relation between age-related decline in intelligence and cerebral white-matter hyperintensities in healthy octogenarians: a longitudinal study. Lancet (London, England) 356(9230):628–634
26. Murray AD, Staff RT, Shenkin SD, Deary IJ, Starr JM, Whalley LJ (2005) Brain white matter hyperintensities: relative importance of vascular risk factors in nondemented elderly people. Radiology 237(1):251–257
27. Salarirad S, Staff RT, Fox HC, Deary IJ, Whalley L, Murray AD (2011) Childhood intelligence and brain white matter hyperintensities predict fluid intelligence age 78-81 years: a 1921 Aberdeen birth cohort study. Age Ageing 40(5):562–567
28. de Groot M, Ikram MA, Akoudad S, Krestin GP, Hofman A, van der Lugt A et al (2015) Tract-specific white matter degeneration in aging: the Rotterdam study. Alzheimers Dement 11(3):321–330
29. Ziegler DA, Piguet O, Salat DH, Prince K, Connally E, Corkin S (2010) Cognition in healthy aging is related to regional white matter integrity, but not cortical thickness. Neurobiol Aging 31(11):1912–1926
30. Allen JS, Bruss J, Brown CK, Damasio H (2005) Normal neuroanatomical variation due to age: the major lobes and a parcellation of the temporal region. Neurobiol Aging 26(9):1245–1260
31. Cees De Groot J, De Leeuw FE, Oudkerk M, Van Gijn J, Hofman A, Jolles J et al (2000) Cerebral white matter lesions and cognitive function: the Rotterdam scan study. Ann Neurol 47(2):145–151
32. Head D, Buckner RL, Shimony JS, Williams LE, Akbudak E, Conturo TE et al (2004) Differential vulnerability of anterior white matter in nondemented aging with minimal acceleration in dementia of the Alzheimer type: evidence from diffusion tensor imaging. Cereb Cortex 14(4):410–423
33. Ardekani S, Kumar A, Bartzokis G, Sinha U (2007) Exploratory voxel-based analysis of diffusion indices and hemispheric asymmetry in normal aging. Magn Reson Imaging 25(2):154–167
34. Salat DH, Buckner RL, Snyder AZ, Greve DN, Desikan RS, Busa E et al (2004) Thinning of the cerebral cortex in aging. Cereb Cortex 14(7):721–730
35. Bartres-Faz D, Clemente IC, Junque C (2001) White matter changes and cognitive performance in aging. Rev Neurol 33(4):347–353
36. Kim D, Moon Y (2013) The relationship between depressive symptoms and white matter lesions in elderly individuals with cognitive impairment. Alzheimer's & Dementia
37. Babiloni C, Binetti G, Cassarino A, Dal Forno G, Del Percio C, Ferreri F et al (2006) Sources of cortical rhythms in adults during physiological aging: a multicentric EEG study. Hum Brain Mapp 27(2):162–172
38. Babiloni C, Binetti G, Cassetta E, Dal Forno G, Del Percio C, Ferreri F et al (2006) Sources of cortical rhythms change as a function of cognitive impair-

ment in pathological aging: a multicenter study. Clin Neurophysiol 117(2):252–268
39. Choi J, Ku B, You YG, Jo M, Kwon M, Choi Y et al (2019) Resting-state prefrontal EEG biomarkers in correlation with MMSE scores in elderly individuals. Sci Rep 9(1):1–15
40. Garcés P, Vicente R, Wibral M, Pineda Pardo JA, López ME, Aurtenetxe S et al (2013) Brain-wide slowing of spontaneous alpha rhythms in mild cognitive impairment. Front Aging Neurosci 5:100
41. Babiloni C, Visser PJ, Frisoni G, De Deyn PP, Bresciani L, Jelic V et al (2010) Cortical sources of resting EEG rhythms in mild cognitive impairment and subjective memory complaint. Neurobiol Aging 31(10):1787–1798
42. Babiloni C, Triggiani AI, Lizio R, Cordone S, Tattoli G, Bevilacqua V et al (2016) Classification of single normal and Alzheimer's disease individuals from cortical sources of resting state EEG rhythms. Front Neurosci 10:47
43. Stomrud E, Hansson O, Minthon L, Blennow K, Rosén I, Londos E (2010) Slowing of EEG correlates with CSF biomarkers and reduced cognitive speed in elderly with normal cognition over 4 years. Neurobiol Aging 31(2):215–223
44. Hot P, Rauchs G, Bertran F, Denise P, Desgranges B, Clochon P et al (2011) Changes in sleep theta rhythm are related to episodic memory impairment in early Alzheimer's disease. Biol Psychol 87(3):334–339
45. Reinhart RM, Nguyen JA (2019) Working memory revived in older adults by synchronizing rhythmic brain circuits. Nat Neurosci 22(5):820
46. Solesio-Jofre E, Lorenzo-López L, Gutiérrez R, López-Frutos JM, Ruiz-Vargas JM, Maestú F (2011) Age effects on retroactive interference during working memory maintenance. Biol Psychol 88(1):72–82
47. Solesio-Jofre E, Lorenzo-López L, Gutiérrez R, López-Frutos JM, Ruiz-Vargas JM, Maestú F et al (2012) Age-related effects in working memory recognition modulated by retroactive interference. J Gerontol A Biomed Sci Med Sci 67(6):565–572
48. García-Pacios J, Gutiérrez R, Solesio E, Moratti S, Ruiz-Vargas JM, López-Frutos JM et al (2013) Early prefrontal activation as a mechanism to prevent forgetting in the context of interference. Am J Geriatr Psychiatry 21(6):580–588
49. Fernández A, Maestú F, Amo C, Gil P, Fehr T, Wienbruch C et al (2002) Focal temporoparietal slow activity in Alzheimer's disease revealed by magnetoencephalography. Biol Psychiatry 52(7):764–770
50. Fernández A, Arrazola J, Maestú F, Amo C, Gil-Gregorio P, Wienbruch C et al (2003) Correlations of hippocampal atrophy and focal low-frequency magnetic activity in Alzheimer disease: volumetric MR imaging-magnetoencephalographic study. Am J Neuroradiol 24(3):481–487
51. Fakhri M, Sikaroodi H, Maleki F, Oghabian MA, Ghanaati H (2012) Age-related frontal hyperactivation observed across different working memory tasks: An fMRI study. Behav Neurol 25(4):351–361
52. Hedden T, Gabrieli JDE (2004) Insights into the ageing mind: a view from cognitive neuroscience. Nat Rev Neurosci 5(2):87–96
53. Grady CL, Craik FI (2000) Changes in memory processing with age. Curr Opin Neurobiol 10(2):224–231
54. Rypma B (2000) Isolating the neural mechanisms of age-related changes in human working memory. Nat Neurosci 3:509–515
55. Stebbins GT, Carrillo MC, Dorfman J, Dirksen C, Desmond JE, Turner DA et al (2002) Aging effects on memory encoding in the frontal lobes. Psychol Aging 17(1):44–55
56. Bergerbest D, JDE G, Whitfield-Gabrieli S, Kim H, Stebbins GT, Bennett DA et al (2009) Age-associated reduction of asymmetry in prefrontal function and preservation of conceptual repetition priming. NeuroImage 45(1):237–246
57. Buckner RL (2004) Memory and executive function in aging and AD: multiple factors that cause decline and reserve factors that compensate. Neuron 44(1):195–208
58. Cabeza R, Daselaar SM, Dolcos F, Prince SE, Budde M, Nyberg L (2004) Task-independent and task-specific age effects on brain activity during working memory, visual attention and episodic retrieval. Cereb Cortex 14(4):364–375
59. Rajah MN, D'Esposito M (2005) Region-specific changes in prefrontal function with age: a review of PET and fMRI studies on working and episodic memory. Brain 128(Pt 9):1964–1983
60. Cabeza R, Anderson ND, Locantore JK, McIntosh AR (2002) Aging gracefully: compensatory brain activity in high-performing older adults. NeuroImage 17(3):1394–1402
61. Beeri MS, Lee H, Cheng H, Wollman D, Silverman JM, Prohovnik I (2011) Memory activation in healthy nonagenarians. Neurobiol Aging 32(3):515–523
62. Hutchinson S, Kobayashi M, Horkan CM, Pascual-Leone A, Alexander MP, Schlaug G (2002) Age-related differences in movement representation. NeuroImage 17(4):1720–1728
63. Ward NS, Frackowiak RS (2003) Age-related changes in the neural correlates of motor performance. Brain 126(Pt 4):873–888
64. Smith CD, Umberger GH, Manning EL, Slevin JT, Wekstein DR, Schmitt FA et al (1999) Critical decline in fine motor hand movements in human aging. Neurology 53(7):1458–1461
65. Verstynen T, Diedrichsen J, Albert N, Aparicio P, Ivry RB (2005) Ipsilateral motor cortex activity during unimanual hand movements relates to task complexity. J Neurophysiol 93(3):1209–1222
66. Naccarato M, Calautti C, Jones PS, Day DJ, Carpenter TA, Baron JC (2006) Does healthy aging affect the hemispheric activation balance during paced index-to-thumb opposition task? An fMRI study. Neuroimage 32(3):1250–1256
67. Mattay VS, Fera F, Tessitore A, Hariri AR, Das S, Callicott JH et al (2002) Neurophysiological corre-

lates of age-related changes in human motor function. Neurology 58(4):630–635
68. Riecker A, Groschel K, Ackermann H, Steinbrink C, Witte O, Kastrup A (2006) Functional significance of age-related differences in motor activation patterns. NeuroImage 32(3):1345–1354
69. Li HJ, Hou XH, Liu HH, Yue CL, Lu GM, Zuo XN (2015) Putting age-related task activation into large-scale brain networks: a meta-analysis of 114 fMRI studies on healthy aging. Neurosci Biobehav Rev 57:156–174
70. Spreng RN, Wojtowicz M, Grady CL (2010) Reliable differences in brain activity between young and old adults: a quantitative meta-analysis across multiple cognitive domains. Neurosci Biobehav Rev 34(8):1178–1194
71. Grady CL, Maisog JM, Horwitz B, Ungerleider LG, Mentis MJ, Salerno JA et al (1994) Age-related changes in cortical blood flow activation during visual processing of faces and location. J Neurosci 14(3 Pt 2):1450–1462
72. Cappell RLA (2008) Neurocognitive aging and the compensation hypothesis. Curr Dir Psychol Sci 17(3):177–182
73. Huettel SA, Singerman JD, McCarthy G (2001) The effects of aging upon the hemodynamic response measured by functional MRI. NeuroImage 13(1):161–175
74. Iidaka T, Sadato N, Yamada H, Murata T, Omori M, Yonekura Y (2001) An fMRI study of the functional neuroanatomy of picture encoding in younger and older adults. Brain Res Cogn Brain Res 11(1):1–11
75. Madden DJ, Turkington TG, Provenzale JM, Denny LL, Langley LK, Hawk TC et al (2002) Aging and attentional guidance during visual search: functional neuroanatomy by positron emission tomography. Psychol Aging 17(1):24–43
76. Nyberg L, Sandblom J, Jones S, Neely AS, Petersson KM, Ingvar M et al (2003) Neural correlates of training-related memory improvement in adulthood and aging. Proc Natl Acad Sci U S A 100(23):13728–13733
77. Meulenbroek O, Petersson KM, Voermans N, Weber B, Fernandez G (2004) Age differences in neural correlates of route encoding and route recognition. NeuroImage 22(4):1503–1514
78. Grossman M, Cooke A, DeVita C, Alsop D, Detre J, Chen W et al (2002) Age-related changes in working memory during sentence comprehension: an fMRI study. NeuroImage 15(2):302–317
79. Dennis NA, Daselaar S, Cabeza R (2007) Effects of aging on transient and sustained successful memory encoding activity. Neurobiol Aging 28(11):1749–1758
80. Gutchess AH, Welsh RC, Hedden T, Bangert A, Minear M, Liu LL et al (2005) Aging and the neural correlates of successful picture encoding: frontal activations compensate for decreased medial-temporal activity. J Cogn Neurosci 17(1):84–96
81. Daselaar SM, Veltman DJ, Rombouts SA, Raaijmakers JG, Jonker C (2003) Neuroanatomical correlates of episodic encoding and retrieval in young and elderly subjects. Brain 126(Pt 1):43–56
82. Milham MP, Erickson KI, Banich MT, Kramer AF, Webb A, Wszalek T et al (2002) Attentional control in the aging brain: insights from an fMRI study of the stroop task. Brain Cogn 49(3):277–296
83. Ford JH, Kensinger EA (2017) Age-related reversals in neural recruitment across memory retrieval phases. J Neurosci 37(20):5172–5182
84. Dew IT, Buchler N, Dobbins IG, Cabeza R (2012) Where is ELSA? The early to late shift in aging. Cereb Cortex 22(11):2542–2553
85. Jack CR Jr, Bennett DA, Blennow K, Carrillo MC, Dunn B, Haeberlein SB et al (2018) NIA-AA research framework: toward a biological definition of Alzheimer's disease. Alzheimers Dement 14(4):535–562
86. Jack CR, Knopman DS, Jagust WJ, Petersen RC, Weiner MW, Aisen PS et al (2013) Tracking pathophysiological processes in Alzheimer's disease: an updated hypothetical model of dynamic biomarkers. Lancet Neurol 12(2):207–216
87. Jack CR Jr, Knopman DS, Jagust WJ, Shaw LM, Aisen PS, Weiner MW et al (2010) Hypothetical model of dynamic biomarkers of the Alzheimer's pathological cascade. Lancet Neurol 9(1):119–128
88. Chetelat G, La Joie R, Villain N, Perrotin A, de La Sayette V, Eustache F et al (2013) Amyloid imaging in cognitively normal individuals, at-risk populations and preclinical Alzheimer's disease. Neuroimage Clin 2:356–365
89. Storandt M, Mintun MA, Head D, Morris JCJA, o. n. (2009) Cognitive decline and brain volume loss as signatures of cerebral amyloid-β peptide deposition identified with Pittsburgh compound B: cognitive decline associated with Aβ deposition. Arch Neurol 66(12):1476–1481
90. Fjell AM, Walhovd KB, Fennema-Notestine C, McEvoy LK, Hagler DJ, Holland D et al (2010) Brain atrophy in healthy aging is related to CSF levels of Aβ1-42. Cereb Cortex 20(9):2069–2079
91. Tosun D, Schuff N, Truran-Sacrey D, Shaw LM, Trojanowski JQ, Aisen P et al (2010) Relations between brain tissue loss, CSF biomarkers, and the ApoE genetic profile: a longitudinal MRI study. Neurobiol Aging 31(8):1340–1354
92. Becker JA, Hedden T, Carmasin J, Maye J, Rentz DM, Putcha D et al (2011) Amyloid-β associated cortical thinning in clinically normal elderly. Ann Neurol 69(6):1032–1042
93. De Leon M, DeSanti S, Zinkowski R, Mehta P, Pratico D, Segal S et al (2006) Longitudinal CSF and MRI biomarkers improve the diagnosis of mild cognitive impairment. Neurobiol Aging 27(3):394–401
94. Fjell AM, Walhovd KB, Fennema-Notestine C, McEvoy LK, Hagler DJ, Holland D et al (2010) CSF biomarkers in prediction of cerebral and

clinical change in mild cognitive impairment and Alzheimer's disease. J Neurosci 30(6):2088–2101
95. Hedden T, Oh H, Younger AP, Patel TAJN (2013) Meta-analysis of amyloid-cognition relations in cognitively normal older adults. Neurology 80(14):1341–1348
96. Aizenstein HJ, Nebes RD, Saxton JA, Price JC, Mathis CA, Tsopelas ND et al (2008) Frequent amyloid deposition without significant cognitive impairment among the elderly. Arch Neurol 65(11):1509–1517
97. Jansen WJ, Ossenkoppele R, Tijms BM, Fagan AM, Hansson O, Klunk WE et al (2018) Association of cerebral amyloid-β aggregation with cognitive functioning in persons without dementia. JAMA Psychiatry 75(1):84–95
98. Donohue MC, Sperling RA, Petersen R, Sun C-K, Weiner MW, Aisen PSJJ (2017) Association between elevated brain amyloid and subsequent cognitive decline among cognitively normal persons. JAMA 317(22):2305–2316
99. Villeneuve S, Rabinovici GD, Cohn-Sheehy BI, Madison C, Ayakta N, Ghosh PM et al (2015) Existing Pittsburgh compound-B positron emission tomography thresholds are too high: statistical and pathological evaluation. Brain 138(Pt 7):2020–2033
100. Rowe CC, Ellis KA, Rimajova M, Bourgeat P, Pike KE, Jones G et al (2010) Amyloid imaging results from the Australian imaging, Biomarkers and Lifestyle (AIBL) study of aging. Neurobiol Aging 31(8):1275–1283
101. Vandenberghe R, Van Laere K, Ivanoiu A, Salmon E, Bastin C, Triau E et al (2010) 18F-flutemetamol amyloid imaging in Alzheimer disease and mild cognitive impairment: a phase 2 trial. Ann Neurol 68(3):319–329
102. Mielke MM, Wiste HJ, Weigand SD, Knopman DS, Lowe VJ, Roberts RO et al (2012) Indicators of amyloid burden in a population-based study of cognitively normal elderly. Neurology 79(15):1570–1577
103. Reiman EM, Chen K, Liu X, Bandy D, Yu M, Lee W et al (2009) Fibrillar amyloid-β burden in cognitively normal people at 3 levels of genetic risk for Alzheimer's disease. Proc Natl Acad Sci USA 106(16):6820–6825
104. Rowe CC, Ng S, Ackermann U, Gong SJ, Pike K, Savage G et al (2007) Imaging β-amyloid burden in aging and dementia. Neurology 68(20):1718–1725
105. Fleisher AS, Chen K, Liu X, Ayutyanont N, Roontiva A, Thiyyagura P et al (2013) Apolipoprotein E ε4 and age effects on florbetapir positron emission tomography in healthy aging and Alzheimer disease. Neurobiol Aging 34(1):1–12
106. Mosconi L, Rinne JO, Tsui WH, Berti V, Li Y, Wang H et al (2010) Increased fibrillar amyloid-β burden in normal individuals with a family history of late-onset Alzheimer's. Proc Natl Acad Sci 107(13):5949–5954
107. Rentz DM, Locascio JJ, Becker JA, Moran EK, Eng E, Buckner RL et al (2010) Cognition, reserve, and amyloid deposition in normal aging. Ann Neurol 67(3):353–364
108. Head D, Bugg JM, Goate AM, Fagan AM, Mintun MA, Benzinger T et al (2012) Exercise engagement as a moderator of the effects of APOE genotype on amyloid deposition. Arch Neurol 69(5):636–643
109. Braak H, Thal DR, Ghebremedhin E, Del Tredici K (2011) Stages of the pathologic process in Alzheimer disease: age categories from 1 to 100 years. J Neuropathol Exp Neurol 70(11):960–969
110. Crary JF, Trojanowski JQ, Schneider JA, Abisambra JF, Abner EL, Alafuzoff I et al (2014) Primary age-related tauopathy (PART): a common pathology associated with human aging. Acta Neuropathol 128(6):755–766
111. Johnson KA, Schultz A, Betensky RA, Becker JA, Sepulcre J, Rentz D et al (2016) Tau positron emission tomographic imaging in aging and early Alzheimer disease. Ann Neurol 79(1):110–119
112. Scholl M, Lockhart SN, Schonhaut DR, O'Neil JP, Janabi M, Ossenkoppele R et al (2016) PET imaging of tau deposition in the aging human brain. Neuron 89(5):971–982
113. Sonnen JA, Santa Cruz K, Hemmy LS, Woltjer R, Leverenz JB, Montine KS et al (2011) Ecology of the aging human brain. Arch Neurol 68(8):1049–1056
114. Maass A, Lockhart SN, Harrison TM, Bell RK, Mellinger T, Swinnerton K et al (2018) Entorhinal tau pathology, episodic memory decline, and neurodegeneration in aging. J Neurosci 38(3):530–543
115. Hanseeuw BJ, Betensky RA, Schultz AP, Papp KV, Mormino EC, Sepulcre J et al (2017) Fluorodeoxyglucose metabolism associated with tau-amyloid interaction predicts memory decline. Ann Neurol 81(4):583–596
116. Sepulcre J, Sabuncu MR, Li Q, El Fakhri G, Sperling R, Johnson KA et al (2017) Tau and amyloid β proteins distinctively associate to functional network changes in the aging brain. Alzheimers Dement 13(11):1261–1269
117. Jellinger K, Attems J (2007) Neurofibrillary tangle-predominant dementia: comparison with classical Alzheimer disease. Acta Neuropathol 113(2):107–117
118. Yamada MJN (2003) Senile dementia of the neurofibrillary tangle type (tangle-only dementia): neuropathological criteria and clinical guidelines for diagnosis. Neuropathology 23(4):311–317

Brain Network Organization and Aging

8

Feng Sang, Kai Xu, and Yaojing Chen

Abstract

Despite recent substantial progress in neuroscience, the mechanisms and principles of the complex structure, functions, and the relationship between the brain and cognitive functions have not been fully understood. The modeling method of brain network can provide a new perspective for neuroscience research, and it is possible to provide new solutions to the related research problems. On this basis, the researchers define the concept of human brain connectome to highlight and emphasize the importance of network modeling methods in neuroscience. For example, using diffusion-weighted magnetic resonance imaging (dMRI) technology and fiber tractography methods, a white matter connection network of the whole brain can be constructed. From the perspective of brain function, functional magnetic resonance imaging (fMRI) data can build the brain functional connection network. A structural covariation modeling method is used to obtain a brain structure covariation network, and it appears to reflect developmental coordination or synchronized maturation between areas of the brain. In addition, network modeling and analysis methods can also be applied to other types of image data, such as positron emission tomography (PET), electroencephalogram (EEG), and magnetoencephalography (MEG). This chapter mainly reviews the research progress of researchers on brain structure, function, and other aspects at the network level in recent years.

Keywords

Brain connectome · Cognitive aging · White matter network · Default mode network

8.1 Brain Structural Morphology Network Based on Magnetic Resonance Imaging (MRI)

Brain structural networks have been determined based on the similarity in gray matter morphology between brain areas measurement using structural MRI and structural covariance networks. The vertex of a structured covariance network is usually defined by the region of interest (ROI) in a priori brain parcellation atlases (such as the Desikan Killiany atlas) [1]. The edge of the network is defined by the correlation between the measures (such as cortical thickness) of two ROIs obtained from MRI data processing.

F. Sang · K. Xu · Y. Chen (✉)
State Key Laboratory of Cognitive Neuroscience and Learning, Faculty of Psychology, Beijing Normal University, Beijing, China

Beijing Aging Brain Rejuvenation Initiative (BABRI) Centre, Beijing Normal University, Beijing, China
e-mail: chenyaojing@bnu.edu.cn

Z. Zhang (ed.), *Cognitive Aging and Brain Health*, Advances in Experimental Medicine and Biology 1419, https://doi.org/10.1007/978-981-99-1627-6_8

In 2007, He et al. [2] used the cortical thickness of gray matter as a measurement tool to construct a structural network of the brain and found for the first time that the structural covariance network of humans is a "small-world" network [3]. Modularity was also identified in the network in subsequent studies [4]. In addition to the cortical thickness, researchers also used other gray matter measures to construct the corresponding gray matter structure covariance network. Bassett and colleagues [5] constructed a gray matter volume covariance network of 259 subjects. The network had hierarchical structural characteristics, while the network topology pattern of patients with schizophrenia was abnormal and was characterized by a hierarchical decrease. Sanabria-Diaz and colleagues [6] used the surface area of gray matter to construct a structural covariance network of the brain and found small-world attribution of the network. Additionally, researchers [7] constructed a structural covariance network of the brain using the annual rate of change in cortical thickness in a longitudinal cohort of 108 subjects. This network has similar global and node topology properties to the other structural covariance networks.

Changes in the gray matter structural covariance network might be an early marker and diagnostic marker of some cognitive diseases. In elderly individuals who complain about subjective memory complaints (SMC), the amyloid aggregation is associated with gray matter morphology covariance network measures (such as the clustering coefficient and small-world properties). Even among the young to middle-aged adult group, gray matter covariance network disruption has been linked to a risk of developing Alzheimer's disease (AD) in the future [8]. This evidence suggests that gray matter structural covariance networks differ decades before the potential onset of dementia or cognitive aging. A longitudinal study showed lower small-world coefficient values were associated with a steeper decline in most cognitive domains, including memory, attention, executive function, visuospatial function, and language. Lower betweenness centrality values correlated with a faster reduction in the Mini-Mental State Examination (MMSE) score and memory at a regional level. These associations were specific to the precuneus, medial frontal, and temporal cortex [9]. In contrast to anatomical studies of specific areas of the brain, imaging-related structures between the regions have the potential to reveal the pathology of cognitive impairment in elderly individuals at the network level. For example, atrophy in AD may target areas of the default-mode network, a group of anatomical areas characterized by their relatively intrinsic solid activity when the brain is in a resting state, since these regions may be vulnerable due to their high metabolic load [10].

Besides, inter-regional correlation patterns in SMC or patient populations are also altered compared with inter-regional correlation patterns in healthy people. The most remarkable associations were in orbito- and dorsolateral frontal and parieto-occipital regions in SMC [11]. For AD individuals, the researchers found the correlation of cortical thickness was decreased between bilateral parietal regions and was increased between bilateral temporal, parietal cortex, cingulate, and medial frontal cortex [12]. They also found the nodal centrality reduced in the temporal area and raised in the parietal heterometal association cortex and the occipital cortex regions. Thus, the extensively altered covariance pattern of cortical morphology provides structural evidence for the destruction of the integrity of the cognitive brain networks in patients with AD.

Although structural covariance networks have a certain consistency in some aspects [7], many studies have suggested that structural covariance networks constructed with different gray matter structure measures have different attribution [6]. Thus, different morphology measures reflect the limited characteristics of the gray matter structure. Therefore, more comprehensive studies of structural covariance networks may be required in the future.

8.2 Changes in White Matter Structural Networks in the Brain During Cognitive Aging

The microstructural changes of cerebral white matter (WM) can be quantified with diffusion tensor imaging [13]. As mentioned before in

Chap. 7, the volume and integrity of WM decrease with aging, which is also associated with cognitive declines in cognitive domains such as processing speed and memory. WM integrity is associated with cerebrovascular disease and neurodegenerative disorders. In this section, we will focus on the age-related WM changes in mesoscale, WM structural networks, as well as the impact of WM alterations on brain function.

8.2.1 White Matter Structural Networks in the Brain Change with Aging

With our understanding of cognition and the brain, people have realized that the integrity of brain function depends not only on local brain areas but also on the integration and communication between them. The anatomical organization of the brain can also be approached from the perspective of complex networks. One can use the topological properties of this network to describe and evaluate the shape and organization patterns of brain networks, such as efficiency, "small-world," and rich-club organization [14]. In WM structural networks, the nodes are often defined as brain regions, while the edges are defined as WM connections between nodes.

In general, aging is related to atrophy in both gray matter and WM, as well as deterioration of WM fiber pathways. Reduced communication between regions in the brain network indicates disrupted WM pathways and a reduced ability to transfer information effectively, which affects the efficiency of communication within and between networks [15]. For example, Gong and colleagues [16] investigated WM connectivity patterns in a sample of 95 cognitively normal subjects ranging from 19 to 85 years old and found that WM structural networks exhibited a "small-world" character that allows efficient information transfer at both the local and global scales. When considering the age effect, changes in the underlying network organization resulted in decreased local efficiency (e.g., how well the information is communicated within the cerebral cortex), but preserved overall global efficiency. This result suggests the differential decline and relative preservation of specific regional anatomical connections in the aging brain. For regional efficiency, the results exhibited a clear shift from the parietal and occipital neocortices to the frontal and temporal neocortices, which indicated a putative compensation mechanism for cortical network reorganization in aging. Similar results have also been found by Zhao and colleagues [17] when the age-related alterations in the topology of WM structural networks in a cohort of 137 healthy subjects ranging from 9 to 85 years old were examined. Their results showed that there was an inverted U-shaped trajectory effect of age on connectivity, with the efficiency peak at around 33 years old and decreasing in older age. In addition, they also found that lower rich clubs (e.g., high-degree nodes of a network tend to be more densely connected among themselves than expected by chance) organization in older adults, which is characterized by fewer hub regions and connections in the frontal lobe. In a larger sample of 342 healthy elderly with a narrow age range from 72 to 92 years old, Wen and colleagues [18] also confirmed these age-related connectivity changes, as the brain global structural network efficiency and regional efficiency of many cortical regions were negatively associated with age.

8.2.2 The Effect of Brain Changes in White Matter Structural Networks Changes and Brain Function

The effects of topological properties on brain function also attracted scientists' interest. As reflected in WM structural networks, the organization of brain regions, which is the basis of information communication, is associated with cognitive performance. For example, the higher efficiency of the organization of the brain structure is significantly correlated with intelligence [19]. Wen and colleagues [18] found a positive association between global effectivity and cognitive performance in healthy elderly individuals, including processing speed, visuospatial ability, and executive function, regardless of age, sex, and years of education. Furthermore, neuroanatomical connectivity has been segregated to explain specific cognitive functions. Higher regional connectivity efficiency in almost all cortical regions is associated with higher processing speed. For the visuospatial domain, fewer core regions have been identified, including the frontal lobe, cingulate cortex, insular cortex, and left parietal region. Executive performance is significantly correlated with more excellent regional connectivity of the superior frontal gyrus, posterior cingulate cortex, and left superior parietal cortex.

Patients with AD [20] exhibit an increased shortest path length, decreased global efficiency, and reduced nodal efficiency predominantly located in the frontal regions compared with controls, implying an abnormal topological organization. These alterations in network properties are significantly correlated with behavioral performance. Further studies also confirmed that the abnormal topological organization had already emerged in the preclinical stage of AD [21, 22] and in individuals with subjective cognitive decline [23].

A disrupted brain connectivity architecture has also been observed in individuals with many other psychiatric and neurological disorders. Recently, network analysis revealed connections between central to global network communication and integration to display more significant disturbances across disorders, including schizophrenia, bipolar disorder, attention deficit hyperactivity disorder, autism spectrum disorder, major depressive disorder, obsessive-compulsive disorder, posttraumatic stress disorder, amyotrophic lateral sclerosis, primary lateral sclerosis, AD, and mild cognitive impairment (MCI) [24]. This result suggested general cross-disorder involvement and the importance of these pathways in normal function. Each disorder may involve a balance between disorder-specific and disorder-shared dysconnectivity. Thus, WM connectivity may serve as a potential imaging biomarker for the early detection and progression of the aforementioned diseases.

8.3 Functional Brain Network

8.3.1 Functional Network Studies Based on fMRI

The human brain, a complex network system, comprises different regions. Each region has its own missions and functions. Examining aging in functional networks, as opposed to structured brain networks, might provide new insights into the exchange of information in the brain and closer associations with cognition. The application of the functional magnetic resonance imaging (fMRI) technique provided a powerful tool for the study of functional brain networks and indicated how information integration related to human behavior and how neurodegenerative diseases [25, 26] and aging [27, 28] might alter this organization.

8.3.1.1 Resting-State fMRI and Functional Brain Network

The majority of fMRI techniques use paramagnetic deoxyhemoglobin as an endogenous contrast agent based on blood oxygenation level-dependent (BOLD) contrast, including resting-state fMRI [29]. The degree of temporal correlation produced by the time series of BOLD signals from some brain regions is called "functional connectivity". Functional connectivity is defined as the temporal dependency of neuronal activation patterns of anatomically separated brain regions [30].

To date, neuroimaging studies have identified eight main functionally linked networks [31, 32], including the parietal–frontal network, default mode network, salience network, executive control network, motor network, visual network, and auditory network. Functional intrinsic connectivity networks (ICNs) [33, 34] are considered as the functional foundation of advanced cognitive function and neurodegenerative disease [32, 35]. Raichle, M.E. et al. [10] illustrated the most frequently reported resting-state networks, describing networks that show a high level of functional connectivity during rest. Changes in the functional connectivity of these ICNs are important neuroimaging characteristics in an elderly person [36].

8.3.1.2 Aging of the Default Mode Network

The default mode network (DMN) is a unique network that prompted neuroimaging researchers to examine connections in resting-state brains. The cognitive neuroscientist who first began to use neuroimaging to investigate human brain function reported a serendipitous discovery: distributed regions of the association cortex showed reduced activity when participants performed active attention-demanding tasks. The result is presented relative to their activity levels during passive conditions. The important discovery of the default network was reported in 2001 in a paper titled "*A default mode of brain function*" [37] that conducted a wide range of studies of unique networks of frontal and parietal lobes in the brain, which was subsequently defined as the default network. By 2020, over 3000 publications had been reported describing the DMN.

Anatomically, the DMN consists of brain regions including the posterior cingulate cortex, precuneus, medial prefrontal, inferolateral parietal cortices, and medial temporal cortex. Reduced activity in this network, which was discovered using PET and generalized across numerous fMRI studies, was robustly detected within individuals, and parallels were observed in monkeys and rodents. The correlated spontaneous activity fluctuations across regions within the DMN suggested that they formed an interacting network. Furthermore, researchers found increased activity in DMN regions during the internally constructed representations task, including remembering, envisioning the future, and masking social inferences. These functional observations suggested that the DMN is activated, especially during advanced forms of human thought.

Increasing evidence suggests a role for coherent intrinsic brain activity in brain function. The DMN is mainly responsible for autobiographical memory retrieval, envisioning the future, and conceiving the perspectives of others [38, 39]. It is the most large-scale system generally recognized to be disrupted in individuals with AD [40]. The DMN also appears to be affected by aging. Previous studies have documented a less pronounced reduction of DMN activity during externally driven cognitive tasks in old adults than in young adults [41, 42]. Reduced resting-state brain activity [43, 44] and functional connectivity [45] of the DMN were observed in normal aging. Detailed high-resolution analyses suggest that the DMN comprises multiple interwoven networks [46].

8.3.1.3 Other Networks Involved in Aging

The salience network (SN) is another network that is closely related to aging. SN-related cognitive functions, such as attention and cognitive control, decline rapidly during aging. The SN mainly consists of the dorsal anterior cingulate (ventral subsystem) and frontoinsular (dorsal subsystem) cortices [34, 47]. A resting-state fMRI study showed that connectivity within the ventral subsystem of the SN increased with age, whereas connectivity within the dorsal subsystem decreased with age [48].

The executive control network (ECN) mainly consists of the dorsolateral prefrontal cortex and posterior parietal cortex and supports goal-oriented (or externally directed) cognition [49]. Dissociation between the ECN and DMN is observed in common forms of aging, and a compensatory effect on activation in ECN regions was observed [50].

Internetwork disconnection between the SN, ECN, and DMN occurs during aging, and this disconnection is associated with cognitive decline [51]. The SN mediates the "switching" between activation of the DMN and the ECN and guides appropriate responses to external stimuli [47]. Another study [52] used dynamic causal modeling to assess directed connectivity within and between these three critical networks. A canonical correlation analysis revealed that the relationship between network connectivity and cognitive function is an age-dependent cognitive performance that relies on neural dynamics, and the correlation is more robust in older adults.

8.3.1.4 Functional Brain Network Studies Using Graph Theory

Graph theory (GT), which was originally derived from network science, has emerged as an effective tool for investigating the changes in functional connectivity of the brain. The whole brain network can be abstracted as a set of nodes and edges. Through the combination of resting-state fMRI and GT techniques, functional networks of the human brain have been demonstrated to have robust topological organization over different scales and types of measurements, such as small-world and modular structures. In 2005, Salvador and colleagues [53], for the first time, built a brain functional network using GT with resting-state fMRI data from normal subjects (age: 23–48 years). They used an a priori brain atlas (the automated anatomical atlas, AAL) to divide the brain into 90 regions and then calculated the BOLD signal of each participant to obtain the partial correlation coefficient between different brain regions.

GT approaches have been used to explore the organization of the aging brains. More importantly, some of these features exhibit specific changes associated with aging and various pathological conditions, which indicate the potential value of capturing and monitoring the brain organization throughout a lifetime. Comparing groups of young and old participants, Achard and Bullmore [54] found that healthy aging was associated with a less efficient global network, whereas Meunier et al. [55] reported the restructuring of the overall modular organization of the aging brain. Roser Sala-Llonch et al. [56] reported that older participants showed lower connectivity strength for long-range connections, in other words, higher functional segregation, suggesting a more localized information processing. Also, the higher localization was related to the worse memory performance. Song et al. [57] demonstrated the consistent regional alterations of network topological properties in elderly individuals, especially in the DMN and the sensorimotor network. Specifically, they found that regional connection strength, local and global efficiency of functional connectivity network was increased in the sensorimotor network with aging, while it is decreased in the DMN.

8.3.2 Functional Brain Network Based on Other Neuroimaging Techniques

Fluorodeoxyglucose PET (FDG–PET) measures glucose metabolism values in brain regions. The signals, such as the standard uptake value ratio (SUVr) of each region, are then converted into three-dimensional images. The connectivity between brain regions is analyzed using covariant network analysis. Metabolic network analyses have an advantage since they directly reflect neuronal activity, and thus metabolism is viewed as "the final fingerprint of functionality" [58]. However, only a few studies have investigated the association of metabolic connectivity and aging because of the limited availability of PET imaging data from healthy participants. Limited relevant research suggests that normal aging leads to a widespread loss of independent metabolic function across the brain [59]. The fractal dimensions and heterogeneity are higher in most brain areas of elderly individuals [60]. In terms of network topology properties, older subjects have increased clustering and decreased efficiency and become more vulnerable to targeted attacks in the PET network [61]. Biochemical markers from FDG–PET, such as oxidative phosphorylation functions and efficiency and the Adenosine triphosphate (ATP)/Adenosine monophosphate (AMP) ratio, may predict the age of an individual with a low error rate [62].

EEG is another commonly used method to build a brain network through scalp surface electrodes that record the electrical activity of neuronal discharge in the brain. In an EEG network, electrodes are treated as network nodes, and correlations between signals recorded at these electrodes are treated as network edges. EEG has a high temporal resolution, but the spatial resolution of EEG is relatively poor, often on the centimeter or even decimeter level. EEG has been widely used to study aging and neuropsychological diseases. Through network analyses, Miraglia et al. [63] reviewed studies assessing signs of aging and dementia in EEG. For example, Smit et al. [64] found that the connectivity revealed by EEG was more random in adolescence and old age. A decrease in small-world properties in older adults was also detected based on connectivity patterns [65]. Another recent study [66] found that the normalized characteristic path length showed the pattern "Young >Adult > Elderly" in the higher alpha band, and the elderly group also showed an increase in delta and theta bands.

MEG measures the neuron post-synaptic potential produced by changes in the magnetic field, and the space positioning accuracy is up to 2 mm. MEG has a substantial advantage in terms of a high temporal resolution (milliseconds), and the dynamic activity of brain function can be observed in real-time. MEG provides a direct measure of neuronal activity and bypasses these neurovascular coupling issues. Networks were revealed from large-scale correlation patterns in the slow fluctuations of the band-limited source envelope (i.e., power envelope correlation), particularly in the α and β frequency bands, using MEG. In 2004, Stam et al. [67] used MEG for the first time to establish an undirected functional network in the brains of healthy subjects at different frequency bands under the condition of no task. In both the low frequency (<8 Hz) and high frequency band (>30 Hz) network structures, "small world" features exist, and intermediate frequency (8–30 Hz) rules of the network structure are similar to the network. The differences between different frequency band network structures show that the brain activity associated with different frequency bands may correspond to different brain functions. The high-frequency and low-frequency bands of the "small world" network structure reflect the corresponding frequencies activated during the brain information processing process of optimization. Coquelet et al. [68] used resting-state MEG and a connectome approach in carefully selected healthy young adults and elderly individuals. However, their results failed to support significant age-related differences in the α and β bands after the correction for multiple comparisons, both for static and dynamic resting-state functional connectivity. This result suggests that the electrophysiological connectome is maintained during healthy aging. Mandal and colleagues [69] recently reviewed MEG studies for brain functionality in healthy aging and AD. This review emphasized that MEG is helpful in bridging other electrophysiological measures, such as EEG, local field potential (LFP), fMRI, PET, and brain stimulation.

At the end of this section, some essential information must be emphasized. As Miraglia et al. indicated in their review, "Further advancement is essential in each modality of imaging methods, but combinations of the different modalities could revolutionize our understanding of both structural and functional connectome." [63] A variety of studies has provided different perspectives in this field, and only studies considering and combining all these findings will provide a complete picture for understanding the underlying mechanism.

References

1. Desikan RS, Segonne F, Fischl B, Quinn BT, Dickerson BC, Blacker D et al (2006) An automated labeling system for subdividing the human cerebral cortex on MRI scans into gyral based regions of interest. NeuroImage 31(3):968–980
2. He Y, Chen ZJ, Evans AC (2007) Small-world anatomical networks in the human brain revealed by cortical thickness from MRI. Cereb Cortex 17(10):2407–2419
3. Watts DJ, Strogatz SH (1998) Collective dynamics of 'small-world' networks. Nature 393(6684):440–442
4. Chen ZJ, He Y, Rosa-Neto P, Germann J, Evans AC (2008) Revealing modular architecture of human brain structural networks by using cortical thickness from MRI. Cereb Cortex 18(10):2374–2381
5. Bassett DS, Bullmore E, Verchinski BA, Mattay VS, Weinberger DR, Meyer-Lindenberg A (2008) Hierarchical organization of human cortical networks in health and schizophrenia. J Neurosci 28(37):9239–9248

6. Sanabria-Diaz G, Melie-Garcia L, Iturria-Medina Y, Aleman-Gomez Y, Hernandez-Gonzalez G, Valdes-Urrutia L et al (2010) Surface area and cortical thickness descriptors reveal different attributes of the structural human brain networks. NeuroImage 50(4):1497–1510
7. Alexander-Bloch A, Raznahan A, Bullmore E, Giedd J (2013) The convergence of maturational change and structural covariance in human cortical networks. J Neurosci 33(7):2889–2899
8. Alexander GE, Bergfield KL, Chen K, Reiman EM, Hanson KD, Lin L et al (2012) Gray matter network associated with risk for Alzheimer's disease in young to middle-aged adults. Neurobiol Aging 33(12):2723–2732
9. Dicks E, Tijms BM, ten Kate M, Gouw AA, Benedictus MR, Teunissen CE et al (2018) Gray matter network measures are associated with cognitive decline in mild cognitive impairment. Neurobiol Aging 61:198–206
10. Raichle ME (2011) The restless brain. Brain Conn 1(1):3–12
11. Ten Kate M, Visser PJ, Bakardjian H, Barkhof F, Sikkes SA, Van Der Flier WM et al (2018) Gray matter network disruptions and regional amyloid beta in cognitively normal adults. Front Aging Neurosci 10:67
12. He Y, Chen Z, Evans A (2008) Structural insights into aberrant topological patterns of large-scale cortical networks in Alzheimer's disease. J Neurosci 28(18):4756–4766
13. Basser PJ, Mattiello J, LeBihan D (1994) MR diffusion tensor spectroscopy and imaging. Biophys J 66(1):259–267
14. van den Heuvel MP, Sporns O (2013) Network hubs in the human brain. Trends Cogn Sci 17(12):683–696
15. Antonenko D, Floel A (2014) Healthy aging by staying selectively connected: a mini-review. Gerontology 60(1):3–9
16. Gong G, Rosa-Neto P, Carbonell F, Chen ZJ, He Y, Evans AC (2009) Age- and gender-related differences in the cortical anatomical network. J Neurosci 29(50):15684–15693
17. Zhao T, Cao M, Niu H, Zuo X-N, Evans A, He Y et al (2015) Age-related changes in the topological organization of the white matter structural connectome across the human lifespan. Hum Brain Mapp 36(10):3777–3792
18. Wen W, Zhu W, He Y, Kochan NA, Reppermund S, Slavin MJ et al (2011) Discrete neuroanatomical networks are associated with specific cognitive abilities in old age. J Neurosci 31(4):1204–1212
19. Li Y, Liu Y, Li J, Qin W, Li K, Yu C et al (2009) Brain anatomical network and intelligence. PLoS Comput Biol 5(5):e1000395
20. Lo CY, Wang PN, Chou KH, Wang J, He Y, Lin CP (2010) Diffusion tensor tractography reveals abnormal topological organization in structural cortical networks in Alzheimer's disease. J Neurosci 30(50):16876–16885
21. Shu N, Liang Y, Li H, Zhang J, Li X, Wang L et al (2012) Disrupted topological organization in white matter structural networks in amnestic mild cognitive impairment: relationship to subtype. Radiology 265(2):518–527
22. Zhao TD, Sheng C, Bi QH, Niu WL, Shu N, Han Y (2017) Age-related differences in the topological efficiency of the brain structural connectome in amnestic mild cognitive impairment. Neurobiol Aging 59:144–155
23. Shu N, Wang X, Bi Q, Zhao T, Han Y (2018) Disrupted topologic efficiency of white matter structural connectome in individuals with subjective cognitive decline. Radiology 286(1):229–238
24. de Lange SC, Scholtens LH, van den Berg LH, Boks MP, Bozzali M, Cahn W et al (2019) Shared vulnerability for connectome alterations across psychiatric and neurological brain disorders. Nat Hum Behav 3(9):988–998
25. Bullmore E, Sporns O (2009) Complex brain networks: graph theoretical analysis of structural and functional systems. Nat Rev Neurosci 10(3):186–198
26. Greicius M (2008) Resting-state functional connectivity in neuropsychiatric disorders. Curr Opin Neurol 21(4):424–430
27. Dennis EL, Thompson PM (2014) Functional brain connectivity using fMRI in aging and Alzheimer's disease. Neuropsychol Rev 24(1):49–62
28. Furst AJ, Mormino EC (2010) A BOLD move: clinical application of fMRI in aging. Neurology 74(24):1940–1941
29. Ogawa S, Lee TM, Kay AR, Tank DW (1990) Brain magnetic resonance imaging with contrast dependent on blood oxygenation. Proc Natl Acad Sci U S A 87(24):9868–9872
30. van den Heuvel MP, Hulshoff Pol HE (2011) Exploración de la red cerebral: una revisión de la conectividad funcional en la RMf en estado de reposo. Psiquiatría Biológica 18(1):28–41
31. Buckner RL, Krienen FM, Yeo BTT (2013) Opportunities and limitations of intrinsic functional connectivity MRI. Nat Neurosci 16(7):832–837
32. Yeo BT, Krienen FM, Sepulcre J, Sabuncu MR, Lashkari D, Hollinshead M et al (2011) The organization of the human cerebral cortex estimated by intrinsic functional connectivity. J Neurophysiol 106(3):1125–1165
33. Sorg C, Riedl V, Muhlau M, Calhoun VD, Eichele T, Laer L et al (2007) Selective changes of resting-state networks in individuals at risk for Alzheimer's disease. [Research support, non-U.S. Govt]. Proc Natl Acad Sci U S A 104(47):18760–18765
34. Seeley WW, Menon V, Schatzberg AF, Keller J, Glover GH, Kenna H et al (2007) Dissociable intrinsic connectivity networks for salience processing and executive control. J Neurosci 27(9):2349–2356
35. Raichle ME (2015) The restless brain: how intrinsic activity organizes brain function. Philos Trans R Soc B Biol Sci 370(1668):20140172

36. Tomasi D, Volkow ND (2012) Aging and functional brain networks. Mol Psychiatry 17(5):471, 549–458
37. Raichle ME, MacLeod AM, Snyder AZ, Powers WJ, Gusnard DA, Shulman GL (2001) A default mode of brain function. Proc Natl Acad Sci 98(2):676–682
38. Buckner RL, Andrews-Hanna JR, Schacter DL (2008) The brain's default network - anatomy, function, and relevance to disease. In: Kingstone A, Miller MB (eds) Year in cognitive neuroscience 2008, vol 1124. Annals of the New York Academy of Sciences, pp 1–38
39. Tsioufis C, Dimitriadis K, Taxiarchou E, Vasiliadou C, Chartzoulakis G, Tousoulis D et al (2006) Diverse associations of microalbuminuria with C-reactive protein, interleukin-18 and soluble CD 40 ligand in male essential hypertensive subjects. [comparative study]. Am J Hypertens 19(5):462–466
40. Greicius MD, Srivastava G, Reiss AL, Menon V (2004) Default-mode network activity distinguishes Alzheimer's disease from healthy aging: evidence from functional MRI. Proc Natl Acad Sci U S A 101(13):4637–4642
41. Grady CL, Springer MV, Hongwanishkul D, McIntosh AR, Winocur G (2006) Age-related changes in brain activity across the adult lifespan. J Cogn Neurosci 18(2):227–241
42. Grady CL, Protzner AB, Natasa K, Strother SC, Babak AP, Magda W et al (2009) A multivariate analysis of age-related differences in default mode and task-positive networks across multiple cognitive domains. Cereb Cortex 6:6
43. Damoiseaux JS, Beckmann CF, Arigita EJS, Barkhof F, Scheltens P, Stam CJ et al (2008) Reduced resting-state brain activity in the "default network" in normal aging. Cereb Cortex 18(8):1856–1864
44. Mowinckel AM, Espeseth T, Westlye LT (2012) Network-specific effects of age and in-scanner subject motion: a resting-state fMRI study of 238 healthy adults. NeuroImage 63(3):1364–1373
45. Andrewshanna JR, Snyder AZ, Vincent JL, Lustig C, Head D, Raichle ME et al (2007) Disruption of large-scale brain systems in advanced aging. Neuron 56(5):924–935
46. Buckner RL, DiNicola LM (2019) The brain's default network: updated anatomy, physiology and evolving insights. Nat Rev Neurosci 20(10):593–608
47. Menon V, Uddin LQ (2010) Saliency, switching, attention and control: a network model of insula function. Brain Struct Funct 214(5–6):655–667
48. Alexandra T, Jiahe Z, Andreano JM et al (2018) Dissociable effects of aging on salience subnetwork connectivity mediate age-related changes in executive function and affect. Front Aging Neuroscience 10:410
49. Uddin LQ (2015) Salience processing and insular cortical function and dysfunction. Nat Rev Neurosci 16(1):55–61
50. Buckner RL (2004) Memory and executive function in aging and AD. Neuron 44(1):195–208
51. He X, Qin W, Liu Y, Zhang X, Duan Y, Song J et al (2014) Abnormal salience network in normal aging and in amnestic mild cognitive impairment and Alzheimer's disease. Hum Brain Mapp 35(7):3446–3464
52. Tsvetanov KA, Henson RN, Tyler LK, Razi A, Geerligs L, Ham TE et al (2016) Extrinsic and intrinsic brain network connectivity maintains cognition across the lifespan despite accelerated decay of regional brain activation. J Neurosci 36(11):3115–3126
53. Salvador R, Suckling J, Coleman MR, Pickard JD, Menon D, Bullmore E (2005) Neurophysiological architecture of functional magnetic resonance images of human brain. Cereb Cortex 15(9):1332–1342
54. Achard S, Bullmore E (2007) Efficiency and cost of economical brain functional networks. PLoS Comput Biol 3(2):e17
55. Meunier D, Achard S, Morcom A, Bullmore E (2009) Age-related changes in modular organization of human brain functional networks. NeuroImage 44(3):715–723
56. Sala-Llonch R, Junque C, Arenaza-Urquijo EM, Vidal-Pineiro D, Valls-Pedret C, Palacios EM et al (2014) Changes in whole-brain functional networks and memory performance in aging. Neurobiol Aging 35(10):2193–2202
57. Song J, Birn RM, Boly M, Meier TB, Nair VA, Meyerand ME et al (2014) Age-related reorganizational changes in modularity and functional connectivity of human brain networks. Brain Connect 4(9):662–676
58. Soltow QA, Jones DP, Promislow DE (2010) A network perspective on metabolism and aging. Integr Comp Biol 50(5):844–854
59. Arnemann KL, Stoeber F, Narayan S, Rabinovici GD, Jagust WJ (2018) Metabolic brain networks in aging and preclinical Alzheimer's disease. Neuroimage Clin 17:987–999
60. Lee JS, Lee DS, Park KS, Chung JK, Lee MC (2004) Changes in the heterogeneity of cerebral glucose metabolism with healthy aging: quantitative assessment by fractal analysis. J Neuroimaging 14(4):350–356
61. Liu Z, Ke L, Liu H, Huang W, Hu Z (2014) Changes in topological organization of functional PET brain network with normal aging. PLoS One 9(2):e88690
62. Ravera S, Podesta M, Sabatini F, Dagnino M, Cilloni D, Fiorini S et al (2019) Discrete changes in glucose metabolism define aging. Sci Rep 9(1):10347
63. Miraglia F, Vecchio F, Rossini PM (2017) Searching for signs of aging and dementia in EEG through network analysis. Behav Brain Res 317:292–300
64. Smit DJ, Boersma M, van Beijsterveldt CE, Posthuma D, Boomsma DI, Stam CJ et al (2010) Endophenotypes in a dynamically connected brain. Behav Genet 40(2):167–177
65. Gaál ZA, Boha R, Stam CJ, Molnár M (2010) Age-dependent features of EEG-reactivity—spectral, complexity, and network characteristics. Neurosci Lett 479(1):79–84
66. Vecchio F, Miraglia F, Bramanti P, Rossini PM (2014) Human brain networks in physiological aging: a graph

theoretical analysis of cortical connectivity from EEG data. J Alzheimers Dis 41(4):1239–1249
67. Stam CJ (2004) Functional connectivity patterns of human magnetoencephalographic recordings: a 'small-world' network? Neurosci Lett 355(1–2):25–28
68. Coquelet N, Mary A, Peigneux P, Goldman S, Wens V, De Tiege X (2017) The electrophysiological connectome is maintained in healthy elders: a power envelope correlation MEG study. Sci Rep 7:13984
69. Mandal PK, Banerjee A, Tripathi M, Sharma A (2018) A comprehensive review of magnetoencephalography (MEG) studies for brain functionality in healthy aging and Alzheimer's disease (AD). Front Comput Neurosci 12:60

Part V

Neurobiological Mechanism of Brain Aging

Biomolecular Markers of Brain Aging

9

Min Li, Haiting An, Wenxiao Wang, and Dongfeng Wei

Abstract

Characterized by the gradual loss of physiological integrity, impaired function, and increased susceptibility to death, aging is considered the primary risk factor for major human diseases, such as cancer, diabetes, cardiovascular disorders, and neurodegenerative diseases. The time-dependent accumulation of cellular damage is widely considered the general cause of aging. While the mechanism of normal aging is still unresolved, researchers have identified different markers of aging, including genomic instability, telomere attrition, epigenetic alterations, loss of proteostasis, deregulated nutrient-sensing, mitochondrial dysfunction, cellular senescence, stem cell exhaustion, and altered intercellular communication. Theories of aging can be divided into two categories: (1) aging is a genetically programmed process, and (2) aging is a random process caused by gradual damage to the organism over time as a result of its vital activities. Aging affects the entire human body, and aging of the brain is undoubtedly different from all other organs, as neurons are highly differentiated postmitotic cells, and the lifespan of most neurons in the postnatal period is equal to the lifespan of the brain. In this chapter, we discuss the conserved mechanisms of aging that may underlie the changes observed in the aging brain, with a focus on mitochondrial function and oxidative stress, autophagy and protein turnover, insulin/IGF signaling, target of rapamycin (TOR) signaling, and sirtuin function.

Keywords

Genomic instability · Proteostasis · Calcium homeostasis · Neural regulators · Mitochondrial dysfunction

M. Li
State Key Laboratory of Cognitive Neuroscience and Learning, Faculty of Psychology, Beijing Normal University, Beijing, China

Beijing Aging Brain Rejuvenation Initiative (BABRI) Centre, Beijing Normal University, Beijing, China
e-mail: 201831061049@mail.bnu.edu.cn

H. An
Beijing Neurosurgical Institute, Beijing Tian Tan Hospital, Capital Medical University, Beijing, China

Beijing Aging Brain Rejuvenation Initiative (BABRI) Centre, Beijing Normal University, Beijing, China

W. Wang
School of Systems Science, Beijing Normal University, Beijing, China

Beijing Aging Brain Rejuvenation Initiative (BABRI) Centre, Beijing Normal University, Beijing, China

D. Wei (✉)
Institute of Basic Research in Clinical Medicine, China Academy of Chinese Medical Sciences, Beijing, China

Beijing Aging Brain Rejuvenation Initiative (BABRI) Centre, Beijing Normal University, Beijing, China

Z. Zhang (ed.), *Cognitive Aging and Brain Health*, Advances in Experimental Medicine and Biology 1419, https://doi.org/10.1007/978-981-99-1627-6_9

Characterized by the gradual loss of physiological integrity, impaired function, and increased susceptibility to death, aging is considered the primary risk factor for major human diseases, such as cancer, diabetes, cardiovascular disorders, and neurodegenerative diseases. The time-dependent accumulation of cellular damage is widely considered the general cause of aging. While the mechanism of normal aging is still unresolved, researchers have identified different markers of aging, including genomic instability, telomere attrition, epigenetic alterations, loss of proteostasis, deregulated nutrient-sensing, mitochondrial dysfunction, cellular senescence, stem cell exhaustion, and altered intercellular communication [1]. Theories of aging can be divided into two classes: (1) aging is a genetically programmed process, and (2) aging is a random process caused by gradual damage to the organism over time as a result of its vital activities [2].

Aging affects the entire human body, and aging of the brain is undoubtedly different from all other organs, as neurons are highly differentiated postmitotic cells, and the lifespan of most neurons in the postnatal period is equal to the lifespan of the brain [3]. In this chapter, we discuss the conserved mechanisms of aging that may underlie the changes observed in the aging brain, with a focus on mitochondrial function and oxidative stress, autophagy and protein turnover, insulin/IGF signaling, target of rapamycin (TOR) signaling, and sirtuin function.

9.1 Genomic Instability

One common characteristic of aging is the accumulation of genetic damage over a lifetime. The integrity and stability of DNA are constantly threatened by exogenous chemical, physical, and biological agents, as well as endogenous intimidations, including DNA replication errors, reactive oxygen species (ROS), and spontaneous hydrolytic reactions.

Genetic damage caused by external or internal damage is highly diverse. This damage includes point mutations, translocations, gains and losses of chromosomes, shortening of telomeres, and gene destruction due to virus or transposon integration. In addition, in response to direct damage to DNA, defects in the nuclear architecture, which are also called laminopathies, can cause genome instability and lead to premature aging syndromes [4]. To minimize this damage, organisms have evolved an advanced DNA repair system or stability system that can selectively deal with most of the lesions that occur on nuclear DNA. Genomic stability systems also include specific mechanisms that maintain the proper length and function of telomeres and ensure the integrity of mitochondrial DNA (mtDNA).

9.1.1 DNA Damages

Comparisons between young and old tissues from several species based on transcriptional microarrays have identified age-related genetic changes that encode key components of inflammatory, mitochondrial, and lysosomal degradation pathways. The age-dependent accumulation of somatic DNA mutations has been noted in single human postmitotic neurons from the prefrontal cortex and hippocampus [5].

mtDNA has been considered the main target of aging-associated somatic mutations. Mutations and deletions in mtDNA in elderly individuals can cause diverse human mitochondrial diseases and are also heavily involved in aging and age-associated diseases. mtDNA damage is also one of the signs of brain aging as the result of the oxidative microenvironment in mitochondria. mtDNA lacks protective histones, and the mtDNA repair mechanism is less efficient than that of nuclear DNA. Owing to the diversity of mitochondrial genomes, the causal relationship of mtDNA mutations in aging has been contentious. The mitochondrial genome allows mutant and wild-type genomes to coexist in the same cell, which is called "heteroplasmy." Although the overall level of mtDNA mutations is low, the mutation load of individual senescent cells is still significant and may be in a homoplasmy state, where the mutant genome dominates the normal genome [6]. Interestingly, recent studies in mice have indicated that most somatic mtDNA mutations in senescent cells appear to be caused by replication errors during the early developmental period rather than by accumulated damage, such as oxidative damage, in adult life. These mutations may undergo polyclonal expansion and induce respiratory chain dysfunction in different tissues [7]. Studies on accelerated aging in HIV-infected patients treated with antiretroviral drugs may affect mtDNA replication, which has supported the phenomenon of clonal amplification of mtDNA mutations that originate early in life [8]. The first evidence that mtDNA damage might be important for aging and age-related diseases was obtained by identifying multisystem disorders in humans, that is, phenotypic aging partially caused by mtDNA mutations [9]. Further causal evidence came from studies of mitochondrial DNA polymerase γ-deficient mice, which exhibited premature aging and reduced lifespan in association with cumulative random point mutations and deletions in mtDNA [10, 11]. Cells from these mice exhibited impaired mitochondrial function, but surprisingly, it was not accompanied by an increase in ROS production [12]. Future studies are necessary to determine whether genetic manipulations that can reduce mtDNA mutational load are able to prolong lifespan.

9.1.2 RNA Alterations

Since aging is associated with aberrant production and maturation of many mRNAs, the spotlighted nature of age-related gene expression may uncover potential biomarkers for aging. In a large-scale transcriptomic microarray study on human peripheral blood leukocytes to detect in vivo RNA, the mRNA levels of *CCR6* (chemokine receptor 6), *CCR7* (chemokine receptor 7), and *CD27* genes were found to be associated with age. This was supported by the observation of age-associated disruption to the balance of alternatively expressed isoforms in selected genes, suggesting that modified mRNA processing may be a characteristic of human aging [13]. A-to-I editing is an adenosine-to-inosine modification of mRNA particularly widespread in the human brain, where it affects thousands of genes. A growing amount of evidence suggests that A-to-I RNA editing is necessary for normal development and maintenance in mammals and that deficiencies in A-to-I RNA editing contribute to a number of pathological states. The mRNA editing levels of CYFIP2, which is implicated in synaptic maintenance, exhibit an age-dependent statistically significant decrease [14].

Several common genetic signatures of aging, including 56 genes, are consistently overexpressed with aging. The most well-known gene is apolipoprotein D (*APOD*), and another 17 genes are under-expressed with aging, characterizing the biological processes associated with these signatures and changes in age-related gene expression. The most notable changes included the overexpression of inflammation and immune response genes and genes associated with the lysosome. An under-expression of collagen genes and genes associated with energy metabolism, particularly mitochondrial genes, as well as alterations in the expression of genes related to apoptosis, cell cycle, and cellular senescence biomarkers, were also observed. By employing a new method that emphasizes the sensitivity of genome detection, more research work will further reveal transcriptional changes that occur with age in many genes, processes, and functions that were previously unknown. These molecular signatures may reflect a combination of degenerative processes and transcriptional responses to the process of aging. Altogether, they may reveal how transcriptional changes relate to the process of aging and could serve as targets for future studies [15].

Nucleolar shrinkage during aging may represent the decreased transcription of ribosomal RNA (rRNA) cistrons and the assembly of ribosomes as a change in gene regulation [16]. A different mechanism suggests that there could be a 30–50% age-related loss of rRNA genes in DNA [17]. Tandemly repeated sequences, such as rRNA cistrons, are hypothesized to be at risk for excision, as classically observed for the Bar locus of Drosophila. Thus far, little is known about the roles in which noncoding RNAs (ncRNAs) play in age-dependent memory decline. There are several possible scenarios for age-dependent control of memory by ncRNAs. MicroRNAs (miRNAs) have the potential to affect complex networks of genes, which provide translational control and affect mRNA stability. Since many miRNAs exhibit developmental patterns of expression, it is reasonable to imagine that age-dependent changes occur in the miRNA constitution of cells, which in turn affect the protein environment. The targeting of sarco/endoplasmic reticulum Ca^{2+} ATPase type 2 (Serca2) protein levels represents one of the first known examples of age-dependent regulation of a target by miRNAs [18]. The control of miRNAs by aging genes, such as the deacetylase gene *SIRT1,* and interactions of miRNAs with the age-dependent *N*-methyl-D-aspartic acid receptor (NMDAR) pathway suggest that many more miRNA pathways may be critical in the aging brain. In addition, the complement of miRNAs may differ not only with aging in individual cells level but also subcellular regions, which adds more variables for future research. Antisense long noncoding RNAs (lncRNAs) might target the transcription of specific genes that are important for memory in an age-dependent model [19]. Discoveries of ncRNAs in aging enlighten the future of the molecular neuroscience of aging.

9.2 Proteostasis

Proteostasis is the compilation of cellular mechanisms that are involved in maintaining the homeostasis of the proteome, which includes the building and recycling of human proteins to avoid misfolding and aggregation of damaged and dysfunctional polypeptides that may compromise cellular resilience [20]. The proteostasis network consists of all types of autophagy, the ubiquitin–proteasome system (UPS), and the unfolded protein response (UPR) in the endoplasmic reticulum (ER) [21]. The loss of any components in this network may lead to protein aggregation and proteotoxic effects. Studies have already demonstrated that the ability of cells to maintain proteostasis deteriorates with aging, and non-native protein aggregates and inclusions have been found in almost all tissues in aged organisms [22]. This aggregation and accumulation of misfolded proteins is the principal cause of some common age-associated diseases in humans, which primarily include neurodegenerative diseases, such as Alzheimer's disease (AD) and Parkinson's disease (PD). The hallmark of all these diseases is the consistent occurrence of detergent-insoluble protein inclusions and aggregates in the nucleus and cytoplasm of neurons [23].

9.2.1 Autophagy

Autophagic degradation of aggregated proteins or organelles begins with the formation of a double membrane enveloping the damaged, aggregated protein or organelle to form an autophagosome, which is transported to the lysosome, where membrane fusion leads to the formation of the autophagolysosome. Endocytosis of the contents in the autophagosome into the acidic interior of the lysosome leads to proteolytic degradation of those proteins. Autophagolysosomes release degraded peptides, amino acids, and other biomolecules for reuse [24]. Proteostasis, including autophagy and proteasome activity, decreases with aging [25]. Various types of interventions support the idea that reducing the proteotoxic load during aging can prolong lifespan. In mammals, studies are just beginning to appear to promote proteasome or autophagy activity by overexpressing proteasome subunits or essential autophagy genes, thereby extending lifespan and conferring resistance to stress. For instance, overexpression of the essential autophagy gene *Atg5* in the whole body of mice revealed an anti-aging phenotype and a lifespan extension of approximately 20% [26]. Interestingly, in experimental models, most of the interventions that slow aging are associated with improved proteostasis, and these interventions demonstrated autophagy activation in many cases. For example, calorie restriction, metformin, rapamycin, resveratrol, and spermidine, which are well known for their ability to prolong lifespan or health span, have all been proven to activate autophagy directly [27]. Because at least some of these interventions can also influence proteostasis through their effects on chaperone levels and protein synthesis, the contribution of autophagy activation to their overall impact on longevity is still being investigated [28].

9.2.2 The Ubiquitin–Proteasome System (UPS)

Damaged proteins are polyubiquitinylated by ubiquitin ligase enzymes E1, E2, and E3. An initial ubiquitin molecule binds to the damaged protein or organelle through this process and is repeated to form a polyubiquitinylated chain. Polyubiquitinylated damaged proteins are selected for degradation by the 26S proteasome prior to entering the 19S cap. These proteins must be deubiquitinylated by the enzyme ubiquitin carboxy-terminal hydrolase L1 (UCHL1) first, one ubiquitin residue at a time. The deubiquitinylated, damaged protein is degraded by proteinases in the 20S portion of the 26S proteasome, while small peptides are ejected by the bottom 19S portion of the 26S proteasome and then degraded by soluble peptidases into individual amino acids for reuse [29].

Proteasomal activity in mammals declines during aging. With the declined proteasomal function, oxidatively damaged proteins (20S substrates) and polyubiquitinated proteins (26S substrates) increase [30]. The enrichment of undegraded substrates may be due to the lowered activity of the UPS or intracellular accumulation of severely oxidized and covalently cross-linked protein aggregates (lipofuscin) [31]. These structures are still recognized as proteasomal substrates and are found to be massively polyubiquitinated, but they are very resistant to proteolytic degradation and, therefore, divert proteasomal activity from substrates which are still degradable. Furthermore, the decline of proteolysis and the improved production of ROS in aged cells lead to increased formation of lipofuscin [32].

Molecular chaperones are small proteins that assist natural polypeptide chains in folding into functional protein structures. The most important chaperones are heat-shock family (HSF) proteins, which play a central role in the proteostasis network (PN) and can be broadly grouped into five families: heat-shock protein 70 (HSP70), HSP90, Hsp70/Hsp40 (also called DnaJ proteins), chaperonin/HSP60, and small HSP (sHSP) families. Chaperones can act alone or in combination with different cochaperones to regulate protein interactions, folding, disaggregation, degradation, and trafficking within the cell [33]. The transcriptional expression of these chaperones is upregulated by cellular stress, including heat shock.

A number of studies on several animal models support the causative impact of chaperone decline on lifetime. Transgenic worms and flies overexpressed Hsp22 chaperones display longer longevity [34]. Moreover, mutant mice deficient in heat-shock family cochaperones exhibit accelerated aging phenotypes, while long-lived mouse strains show that some heat-shock proteins markedly upregulated [35]. The transcription factor HSF1 is considered a master regulator of the heat-shock response in mammalian cells, which is transcriptionally increased in response to increased levels of unfolded proteins [23]. Furthermore, the downregulated expression level of molecular chaperones, alteration of proteasome activity, and disruption of stress responses have been observed in senescent rodent tissues and aged human cells [36–38]. Additionally, an investigation of chaperone and cochaperone gene expression level in young (36 ± 4 years of age) and aged (73 ± 4 years of age) human brain tissue revealed that among 332 genes examined, 101 are significantly suppressed by age, including *HSP70*, *HSP40*, *HSP90*, and *TRiC* genes. Moreover, 62 chaperone genes, including several small HSPs, were found to be significantly induced, likely resulting from the cellular response to protein damage accumulated with aging [39].

9.2.3 The Unfolded Protein Response (UPR) in the Endoplasmic Reticulum (ER)

The ER is a multifunctional organelle with many functions, such as protein folding, lipid synthesis, calcium storage, and calcium release. The ER is directly involved in protein synthesis because its lumen contains a high concentration of chaperones that can detect unfolded protein products and promote their folding. The number and activity of key molecular chaperones and folding enzymes decrease with age, which compromises appropriate protein folding and the adaptive response of the UPR [40]. Interference with ER homeostasis leads to the misfolding of proteins, which could become a type of ER stress and lead to the upregulation of a signaling pathway called the ER stress response. ER stress can be provoked by various physiological conditions, including disturbances in calcium homeostasis, glucose deprivation, redox changes, hyperhomocystinemia, ischemia, viral infections, and mutations, which can impair protein folding.

The ER stress is also known as UPR, which is relevant to chaperones associated with ER stress and is usually induced by an elevation in the levels of misfolded proteins or Ca^{2+} in the ER lumen. The primary features of the UPR are activation of chaperones, degradation of misfolded proteins, and reduction of protein translation. Misfolded proteins are toxic and interfere with normal cellular functions, especially under proteotoxic stress. Thus, molecular chaperones, UPS, and autophagy together promote misfolded proteins to be refolded or cleared [41].

Cells have evolved an adaptive coordinated response to limit the accumulation of unfolded proteins in the ER. On the cellular level, the adaptive mechanism of the UPR triggers three kinds of protective cellular responses to limit protein load and alleviate ER stress: (1) upregulation of ER chaperones, such as immunoglobulin binding protein (BiP)/glucose-regulated proteins 78 (GRP78), to assist in protein refolding [42]; (2) degradation of misfolded proteins by the proteasome in a process called ER-associated degradation (ERAD) [43]; and (3) attenuation of protein translation mediated by the serine-threonine kinase PERK (PKR-like endoplasmic reticulum kinase), which can phosphorylate translational initiation factor eIF2a, accordingly reducing translation [44]. When these strategies unsuccessfully adapt, ER-initiated pathway signaling is activated by activating NF-κB, a transcription factor that induces genes that encode mediators of host defense [45]. Excessive and prolonged stress leads to a maladjustment response and apoptosis [46].

As described above, the BiP level is critical for proper folding and protein quality control (PQC), and BiP expression decreases with age. BiP protein levels exhibit an age-associated decline in brain tissue; there is a 30% reduction in BiP in the cerebral cortex of aged C57/B6 mice (22–24 months old) compared to that in younger mice (3 months old) [47]. Both BiP mRNA and protein expression are decreased in the hippocampus of 23–26-month-old Wistar rats compared to 4–6-month-old rats [48]. The age-related BiP decline at the transcript and protein levels affects the efficiency of protein folding and ER homeostasis.

Formation of the PQC compartments is an active process including chaperones, ubiquitination, and sorting factors that interact with cellular structures, such as nuclear membrane, ER network, and cytoskeleton. ATP-dependent molecular chaperones which belong to the Hsp70, Hsp90, and Hsp110 families have been demonstrated to be important for refolding and degradation of misfolded or stress-denatured proteins [49]. However, these proteins have complex roles that may vary between disease states, and a better understanding of their clinical application is required [50].

9.3 Calcium Homeostasis

Calcium is a ubiquitous second messenger in cells, and calcium levels are strictly regulated in neurons through complex homeostatic mechanisms. Cellular Ca^{2+} homeostasis plays a key regulatory role in many neuronal physiological aspects, including growth and differentiation, action potential properties, synaptic plasticity, and learning and memory [51]. Abnormal Ca^{2+} regulation and altered Ca^{2+} sensitivity in neurons are involved in the aging process. Alterations in Ca^{2+} dynamics may make aging neurons more vulnerable to neuronal death following seizures, stroke, or head trauma. The brains of aged rodents exhibited different levels of age-dependent Ca^{2+} dysregulation, including increased cytoplasmic Ca^{2+} ($[Ca^{2+}]$cyt), elevated Ca^{2+} influx through voltage-gated Ca^{2+} channels (VGCCs), exaggerated agonist-induced ER Ca^{2+} release through inositol 1,4,5-trisphosphate receptors (InsP3R) and ryanodine receptors (RyRs), impaired competence of mitochondrial Ca^{2+} recycling, and attenuated store-operated Ca^{2+} entry (SOCE) [52–55]. Raza first determined that basal $[Ca^{2+}]_i$ levels in middle-aged neurons are significantly higher than those in young neurons. Because of glutamate stimulation and subsequent $[Ca^{2+}]_i$ load, middle-aged neurons took longer to remove the excess $[Ca^{2+}]_i$ than young neurons, which directly indicates that altered Ca^{2+} homeostasis might be present in animals at significantly younger ages (12–16 months) than commonly considered ages (> or = 24 months). Elucidating the function of the Ca^{2+} homeostasis mechanism may provide insights into the neuronal loss that increases with aging and allow the development of new therapeutic agents that target the reduction in $[Ca^{2+}]_i$ levels, repairing and maintaining a steady state of Ca^{2+} homeostasis in the aged neurons [54].

Ca^{2+} influx via L-type voltage-gated Ca^{2+} channels (LTCCs) is a major source of Ca^{2+} involved in Ca^{2+} dysregulation during aging, and Ca^{2+}-induced Ca^{2+} release (CICR) enhances cytoplasmic Ca^{2+} levels in aging. CICR is mediated by RyRs in the ER, which release increased Ca^{2+} in response to Ca^{2+} influx via plasmalemmal channels. This process also leads to the loss of input specificity, which is an electrophysiological biomarker of aging that can be prevented by blocking LTCC or restraining CICR [56]. CICR enhances the inflow of Ca^{2+}, thereby increasing the concentration of Ca^{2+} to drive Ca^{2+}-dependent functions. As changed precise adjustment leads to Ca^{2+} imbalance, in the aging process, CICR further amplifies the already increased source of LTCC Ca^{2+} and leads to an imbalance of Ca^{2+} [57, 58].

Synaptic plasticity in all kinds of excitatory and inhibitory synapses is dependent on postsynaptic Ca^{2+} release. Accumulating evidence indicates that any dysregulation of Ca^{2+} homeostasis can lead to dramatic changes in synaptic and neuronal integrity [59]. Landfield and Pilter showed for the first time that Ca^{2+} levels are elevated in aged brains, and that in aged brains, Ca^{2+}-dependent K^+ conductance and the resulting afterhyperpolarization (AHP) were increased [60].

Since the increase in Ca^{2+} levels is a consistent characteristic of aging hippocampal neurons, it may be an important factor leading to increased hippocampal vulnerability, AD, and other neurodegenerative diseases. Hypotheses explaining changes in excitatory synaptic function that occur during aging feature the idea that Ca^{2+} dysregulation in aged CA1 neurons destroys the circuit dynamics of aged pyramidal cells. The change in intracellular Ca^{2+} level homeostatic control in old rats is related to changes in gene expression [61], which explains many observed age-associated plasticity deficits of CA3-CA1 Schaffer collateral synapses in vitro and accounts for the decrease in behaviorally induced plasticity in the CA1 of awake-behaving old rats [62]. Moreover, neuronal therapeutic treatments of related gene expression, which can alter intracellular Ca^{2+} levels, improve neuronal plasticity [63, 64] and cognition in aged animals [65, 66]. Aged CA1 pyramidal cells have higher Ca^{2+} conductivity due to the higher density of L-type Ca^{2+} channels [58], which may lead to disrupted Ca^{2+} homeostasis that primarily contributes to age-associated plasticity deficits [67]. Furthermore, it has been speculated that postsynaptic intracellular Ca^{2+} levels are involved in setting the synaptic modification curve, which determines the possibility of synapses being inhibited or potentiated for a given input mode [68]. If Ca^{2+} homeostasis is disrupted in aged animals, then it is likely that the probability for a given synapse to be inhibited or potentiated will also be altered accordingly [58, 67].

Increased cytoplasmic Ca^{2+} has also been found in APP processing and tau phosphorylation. Ca^{2+} disruption may underlie the impaired misfolded protein removal, leading to the accumulation of amyloid plaques and tangles [69, 70]. Calcium ion channels have been proposed to have a significant role in maintaining lysosomal acidification. A previous study implicated the Ca^{2+} pore Mucolipin 1, which contributes to lysosomal pH regulation [71], and loss of function of this channel causes mucolipidosis type IV, a disease in which lysosomal pH is increased and cellular pathology including defective autophagy and prominent autolysosome accumulation [72, 73]. A second lysosomal Ca^{2+} channel, called two pore segment channel 2 (TPC2), may also regulate lysosomal pH via a nicotinic acid adenine dinucleotide phosphate (NAADP)-dependent Ca^{2+}/pH feedback mechanism, whereby calcium regulation and acidification of the lysosome are privately linked through the activity of these two pores, outwardly rectifying ion channels [74].

Altered calcium signaling is connected to the enzymatic activity of γ-secretase, as neurons expressing mutant presenilin, which forms a catalytic subunit of the APP-cleaving enzyme γ-secretase, appear to have weak capacitive calcium entry, which is also known as store-operated calcium entry capabilities [75]. Aβ interaction with the plasma membrane leads to increased intracellular Ca^{2+} concentrations ($[Ca^{2+}]_i$) and elevated vulnerability of neurons to excitotoxicity [76], and oligomeric forms of Aβ42 cause Ca^{2+}-mediated toxicity in cultured cells [77]. Many mutations in presenilin, which cause familial AD (FAD), increase the production of the long aggregation-prone form of Aβ (Aβ42) or reduce the production of a short soluble form (Aβ40); therefore, one way in which presenilin mutations may disturb neuronal Ca^{2+} homeostasis is by increasing the Aβ42:Aβ40 ratio and activating Aβ oligomer-mediated mechanisms [78]. Understanding Ca^{2+} homeostatic mechanisms will allow us to develop novel therapeutic agents that target Ca^{2+} levels and restore plasticity and function by maintaining normal Ca^{2+} homeostasis.

9.4 Neural Regulators

9.4.1 Neural Transmitter

The central nervous system (CNS) is responsible for maintaining the metabolism and function of each organ. Age-related changes in the CNS may affect the regulation of other additional nervous systems and their aging process. Synapses are functional units of neuronal networks in the brain and therefore underlie the functional changes of synapses with aging [79].

With possible mechanisms to be explored, acetylcholine released from synapses was found to decline in the aging brain [80]. During the aging phase, the release of acetylcholine from synaptosomes induced by depolarization is significantly reduced, which may be due to the decreased production of acetylcholine. Gibson and Peterson reported that the synthesis of acetylcholine decreases in brain tissue slices from old rats using radiolabeled precursors [81]. However, their method only focused on labeled tracers and may have missed the unlabeled acetylcholine portion of mitochondria derived from unlabeled acetyl-CoA from mitochondria. It was believed that the decrease in acetylcholine release from aging synapses was due to dysfunction of the release mechanism rather than a decline in acetylcholine synthesis [82]. However, the choline acetyltransferase activity and the content of acetylcholine in synaptosomes may actually remain constant from childhood to old age, as Tanaka's more convincing experiments determined the total amount of acetylcholine using an electrochemical detector and showed that both the production rate of acetylcholine and the content of acetylcholine in synaptosomes remained unchanged across all ages.

Attenuated calcium influx through voltage-gated calcium channels is thought to reduce the release of acetylcholine in elderly synapses [82]. Tanaka further investigated the age-related contribution of each voltage-gated calcium channel to depolarization induced by calcium influx into synapses and found that the density of calcium channels in the synaptic plasma membrane changed. They found that the relative contribution of different types of calcium channels to the calcium influx caused by membrane depolarization was determined by measuring the inhibitory effects of specific blockers on L-, N-, P-, and Q-type channels on calcium influx, with the effects of N- and P-type channels decreasing during aging. The density of calcium channels on the aging synaptic plasma membrane was significantly reduced. These changes in voltage-gated calcium channels may cause an age-related decline in synaptic transmission [83].

The electrical excitability of the synaptic plasma membrane also modulates synaptic transmission. Resting-state membrane potential affects the release of neurotransmitters through voltage-gated calcium channels. The resting membrane potential measured in mouse cortical synaptosomes decreased by 10% in older age, and the decrease in membrane potential was consistent with the change in intrasynaptosomal potassium concentration [84]. Enzyme activity is known to be regulated partially by its lipid microenvironment [85], and among membrane phospholipids, only phosphatidylcholine decreased, and its changes were tightly correlated with changes in the activity of Na^+, K^+-ATPase [84].

A study of synaptophysin, which is located on the outer surface of synaptic vesicles, in brain areas of 3-, 12-, 24- and 33-month-old rats, including the frontal, occipital, temporal, entorhinal cortex, hippocampus, striatum, and cerebellum, suggested that synaptophysin contents decrease in every region of the aged brain, and the largest reduction occurs in the occipital cortex [86]. Furthermore, the age-associated decline in synaptic transmission relates to the decreased density of synaptic vesicles.

An age-associated decline in brain dopamine levels was first discovered in the basal ganglia in post-mortem human samples [87], and a later living dopamine imaging system was developed [88]. Positron emission tomography (PET) and single-photon emission computed tomography (SPECT) studies have mostly verified age-associated dopaminergic loss in autopsies, especially tracers targeting dopamine receptors or transporters. Interneuronal communication in the dopaminergic pathway is accomplished through dopamine. Dopamine receptors are proteins located in neuronal membranes. There are five subtypes of dopamine receptors in the brain, and according to their binding qualities, these receptors can be divided into two groups, named the D1-like and D2-like dopamine receptor families. D1 and D5 receptors belong to the D1-like family, while D2, D3, and D4 receptors belong to the D2-like family [89], and the availabilities of both receptor families decrease with age. The availability of D1-like receptors appears to decline in the human striatum, which is a region with high dopaminergic activity at a rate of 7% per decade of life [90, 91]. For the D2-like family, a consistent decrease with aging has been observed in the same region [92]. The dopamine transporter is a protein located in the membrane of the presynaptic neuronal terminal and is involved in dopamine reuptake. Several PET and SPECT studies have consistently demonstrated the age-related loss of dopamine transporter density in the healthy brain, and the density of dopamine transporters in the striatum decreases by 3–10% per decade [93, 94]. In addition, changes in serotonin (5-HT) signaling have been regarded as another leading factor in aging [95, 96].

9.4.2 Neuropeptide

In the CNS, peripheral nervous system (PNS), and other surrounding tissues, NPY is one of the most abundant peptides [97]. NPY receptors, which are called NPY Y1, Y2, Y4, and Y5, are the same for all members of the NPY family and belong to the classic G-protein coupled receptors (GPCRs) [98]. The NPY system is related to the aging process: transgenic rats overexpressing NPY live longer [99], whereas NPY Y2-receptor knockout mice display memory deterioration, learning deficits, and cognitive function decline during the aging process [100, 101]. In elderly rodents and brain samples from people with neurodegenerative diseases, several specific brain regions were observed to have reduced levels of NPY and NPY receptors [102, 103].

9.4.3 Neurotrophic Factors

Neurotrophic factors (NTFs) are diffusible peptides secreted from neurons and the cells supporting them. They are essential to nervous system function since they regulate the development, maintenance. and survival of neurons and cells, such as glial cells and oligodendrocytes. NTFs belong to several structurally and functionally associated molecule superfamilies: (1) the transforming growth factor (TGF)-superfamily, which includes glial cell line-derived neurotrophic factor (GDNF); (2) neurotrophins, including nerve growth factor (NGF), brain-derived neurotrophic factor (BDNF), neurotrophin-3 (NT-3), and NT-4; (3) the neurokine superfamily; and (4) nonneuronal factors [104]. NTFs promote the survival of specific neuronal populations in terms of brain aging and neurodegenerative diseases [105].

The role of GDNF in aging is based on observations that it can reverse some forms of aging performance, such as cognitive deficits and impaired motor function, in monkeys [106]. BDNF increases until reaching a maximal level after birth [107] and does not seem to decrease with age [108]. However, the role of BDNF in the aging brain still needs to be discussed. Evidence shows that BDNF concentration increases in the dentate gyrus of aged rats [109]. Meanwhile, in some other studies, no changes in BDNF levels were found in the rat hippocampus during the aging process, suggesting that alterations in BDNF receptors may occur during aging [110]. This speculation is strengthened by previous observations that BDNF administration did not prevent spatial learning impairments in aged rats [111].

9.5 Mitochondrial Dysfunction

Considering the high metabolic demands and negligible inherent energy storage of neurons, the brain depends on the constant influx of substrates from the blood. Neurons in the adult brain rely primarily on glucose for energy [112]. To protect the brain from potentially hazardous fluctuations in blood, the blood-brain barrier (BBB) and blood-cerebrospinal fluid barrier (BCSFB) regulate molecular exchange between the blood and cerebral fluid. The primary function of these barriers is to restrict the free diffusion of solutes in blood or cerebral fluid, selectively transport essential ions, nutrients, and signal molecules, and simultaneously remove metabolic wastes.

The metabolic fate of glucose in the brain relies on the cell type and the selective expression of metabolic enzymes. Neurons are primarily oxidative, while astrocytes are primarily glycolytic [113, 114]. In addition to producing adenosine-5′-triphosphate (ATP), glucose can also generate metabolic intermediates that synthesize fatty acids and other lipids required for membrane and myelin synthesis [115, 116]. In neurons, each glucose molecule is oxidized to produce carbon dioxide, water, and 30-36 ATP molecules, which depend on the specific mitochondrial proton leakage rate through glycolysis, the pentose phosphate pathway (PPP), the tricarboxylic acid (TCA) cycle, and oxidative phosphorylation [117]. The glycolysis process metabolizes glucose into pyruvate, which is actively transported to the mitochondria and then converted to acetyl coenzyme A (acetyl-CoA). Acetyl-CoA is complexed with citrate, which undergoes a series of regenerative enzymatic reactions in the TCA cycle to produce reduced nicotinamide adenine dinucleotide (NADH) and flavin adenine dinucleotide (FADH2). NADH and FADH2 produced in glycolysis and the TCA cycle are then reoxidized in the electron transport chain (ETC).

Studies in humans and animals have widely demonstrated that the expression of glucose transporters in the brain decreases during aging [118], and the expression of key enzymes involved in glycolysis and oxidative phosphorylation changes [119, 120]. Studies in mice have shown that ATP levels in white matter decrease during aging, which is related to ultrastructural changes in mitochondria and the reduced association of mitochondria and ER [121]. NAD levels are critical for mitochondrial function and ATP production [122, 123]. Meanwhile, an increased NADH level, with decreased total NAD and NAD^+ levels, has been exhibited in the human brain during normal aging [124].

Low glucose metabolism and mitochondrial dysfunction are early signs of age-associated functional changes during normal aging in the brain [125–127]. PET is used to analyze fluorodeoxyglucose uptake into brain cells in humans at different ages and has demonstrated age-associated degradation in glucose utilization in several different specific brain regions [128], with the most significantly related regions found in the frontal cortex, especially the dorsolateral and medial prefrontal cortex, with the highest positive loads. Furthermore, the posterior parietal and posterior cingulate cortices have higher positive loads, indicating that the metabolism of these areas decreases significantly with age. In addition, regional analysis showed that the temporal, parietal, and cerebral cortex exhibited an age-related decline in metabolism, and the frontal cortex declined particularly rapidly [129]. In rats, an age-related reduction in brain cellular energy metabolism in the hippocampus and prefrontal cortex is also associated with impaired learning and memory test performance [130].

During human evolution, the increase in mitochondrial capacity and oxidative metabolism seems to drive the expansion of the cerebral cortex [131–133], which may have rendered the brain resistance to cognitive decline in aging. Synaptic spine is the site of neurotransmission; therefore, it is essential for forms of synaptic plasticity such as long-term potentiation or long-term inhibition [134]. The excitatory synapse is a subcellular site with a high energy expenditure because it requires large amounts of ATP to support the activity of neurotransmitter transporters. Membrane Na^+ and Ca^{2+} pumps can rapidly restore gradients of these ions after synapse activation [135, 136]. Therefore, when the ability of neurons to produce enough ATP is compromised during aging, synapses are prone to dysfunction and degeneration [135].

Many factors may cause age-dependent hypometabolism in the brain. Many clinical studies have shown that there is a negative correlation between cerebral blood flow and age [137, 138]. In addition, the permeability of the BBB and BCSFB is higher in elderly individuals than in younger individuals [139]. Brain hypoperfusion and loss of BBB integrity can lead to decreased import of nutrients and removal of toxins. Furthermore, the damaged BBB allows parenchymal accumulation of blood-derived proteins, such as fibrinogen, immunoglobulins, albumin, thrombin, and hemoglobin, and even immune cells, which can cause inflammation to invade the brain parenchyma [140].

Age-related metabolic decline is a "malfunction" of the brain, as brain functions deteriorate in several energy metabolism-related pathways in brain cells, including glucose transport, mitochondrial electron transport, DNA repair, and neurotrophic factor signaling. Many epidemiological, experimental, and clinical studies indicate impaired neuronal bioenergetics and decreased adaptation to stress as important roles in normal aging or even preclinical stages of neurodegenerative disorders such as AD and PD [117].

References

1. Lopez-Otin C, Blasco MA, Partridge L, Serrano M, Kroemer G (2013) The hallmarks of aging. Cell 153(6):1194–1217
2. Isaev NK, Stelmashook EV, Genrikhs EE (2019) Neurogenesis and brain aging. Rev Neurosci 30(6):573–580
3. Isaev NK, Stelmashook EV, Stelmashook NN, Sharonova IN, Skrebitsky VG (2013) Brain aging and mitochondria-targeted plastoquinone antioxidants of SkQ-type. Biochemistry (Mosc) 78(3):295–300
4. Worman HJ (2012) Nuclear lamins and laminopathies. J Pathol 226(2):316–325
5. Lodato MA, Rodin RE, Bohrson CL, Coulter ME, Barton AR, Kwon M et al (2018) Aging and neurodegeneration are associated with increased mutations in single human neurons. Science 359(6375):555–559
6. Khrapko K, Bodyak N, Thilly WG, van Orsouw NJ, Zhang X, Coller HA et al (1999) Cell-by-cell scanning of whole mitochondrial genomes in aged human heart reveals a significant fraction of myocytes with clonally expanded deletions. Nucleic Acids Res 27(11):2434–2441
7. Ameur A, Stewart JB, Freyer C, Hagstrom E, Ingman M, Larsson NG et al (2011) Ultra-deep sequencing of mouse mitochondrial DNA: mutational patterns and their origins. PLoS Genet 7(3):e1002028
8. Payne BAI, Wilson IJ, Hateley CA, Horvath R, Santibanez-Koref M, Samuels DC et al (2011) Mitochondrial aging is accelerated by anti-retroviral therapy through the clonal expansion of mtDNA mutations. Nat Genet 43(8):806–810
9. Wallace DC (2005) A mitochondrial paradigm of metabolic and degenerative diseases, aging, and cancer: a dawn for evolutionary medicine. Annu Rev Genet 39:359–407

10. Kujoth GC, Hiona A, Pugh TD, Someya S, Panzer K, Wohlgemuth SE et al (2005) Mitochondrial DNA mutations, oxidative stress, and apoptosis in mammalian aging. Science 309(5733):481–484
11. Vermulst M, Wanagat J, Kujoth GC, Bielas JH, Rabinovitch PS, Prolla TA et al (2008) DNA deletions and clonal mutations drive premature aging in mitochondrial mutator mice. Nat Genet 40(4):392–394
12. Hiona A, Sanz A, Kujoth GC, Pamplona R, Seo AY, Hofer T et al (2010) Mitochondrial DNA mutations induce mitochondrial dysfunction, apoptosis and sarcopenia in skeletal muscle of mitochondrial DNA mutator mice. PLoS One 5(7):e11468
13. Harries LW, Hernandez D, Henley W, Wood AR, Holly AC, Bradley-Smith RM et al (2011) Human aging is characterized by focused changes in gene expression and deregulation of alternative splicing. Aging Cell 10(5):868–878
14. Nicholas A, de Magalhaes JP, Kraytsberg Y, Richfield EK, Levanon EY, Khrapko K (2010) Age-related gene-specific changes of A-to-I mRNA editing in the human brain. Mech Ageing Dev 131(6):445–447
15. de Magalhães JP, Curado J, Church GM (2009) Meta-analysis of age-related gene expression profiles identifies common signatures of aging. Bioinformatics (Oxford, England) 25(7):875–881
16. Finch CE, Morgan DG (1990) RNA and protein metabolism in the aging brain. Annu Rev Neurosci 13:75–88
17. Strehler BL (1986) Genetic instability as the primary cause of human aging. Exp Gerontol 21(4-5):283–319
18. Earls LR, Fricke RG, Yu J, Berry RB, Baldwin LT, Zakharenko SS (2012) Age-dependent microRNA control of synaptic plasticity in 22q11 deletion syndrome and schizophrenia. J Neurosci 32(41):14132–14144
19. Earls LR, Westmoreland JJ, Zakharenko SS (2014) Non-coding RNA regulation of synaptic plasticity and memory: implications for aging. Ageing Res Rev 17:34–42
20. Aunan JR, Watson MM, Hagland HR, Soreide K (2016) Molecular and biological hallmarks of ageing. Br J Surg 103(2):e29–e46
21. Butterfield DA, Halliwell B (2019) Oxidative stress, dysfunctional glucose metabolism and Alzheimer disease. Nat Rev Neurosci 20(3):148–160
22. Koga H, Kaushik S, Cuervo AM (2011) Protein homeostasis and aging: the importance of exquisite quality control. Ageing Res Rev 10(2):205–215
23. Labbadia J, Morimoto RI (2015) The biology of proteostasis in aging and disease. Annu Rev Biochem 84:435–464
24. Nixon RA (2017) Amyloid precursor protein and endosomal-lysosomal dysfunction in Alzheimer's disease: inseparable partners in a multifactorial disease. FASEB J 31(7):2729–2743
25. Morimoto RI, Cuervo AM (2014) Proteostasis and the aging proteome in health and disease. J Gerontol A Biol Sci Med Sci 69(Suppl 1):S33–S38
26. Pyo J-O, Yoo S-M, Ahn H-H, Nah J, Hong S-H, Kam T-I et al (2013) Overexpression of Atg5 in mice activates autophagy and extends lifespan. Nat Commun 4:2300–2300
27. Madeo F, Zimmermann A, Maiuri MC, Kroemer G (2015) Essential role for autophagy in life span extension. J Clin Invest 125(1):85–93
28. Kaushik S, Cuervo AM (2015) Proteostasis and aging. Nat Med 21(12):1406–1415
29. Di Domenico F, Head E, Butterfield DA, Perluigi M (2014) Oxidative stress and Proteostasis network: culprit and casualty of Alzheimer's-like neurodegeneration. Adv Geriatr 2014:1–14
30. Bulteau AL, Szweda LI, Friguet B (2002) Age-dependent declines in proteasome activity in the heart. Arch Biochem Biophys 397(2):298–304
31. Korovila I, Hugo M, Castro JP, Weber D, Höhn A, Grune T et al (2017) Proteostasis, oxidative stress and aging. Redox Biol 13:550–567
32. König J, Ott C, Hugo M, Jung T, Bulteau A-L, Grune T et al (2017) Mitochondrial contribution to lipofuscin formation. Redox Biol 11:673–681
33. Kim YE, Hipp MS, Bracher A, Hayer-Hartl M, Hartl FU (2013) Molecular chaperone functions in protein folding and proteostasis. Annu Rev Biochem 82:323–355
34. Morrow G, Samson M, Michaud S, Tanguay RM (2004) Overexpression of the small mitochondrial Hsp22 extends drosophila life span and increases resistance to oxidative stress. FASEB J 18(3):598–599
35. Swindell WR, Masternak MM, Kopchick JJ, Conover CA, Bartke A, Miller RA (2009) Endocrine regulation of heat shock protein mRNA levels in long-lived dwarf mice. Mech Ageing Dev 130(6):393–400
36. Blake MJ, Udelsman R, Feulner GJ, Norton DD, Holbrook NJ (1991) Stress-induced heat shock protein 70 expression in adrenal cortex: an adrenocorticotropic hormone-sensitive, age-dependent response. Proc Natl Acad Sci U S A 88(21):9873–9877
37. Carnemolla A, Labbadia JP, Lazell H, Neueder A, Moussaoui S, Bates GP (2014) Contesting the dogma of an age-related heat shock response impairment: implications for cardiac-specific age-related disorders. Hum Mol Genet 23(14):3641–3656
38. Westerheide SD, Anckar J, Stevens SM Jr, Sistonen L, Morimoto RI (2009) Stress-inducible regulation of heat shock factor 1 by the deacetylase SIRT1. Science (New York, N.Y.) 323(5917):1063–1066
39. Brehme M, Voisine C, Rolland T, Wachi S, Soper JH, Zhu Y et al (2014) A chaperome subnetwork safeguards proteostasis in aging and neurodegenerative disease. Cell Rep 9(3):1135–1150
40. Naidoo N (2009) ER and aging-protein folding and the ER stress response. Ageing Res Rev 8(3):150–159
41. Kaufman RJ (2002) Orchestrating the unfolded protein response in health and disease. J Clin Invest 110(10):1389–1398
42. Kaufman RJ (1999) Stress signaling from the lumen of the endoplasmic reticulum: coordination of gene

transcriptional and translational controls. Genes Dev 13(10):1211–1233
43. Hampton RY (2000) ER stress response: getting the UPR hand on misfolded proteins. Curr Biol 10(14):R518–R521
44. Prostko CR, Brostrom MA, Brostrom CO (1993) Reversible phosphorylation of eukaryotic initiation factor 2 alpha in response to endoplasmic reticular signaling. Mol Cell Biochem 127–128:255–265
45. Xu C, Bailly-Maitre B, Reed JC (2005) Endoplasmic reticulum stress: cell life and death decisions. J Clin Invest 115(10):2656–2664
46. Szegezdi E, Logue SE, Gorman AM, Samali A (2006) Mediators of endoplasmic reticulum stress-induced apoptosis. EMBO Rep 7(9):880–885
47. Naidoo N, Ferber M, Master M, Zhu Y, Pack AI (2008) Aging impairs the unfolded protein response to sleep deprivation and leads to proapoptotic signaling. J Neurosci 28(26):6539–6548
48. Paz Gavilan M, Vela J, Castano A, Ramos B, del Rio JC, Vitorica J et al (2006) Cellular environment facilitates protein accumulation in aged rat hippocampus. Neurobiol Aging 27(7):973–982
49. Sontag EM, Vonk WIM, Frydman J (2014) Sorting out the trash: the spatial nature of eukaryotic protein quality control. Curr Opin Cell Biol 26:139–146
50. McConnell KW, Fox AC, Clark AT, Chang N-YN, Dominguez JA, Farris AB et al (2011) The role of heat shock protein 70 in mediating age-dependent mortality in sepsis. J Immunol (Baltimore, Md: 1950) 186(6):3718–3725
51. Tong BC, Wu AJ, Li M, Cheung KH (2018) Calcium signaling in Alzheimer's disease & therapies. Biochim Biophys Acta Mol Cell Res 1865(11 Pt B):1745–1760
52. Mattson MP (2010) ER calcium and Alzheimer's disease: in a state of flux. Sci Signal 3(114):pe10
53. Popugaeva E, Bezprozvanny I (2013) Role of endoplasmic reticulum Ca^{2+} signaling in the pathogenesis of Alzheimer disease. Front Mol Neurosci 6:29–29
54. Raza M, Deshpande LS, Blair RE, Carter DS, Sombati S, DeLorenzo RJ (2007) Aging is associated with elevated intracellular calcium levels and altered calcium homeostatic mechanisms in hippocampal neurons. Neurosci Lett 418(1):77–81
55. Thibault O, Gant JC, Landfield PW (2007) Expansion of the calcium hypothesis of brain aging and Alzheimer's disease: minding the store. Aging Cell 6(3):307–317
56. Ris L, Godaux E (2007) Synapse specificity of long-term potentiation breaks down with aging. Learn Mem 14(3):185–189
57. Gant JC, Sama MM, Landfield PW, Thibault O (2006) Early and simultaneous emergence of multiple hippocampal biomarkers of aging is mediated by Ca^{2+}–induced Ca^{2+} release. J Neurosci 26(13):3482–3490
58. Thibault O, Landfield PW (1996) Increase in single L-type calcium channels in hippocampal neurons during aging. Science 272(5264):1017–1020
59. Navakkode S, Liu C, Soong TW (2018) Altered function of neuronal L-type calcium channels in ageing and neuroinflammation: implications in age-related synaptic dysfunction and cognitive decline. Ageing Res Rev 42:86–99
60. Landfield PW, Pitler TA (1984) Prolonged Ca^{2+}–dependent afterhyperpolarizations in hippocampal neurons of aged rats. Science 226(4678):1089–1092
61. Toescu EC, Verkhratsky A, Landfield PW (2004) Ca^{2+} regulation and gene expression in normal brain aging. Trends Neurosci 27(10):614–620
62. Burke SN, Barnes CA (2010) Senescent synapses and hippocampal circuit dynamics. Trends Neurosci 33(3):153–161
63. Kumar A, Foster TC (2005) Intracellular calcium stores contribute to increased susceptibility to LTD induction during aging. Brain Res 1031(1):125–128
64. Norris CM, Halpain S, Foster TC (1998) Reversal of age-related alterations in synaptic plasticity by blockade of L-type Ca^{2+} channels. J Neurosci 18(9):3171–3179
65. Rose GM, Ong VS, Woodruff-Pak DS (2007) Efficacy of MEM 1003, a novel calcium channel blocker, in delay and trace eyeblink conditioning in older rabbits. Neurobiol Aging 28(5):766–773
66. Zoladz PR, Campbell AM, Park CR, Schaefer D, Danysz W, Diamond DM (2006) Enhancement of long-term spatial memory in adult rats by the non-competitive NMDA receptor antagonists, memantine and neramexane. Pharmacol Biochem Behav 85(2):298–306
67. Foster TC, Norris CM (1997) Age-associated changes in ca(2+)-dependent processes: relation to hippocampal synaptic plasticity. Hippocampus 7(6):602–612
68. Bear MF, Cooper LN, Ebner FF (1987) A physiological basis for a theory of synapse modification. Science 237(4810):42–48
69. Nixon RA (2007) Autophagy, amyloidogenesis and Alzheimer disease. J Cell Sci 120(23):4081
70. Wolfe DM, Lee J-H, Kumar A, Lee S, Orenstein SJ, Nixon RA (2013) Autophagy failure in Alzheimer's disease and the role of defective lysosomal acidification. Eur J Neurosci 37(12):1949–1961
71. Soyombo AA, Tjon-Kon-Sang S, Rbaibi Y, Bashllari E, Bisceglia J, Muallem S et al (2006) TRP-ML1 regulates lysosomal pH and acidic lysosomal lipid hydrolytic activity. J Biol Chem 281(11):7294–7301
72. Curcio-Morelli C, Charles FA, Micsenyi MC, Cao Y, Venugopal B, Browning MF et al (2010) Macroautophagy is defective in mucolipin-1-deficient mouse neurons. Neurobiol Dis 40(2):370–377
73. Vergarajauregui S, Connelly PS, Daniels MP, Puertollano R (2008) Autophagic dysfunction in mucolipidosis type IV patients. Hum Mol Genet 17(17):2723–2737
74. Pitt SJ, Funnell TM, Sitsapesan M, Venturi E, Rietdorf K, Ruas M et al (2010) TPC2 is a novel NAADP-sensitive Ca^{2+} release channel, operating as

a dual sensor of luminal pH and Ca^{2+}. J Biol Chem 285(45):35039–35046
75. Bezprozvanny I (2009) Calcium signaling and neurodegenerative diseases. Trends Mol Med 15(3):89–100
76. Mattson MP, Cheng B, Davis D, Bryant K, Lieberburg I, Rydel RE (1992) Beta-amyloid peptides destabilize calcium homeostasis and render human cortical neurons vulnerable to excitotoxicity. J Neurosci 12(2):376–389
77. Demuro A, Mina E, Kayed R, Milton SC, Parker I, Glabe CG (2005) Calcium dysregulation and membrane disruption as a ubiquitous neurotoxic mechanism of soluble amyloid oligomers. J Biol Chem 280(17):17294–17300
78. Bezprozvanny I, Mattson MP (2008) Neuronal calcium mishandling and the pathogenesis of Alzheimer's disease. Trends Neurosci 31(9):454–463
79. Ando S (2012) Neuronal dysfunction with aging and its amelioration. Proc Jpn Acad B Phys Biol Sci 88(6):266–282
80. Ando S, Tanaka Y (1990) Synaptic membrane aging in the central nervous system. Gerontology 36(Suppl 1):10–14
81. Gibson GE, Peterson C (1981) Aging decreases oxidative metabolism and the release and synthesis of acetylcholine. J Neurochem 37(4):978–984
82. Tanaka Y, Hasegawa A, Ando S (1996) Impaired synaptic functions with aging as characterized by decreased calcium influx and acetylcholine release. J Neurosci Res 43(1):63–76
83. Tanaka Y, Ando S (2001) Age-related changes in the subtypes of voltage-dependent calcium channels in rat brain cortical synapses. Neurosci Res 39(2):213–220
84. Tanaka Y, Ando S (1990) Synaptic aging as revealed by changes in membrane potential and decreased activity of Na^+, K(+)-ATPase. Brain Res 506(1):46–52
85. Marcus MM, Apell HJ, Roudna M, Schwendener RA, Weder HG, Lauger P (1986) (Na^{++} K^+)-ATPase in artificial lipid vesicles: influence of lipid structure on pumping rate. Biochim Biophys Acta 854(2):270–278
86. Saito S, Kobayashi S, Ohashi Y, Igarashi M, Komiya Y, Ando S (1994) Decreased synaptic density in aged brains and its prevention by rearing under enriched environment as revealed by synaptophysin contents. J Neurosci Res 39(1):57–62
87. Carlsson A, Winblad B (1976) Influence of age and time interval between death and autopsy on dopamine and 3-methoxytyramine levels in human basal ganglia. J Neural Transm 38(3-4):271–276
88. Garnett ES, Firnau G, Nahmias C (1983) Dopamine visualized in the basal ganglia of living man. Nature 305(5930):137–138
89. Kaasinen V, Rinne JO (2002) Functional imaging studies of dopamine system and cognition in normal aging and Parkinson's disease. Neurosci Biobehav Rev 26(7):785–793
90. Suhara T, Fukuda H, Inoue O, Itoh T, Suzuki K, Yamasaki T et al (1991) Age-related changes in human D_1 dopamine receptors measured by positron emission tomography. Psychopharmacology 103(1):41–45
91. Wang Y, Chan GL, Holden JE, Dobko T, Mak E, Schulzer M et al (1998) Age-dependent decline of dopamine D_1 receptors in human brain: a PET study. Synapse 30(1):56–61
92. Rinne JO, Hietala J, Ruotsalainen U, Sako E, Laihinen A, Nagren K et al (1993) Decrease in human striatal dopamine D_2 receptor density with age: a PET study with [^{11}C]raclopride. J Cereb Blood Flow Metab 13(2):310–314
93. Rinne JO, Sahlberg N, Ruottinen H, Nagren K, Lehikoinen P (1998) Striatal uptake of the dopamine reuptake ligand [^{11}C]beta-CFT is reduced in Alzheimer's disease assessed by positron emission tomography. Neurology 50(1):152–156
94. Volkow ND, Fowler JS, Wang GJ, Logan J, Schlyer D, MacGregor R et al (1994) Decreased dopamine transporters with age in health human subjects. Ann Neurol 36(2):237–239
95. Lanni C, Govoni S, Lucchelli A, Boselli C (2009) Depression and antidepressants: molecular and cellular aspects. Cell Mol Life Sci 66(18):2985–3008
96. Sibille E, Su J, Leman S, Le Guisquet AM, Ibarguen-Vargas Y, Joeyen-Waldorf J et al (2007) Lack of serotonin$_{1B}$ receptor expression leads to age-related motor dysfunction, early onset of brain molecular aging and reduced longevity. Mol Psychiatry 12(11):1042–1056, 1975
97. Michel MC, Beck-Sickinger A, Cox H, Doods HN, Herzog H, Larhammar D et al (1998) XVI. International Union of Pharmacology Recommendations for the nomenclature of neuropeptide Y, peptide YY, and pancreatic polypeptide receptors. Pharmacol Rev 50(1):143
98. Botelho M, Cavadas C (2015) Neuropeptide Y: an anti-aging player? Trends Neurosci 38(11):701–711
99. Michalkiewicz M, Knestaut KM, Bytchkova EY, Michalkiewicz T (2003) Hypotension and reduced catecholamines in neuropeptide Y transgenic rats. Hypertension 41(5):1056–1062
100. Hedden T, Gabrieli JD (2004) Insights into the aging mind: a view from cognitive neuroscience. Nat Rev Neurosci 5(2):87–96
101. Redrobe JP, Dumont Y, Herzog H, Quirion R (2004) Characterization of neuropeptide Y, Y(2) receptor knockout mice in two animal models of learning and memory processing. J Mol Neurosci 22(3):159–166
102. Cha CI, Lee YI, Lee EY, Park KH, Baik SH (1997) Age-related changes of VIP, NPY and somatostatin-immunoreactive neurons in the cerebral cortex of aged rats. Brain Res 753(2):235–244
103. Kowalski C, Micheau J, Corder R, Gaillard R, Conte-Devolx B (1992) Age-related changes in cortico-releasing factor, somatostatin, neuropeptide Y, methionine enkephalin and beta-endorphin in specific rat brain areas. Brain Res 582(1):38–46

104. Lanni C, Stanga S, Racchi M, Govoni S (2010) The expanding universe of neurotrophic factors: therapeutic potential in aging and age-associated disorders. Curr Pharm Des 16(6):698–717
105. Mattson MP, Barger SW, Furukawa K, Bruce AJ, Wyss-Coray T, Mark RJ et al (1997) Cellular signaling roles of TGF beta, TNF alpha and beta APP in brain injury responses and Alzheimer's disease. Brain Res Brain Res Rev 23(1-2):47–61
106. Ai Y, Markesbery W, Zhang Z, Grondin R, Elseberry D, Gerhardt GA et al (2003) Intraputamenal infusion of GDNF in aged rhesus monkeys: distribution and dopaminergic effects. J Comp Neurol 461(2):250–261
107. Schecterson LC, Bothwell M (1992) Novel roles for neurotrophins are suggested by BDNF and NT-3 mRNA expression in developing neurons. Neuron 9(3):449–463
108. Katoh-Semba R, Takeuchi IK, Semba R, Kato K (1997) Distribution of brain-derived neurotrophic factor in rats and its changes with development in the brain. J Neurochem 69(1):34–42
109. Gooney M, Messaoudi E, Maher FO, Bramham CR, Lynch MA (2004) BDNF-induced LTP in dentate gyrus is impaired with age: analysis of changes in cell signaling events. Neurobiol Aging 25(10):1323–1331
110. Croll SD, Ip NY, Lindsay RM, Wiegand SJ (1998) Expression of BDNF and trkB as a function of age and cognitive performance. Brain Res 812(1-2):200–208
111. Pelleymounter MA, Cullen MJ, Baker MB, Gollub M, Wellman C (1996) The effects of intrahippocampal BDNF and NGF on spatial learning in aged long Evans rats. Mol Chem Neuropathol 29(2-3):211–226
112. Sokoloff L (1992) Chapter 2: the brain as a chemical machine. In: Neuronal-astrocytic interactions - implications for Normal and pathological CNS function. Progress in Brain Research, pp 19–33
113. Hamberger A, Hyden H (1963) Inverse enzymatic changes in neurons and glia during increased function and hypoxia. J Cell Biol 16(3):521–525
114. Hyden H, Lange PW (1962) A kinetic study of the neuronglia relationship. J Cell Biol 13(2):233–237
115. Jones JP, Nicholas HJ, Ramsey RB (1975) Rate of sterol formation by rat brain glia and neurons in vitro and in vivo. J Neurochem 24(1):123–126
116. Ramsey RB, Jones JP, Naqvi SH, Nicholas HJ (1971) The biosynthesis of cholesterol and other sterols by brain tissue. II. A comparison of in vitro and in vivo methods. Lipids 6(4):225–232
117. Camandola S, Mattson MP (2017) Brain metabolism in health, aging, and neurodegeneration. EMBO J 36(11):1474–1492
118. Ding F, Yao J, Rettberg JR, Chen S, Brinton RD (2013) Early decline in glucose transport and metabolism precedes shift to ketogenic system in female aging and Alzheimer's mouse brain: implication for bioenergetic intervention. PLoS One 8(11):e79977
119. Bowling AC, Mutisya EM, Walker LC, Price DL, Cork LC, Beal MF (1993) Age-dependent impairment of mitochondrial function in primate brain. J Neurochem 60(5):1964–1967
120. Meier-Ruge W, Iwangoff P, Reichlmeier K, Sandoz P (1980) Neurochemical findings in the aging brain. Adv Biochem Psychopharmacol 23:323–338
121. Stahon KE, Bastian C, Griffith S, Kidd GJ, Brunet S, Baltan S (2016) Age-related changes in axonal and mitochondrial ultrastructure and function in white matter. J Neurosci 36(39):9990–10001
122. Bai P, Cantó C, Oudart H, Brunyánszki A, Cen Y, Thomas C et al (2011) PARP-1 inhibition increases mitochondrial metabolism through SIRT1 activation. Cell Metab 13(4):461–468
123. Pittelli M, Felici R, Pitozzi V, Giovannelli L, Bigagli E, Cialdai F et al (2011) Pharmacological effects of exogenous NAD on mitochondrial bioenergetics, DNA repair, and apoptosis. Mol Pharmacol 80(6):1136–1146
124. Zhu X-H, Lu M, Lee B-Y, Ugurbil K, Chen W (2015) In vivo NAD assay reveals the intracellular NAD contents and redox state in healthy human brain and their age dependences. Proc Natl Acad Sci U S A 112(9):2876–2881
125. de Leon MJ, Ferris SH, George AE, Reisberg B, Christman DR, Kricheff II et al (1983) Computed tomography and positron emission transaxial tomography evaluations of normal aging and Alzheimer's disease. J Cereb Blood Flow Metab 3(3):391–394
126. Mosconi L, De Santi S, Li J, Tsui WH, Li Y, Boppana M et al (2008) Hippocampal hypometabolism predicts cognitive decline from normal aging. Neurobiol Aging 29(5):676–692
127. Small GW, Ercoli LM, Silverman DH, Huang SC, Komo S, Bookheimer SY et al (2000) Cerebral metabolic and cognitive decline in persons at genetic risk for Alzheimer's disease. Proc Natl Acad Sci U S A 97(11):6037–6042
128. Zuendorf G, Kerrouche N, Herholz K, Baron JC (2003) Efficient principal component analysis for multivariate 3D voxel-based mapping of brain functional imaging data sets as applied to FDG-PET and normal aging. Hum Brain Mapp 18(1):13–21
129. Kuhl DE, Metter EJ, Riege WH, Hawkins RA (1984) The effect of normal aging on patterns of local cerebral glucose utilization. Ann Neurol 15(Suppl):S133–S137
130. Gage FH, Kelly PA, Björklund A (1984) Regional changes in brain glucose metabolism reflect cognitive impairments in aged rats. J Neurosci 4(11):2856–2865
131. Grossman LI, Schmidt TR, Wildman DE, Goodman M (2001) Molecular evolution of aerobic energy metabolism in primates. Mol Phylogenet Evol 18(1):26–36
132. Haygood R, Fedrigo O, Hanson B, Yokoyama KD, Wray GA (2007) Promoter regions of many neural- and nutrition-related genes have experienced posi-

tive selection during human evolution. Nat Genet 39(9):1140–1144

133. Pontzer H, Brown MH, Raichlen DA, Dunsworth H, Hare B, Walker K et al (2016) Metabolic acceleration and the evolution of human brain size and life history. Nature 533(7603):390–392
134. Attwell D, Laughlin SB (2001) An energy budget for signaling in the grey matter of the brain. J Cereb Blood Flow Metab 21(10):1133–1145
135. Harris JJ, Jolivet R, Attwell D (2012) Synaptic energy use and supply. Neuron 75(5):762–777
136. Rangaraju V, Calloway N, Ryan TA (2014) Activity-driven local ATP synthesis is required for synaptic function. Cell 156(4):825–835
137. Fabiani M, Low KA, Tan CH, Zimmerman B, Fletcher MA, Schneider-Garces N et al (2014) Taking the pulse of aging: mapping pulse pressure and elasticity in cerebral arteries with optical methods. Psychophysiology 51(11):1072–1088
138. Schultz SK, O'Leary DS, Boles Ponto LL, Watkins GL, Hichwa RD, Andreasen NC (1999) Age-related changes in regional cerebral blood flow among young to mid-life adults. Neuroreport 10(12):2493–2496
139. Rosenberg GA (2012) Neurological diseases in relation to the blood-brain barrier. J Cereb Blood Flow Metab 32(7):1139–1151
140. Zlokovic BV (2011) Neurovascular pathways to neurodegeneration in Alzheimer's disease and other disorders. Nat Rev Neurosci 12(12):723–738

10 Neurobiological Mechanisms of Cognitive Decline Correlated with Brain Aging

Xiaxia Zhang, Haiting An, Yuan Chen, and Ni Shu

Abstract

Cognitive decline has emerged as one of the greatest health threats of old age. Meanwhile, aging is the primary risk factor for Alzheimer's disease (AD) and other prevalent neurodegenerative disorders. Developing therapeutic interventions for such conditions demands a greater understanding of the processes underlying normal and pathological brain aging. Despite playing an important role in the pathogenesis and incidence of disease, brain aging has not been well understood at a molecular level. Recent advances in the biology of aging in model organisms, together with molecular- and systems-level studies of the brain, are beginning to shed light on these mechanisms and their potential roles in cognitive decline. This chapter seeks to integrate the knowledge about the neurological mechanisms of age-related cognitive changes that underlie aging.

Keywords

AD · Synaptic plasticity · White matter · Demyelination · Neurovascular unit injury

X. Zhang · Y. Chen
State Key Laboratory of Cognitive Neuroscience and Learning, Faculty of Psychology, Beijing Normal University, Beijing, China

Beijing Aging Brain Rejuvenation Initiative (BABRI) Centre, Beijing Normal University, Beijing, China

H. An
Beijing Neurosurgical Institute, Beijing Tian Tan Hospital, Capital Medical University, Beijing, China

Beijing Aging Brain Rejuvenation Initiative (BABRI) Centre, Beijing Normal University, Beijing, China

N. Shu (✉)
State Key Laboratory of Cognitive Neuroscience and Learning, Faculty of Psychology, Beijing Normal University, Beijing, China
e-mail: nshu@bnu.edu.cn

10.1 Loss of Neural Circuits and Synaptic Plasticity

10.1.1 Synaptic Structure Deficit

Cognitive decline is among the most striking features of brain aging, and studies have shown similar cognitive impairment in older people, primates, and rodents [1]. Brain aging leads to serious impairment in fluid intelligence, rendering it difficult for individuals to solve new problems effectively. More importantly, the decline in fluid intelligence is related to a decrease in cognitive function. Combined with the age-related increase in cognitive defects, this situation is referred to as "age-associated memory impairment (AAMI)" [2] or "age-associated cognitive decline (AACD)" [3]. However, the brain's regional aging pattern varies among individuals, and the mechanisms remain unclear.

At the macro-level, the following five main theories explain cognitive aging: working memory theory [4], processing speed theory [5], inhi-

Z. Zhang (ed.), *Cognitive Aging and Brain Health*, Advances in Experimental Medicine and Biology 1419, https://doi.org/10.1007/978-981-99-1627-6_10

bition decline theory [6], sensory function theory [7], and executive decline theory [8]. With the development of brain imaging technology, researchers have proposed the following two cognitive aging neural mechanism models: the aging from back to forwarding model (PASA) [9] and the hemispheric asymmetry reduction in older adults (HAROLD) model [10]. Previous researchers believe that there is a difference in brain activation patterns between elderly individuals and young individuals when performing certain cognitive tasks. Elderly individuals recruit the anterior brain area to compensate for the loss of sensory processing function in the occipitotemporal area. Current research reports that under the same conditions, PFC (prefrontal cortex) activation in elderly adults is lower than that in young adults and tends to present a symmetric pattern. In addition, functional magnetic resonance imaging (fMRI) studies have shown that in normal aging, the metabolic activity of the subiculum and dentate gyrus is decreased, whereas, in pathological aging, the metabolic activity of the entorhinal cortex is decreased [11].

At the micro-level, the aging human brain is associated with memory loss and synaptic connectivity reduction, but the loss of neurons is not obvious. For example, there was no significant change in the number of primary hippocampal glutamatergic neurons in humans, monkeys, or rats [12]. However, over the whole lifespan development, volume loss and functional changes appear in the medial temporal lobe (MTL) [13]. Although the loss of neurons is minimal in most cortical regions of the normal aging brain [14], the changes in the synaptic physiology of aging neurons may contribute to altered connectivity and higher-order integration [15].

Barnes and others believe that the decline in cognitive ability, such as learning and memory, is due to subtle changes in the synaptic structure and function [1], such as axonal degeneration [16], dendritic spine density reduction [17], and synaptic plasticity change. However, synaptic degeneration is not a common phenomenon in the brain and is more obvious in the hippocampus and other areas sensitive to aging. Aging decreases the expression of synaptic-related genes, such as synaptic vesicle release- and transport-related molecules, neurotransmitter receptors, postsynaptic dense areas of scaffold proteins, and cell adhesion molecules regulating synaptic connections [13, 18–21].

With brain aging, there are also physiological changes, such as a decrease in the synaptic interface curvature, synaptic membrane fluidity, and synaptic element content and an increase in widening of the gap width and thinning of postsynaptic dense material [22–24]. Animal studies have shown that normally aging mice had preserved Schaffer collateral-CA1 synapses and excitatory postsynaptic potential (EPSP) amplitudes at CA3-CA1 synapses, whereas cognition-impaired mice had decreased amplitudes in Schaffer collateral-induced field EPSPs and the postsynaptic density area of axospinous synapses, suggesting that the cognitive decline in brain aging may be due to an increase in nonfunctional or silent synapses in the hippocampus, which may lead to a change in the plasticity mechanism but is not related to the alterations in the strength of individual synaptic connections.

Synaptic integrity changes with aging. Transcriptomic analyses of the brain of an elderly human show robust negative associations with the age of genes encoding pre-and postsynaptic proteins, which may be related to changes in synaptic integrity [23, 24]. Some of the strongest effects were observed in the hippocampus, which is consistent with the increased vulnerability of the structure during aging [25].

Abnormal numbers of synapses, dendritic spines, synaptic density, synaptic loss, transmitter transmission, morphology, and function were also found in pathological aging, such as AD, Parkinson's disease (PD), and other neurodegenerative diseases. In AD, the accumulation of amyloid plaques and neurofibrillary tangles reduces the number of hippocampal synapses and the levels of acetylcholine, 5-hydroxytryptamine, and other neurotransmitters. Both neurotransmitter levels were damaged in the early stage of AD, while dopamine and

GABA were damaged in the late stage of AD [26]. In AD, amyloid plaques and neurofibrillary tangles cause information transmission disorder [27], and the degree of synaptic damage is positively correlated with the degree of cognitive impairment [28]. As changes in synaptic function can be used as biomarkers of pathological changes in early AD, AD is also defined as a "synaptic failure disease" [29, 30].

What factors have an effect on synapses? The epigenetic state plays an important role in controlling gene expression in brain neurons underlying synaptic plasticity and memory. The transient expression of p25 in adult mice can lead to neurodegeneration, synaptic loss, and memory loss in the hippocampus [31]. Changing the environment or using deacetylase inhibitors to simulate environmental changes can promote memory recovery in p25 mice. Furthermore, the increase in synaptic plasticity and the activation of histone acetylation markers emphasize the role of epigenetic changes in memory loss, which is related to neurodegeneration [32].

Because the process of aging is also described as inflammation, it is reasonable to consider the roles of astrocytes and microglia [33]. Although the number of astrocytes was unaffected in aged humans [34, 35], the volume of astrocytes in the hippocampus appeared to increase in rats [36]. In mice, aging reduced the expression of Ca^{2+} signals induced by ionophilic and purinergic receptors and neurotransmitters [37]. Importantly, aging is also related to a decreased expression of water channels (aquaporin 4) in the perivascular process of astrocytes and decreased brain parenchyma clearance through the glymphatic pathway, which is the key process to prevent the accumulation of misfolded protein aggregation [38]. Aging also changes microglia as follows: the activation of age-related microglia in rodents, nonhuman primates, and humans [39–41] increases MHCII, CD68, TLRs, and proinflammatory cytokines, such as TNFa, interleukin-1b (IL-1b), and IL-6 [41, 42]. However, other studies debunked the hypothesis that rather than inducing microglial activation, progressive microglial degeneration, and the loss of microglial neuroprotection, the microglial dystrophic/senescent phenotype in aged individuals is related to aging and further leads to the occurrence and progression of neurodegenerative diseases, such as AD [43, 44]. Gu et al. found that in mice, immune microglial cells caused forgetting by clearing synapses and that the complement signaling pathway was involved in microglia-mediated forgetting, which depended on the activity of memory-imprint cells. These results suggest that complement-dependent synapse elimination by microglia is a mechanism underlying the forgetting of remote memories [45].

10.1.2 Decreased Synaptic Plasticity

The hippocampus plays a key role in the formation and consolidation of learning and memory and is closely related to the regulation of emotion, fear, anxiety, and stress. Studies have shown that synaptic plasticity in the hippocampus directly affects hippocampal function. Synaptic plasticity has been identified and confirmed in the hippocampus [46]. Additionally, the performance of the hippocampus in different species decreases with aging and decreased synaptic plasticity, such as long-term potentiation (LTP) and long-term depression (LTD), or cellular alterations, which can be partially explained by directly affecting mechanisms of plasticity [24].

Synaptic plasticity, considered a crucial neurobiological foundation for learning, memory, and behavioral adaptation [47], involves the modification of synaptic number, morphology, and function [48]. This results in physical and chemical changes in proteins, neurotransmitters, receptors, ions, and messenger molecules. It is a common phenomenon in both excitatory and inhibitory synapses. Furthermore, and it has been acknowledged that the reaction to various activities and events are all involved [49]. LTP is defined as the long-lasting enhancement of synaptic transmission between two neurons following continuous and strong stimulation (high-frequency stimulation or theta burst stimulation) and was found in the dentate gyrus of the hippocampus in rabbits in 1973 [50], whereas LTD reflects a long-lasting decrease in the efficiency of synaptic transmission following continuous weak stimulation (LFS: low-frequency stimulation) and was found in a hippocampal section in 1977 [51].

Both LTP and LTD activate the α-amino-3-hydroxy-5-methyl-4-isoxazole propionate receptor (AMPAR) to cause a flow of Na^+, leading to the depolarization of the postsynaptic membrane, which releases the Mg^{2+} and Zn^{2+} block from the NMDAR, leading to the outflow of K^+ and the influx of a large amount of Na^+ and Ca^{2+} [52]. However, the transient elevation in Ca^{2+} is essential for controlling membrane excitability and regulating synaptic plasticity mechanisms, gene transcription, and other major cell functions [53, 54].

In rodents, a characteristic of physiological aging is a decline in the action potential firing rates of CA1 pyramidal cells and the amplitude of the post-burst after-hyperpolarization responsible for spike frequency adaptation [55]; in old rats and primates, the excitatory synaptic transmission also changes during aging [56]. Many studies have shown that age-related memory deficits are related to LTP or LTD damage [57, 58].

In older animals, LTP has been found to be decreased [59–61], but some studies proved that LTP is strengthened [62, 63]. Presumably, differences in synaptic circuits or stimulation schemes result in different results [60]. In general, age-related changes in LTP can be observed only when a weaker stimulus is used, which leads to an increase or decrease in LTP [58, 62, 63]. Changes in synaptic plasticity during aging may be related to changes in AMPAR expression and functioning. Studies have shown that the administration of AMPAR-positive allosteric modulators can restore age-related memory and synaptic enhancement defects, suggesting an increase in silent AMPAR rather than alterations in the expression of synaptic AMPAR [64–66]. Additionally, CaMKII is a key component of the molecular mechanism of LTP because LTP induction is blocked in knockout mice lacking the key CaMKII subunit [67].

In addition to AMPA, studies have found that the NMDA-mediated response time increases in elderly CA1 pyramidal neurons [68]. Elderly animals show an excessive activation of NMDA under glutamate or glycine stimulation despite the decrease in the density of these receptors [69]. However, this change in the NMDA-mediated response may be due to the increase in nonfunctional synapses with aging. In fact, the binding of the NMDAR antagonist MK-801 in animals with learning and memory deficits [70, 71] indicates that the increase in the NMDAR channel opening time occurs as a compensatory mechanism for the apparent decrease in the number of receptors because MK-801 only binds open channels. In addition, LTD is related to the levels of Ca^{2+}, and LTD strongly supports age-related disorders and changes in Ca^{2+}-dependent induction mechanisms. With increases in NMDAR activation and the influence of Ca^{2+}, aging is related to changed plasticity induced by weak stimulation. Furthermore, considering the key role of NMDARs in synaptic plasticity and memory, speculative changes in NMDARs may also cause Ca^{2+} imbalance.

The scaffold protein PSD-95 can regulate LTP and LTD. Overexpressing PSD-95 renders LTP more difficult to induce, while enhancing LTD [72] by the knockout of PSD-95 in rat hippocampi weakens the process of LTD [73], which can be explained by the decreased number of AMPA receptors on the postsynaptic membrane and the decreased EPSC amplitude and frequency caused by the decreased PSD-95 expression [74]. In addition, the connection between PSD-95 and NMDA receptors and its downstream enzymes are very important for synaptic signal transmission and synaptic plasticity. When the connection between PSD-95 and nNOS is blocked, nitric oxide synthesis is reduced, leading to abnormal LTP induction [75]. In PSD-95 knockout mice, the excitatory postsynaptic current in the hippocampus was decreased, and LTP was more easily induced [76].

The overactivation of the adenosine A2A receptor in neurons is enough to cause synaptic plasticity and cognitive dysfunction [77]. The adenosine A2A receptor can play a key role by regulating neurotransmitters, activating signaling pathways, and participating in the inflammatory response. The adenosine A2A receptor regulates synaptic plasticity by regulating the neurotransmitter dopamine D1 receptor and changing the ratio of NMDA/AMDA in the dentate gyrus of the hippocampus, thus affecting cognition [78]. The results reported by Furini et al. suggest that an appropriate amount of dopamine can activate the classic adenylate cyclase (AC)/cAMP/protein kinase A (PKA)/CREB/mTOR and dopamine and cAMP-regulated phosphoprotein of 32 kDa (DARPP-32) pathways by activating the D1 receptor, which can synergistically regulate the phosphorylation of many substrates, thus enhancing and improving the cognitive function of LTP [79, 80].

Moreover, diet, calorie restriction, brain-derived neurotrophic factors (BDNF), environment enrichment, diabetes, and other factors have an impact on synaptic plasticity. For instance, a high-calorie diet can damage the structure and function of the hippocampus. Other studies have shown that chronic caffeine can increase the length of neuron processes, the number of branches, and the density of basal dendritic spines, significantly reducing the risk of neurodegenerative diseases [81]. In addition, in the hippocampus and cerebral cortex, BDNF can increase presynaptic signal transduction, promote the release of excitatory neurotransmitters, enhance postsynaptic receptor activity, and participate in the formation of LTP [82]. In addition, changing the living environment and lifestyle by enriching the environment can increase hippocampal neurogenesis, synaptic plasticity, and cognitive behavior. Chronic hyperglycemia changes hippocampal tissue, which may lead to a decrease in the neuron number, abnormalities in cell body morphology, and disorders of synaptic function, resulting in a decrease in the LTP effect or facilitation of the LTD effect, abnormalities in brain signal transmission, and finally, a decline in cognition [83].

In pathological aging, such as AD, synaptic plasticity is reduced similar to normal aging introduced above. 3xTg-AD model mice have NMDAR-mediated LTP damage and memory decline [84], and in AD patients, neurons show significant degeneration and apoptosis and inhibit LTP induction. Soluble amyloid-beta (Aβ) oligomers can impair synaptic plasticity, which is mainly manifested as an effect on LTP. Reducing the level of Aβ or clearing the aggregation of Aβ in the hippocampus can improve synaptic plasticity, learning, and memory [85, 86], which may be related to NMDAR damage. Damage to NMDARs can promote the transformation of APP to Aβ and the soluble forms of oligomeric Aβ, which are related to a decrease in the NMDAR quantity and function. Moreover, Aβ can regulate AMPAR and NMDAR through GluR2 and GluN2B, respectively. In addition, the abnormal aggregation of Aβ reduces the mobility of mGluR5 in the synaptic space, leading to the release of Ca^{2+} in cells and overload [26]. An abnormal increase in the extracellular glutamate concentration leads to damage to synaptic plasticity. In addition, during the early stage of AD, the levels of synapse-related proteins, such as GAP-43 and PSD-95, decrease earlier than the decrease in the number of synapses and the damage to synaptic functioning [87]. This result suggests that synapse-related proteins in AD may also participate in the regulation of synaptic plasticity [87].

Synaptic plasticity is necessary for selectively modulating neuronal circuitry, thus affecting the aging of brain function, and the coordinated expression of LTP and LTD is the basis for enabling learning and memory to proceed normally. However, synaptic plasticity is not enough for the normal progression of neural circuits. For example, an increase in synaptic strength increases the excitability of postsynaptic cells and makes them easier to activate, which, in turn, increases the likelihood of more LTP. Therefore, positive feedback quickly saturates the system, leading to a hyperactive state of synaptic input saturation. In contrast, too much limited liability also sharply increases, leading to a state of silence in which input is completely suppressed, thereby destroying the activity of the neural network [88]. Therefore, in recent years, researchers have proposed the homeostatic plasticity theory, which indicates that a compensatory stabilizing mechanism regulates neuronal excitability, stabilizes the total strength of synapses, and affects the speed and extent of synapse formation [89] using a negative feedback system to maintain the network activity within a dynamic range [90]. These forms of homeostatic plasticity may be a supplement to LTP and LDP that neural networks can be selectively modified [89], but more research is needed to reveal the mechanism of synaptic plasticity and neural plasticity affecting brain aging.

10.2 White Matter Abnormal Alterations

10.2.1 Vascular Pathology Changes

Older people have smaller brains than younger people [91, 92]; gray matter, mainly in the frontal, insular, and cingulate cortex, is decreased by 3%, whereas white matter, mainly in the corpus callosum and frontal cortex, is decreased by 11%. The white matter volume gradually increases during the first 40 years after birth, reaches a peak at approximately the age of 50 years, and then rapidly decreases from the age of 60 years [93], and a quadratic regression relationship has been proposed between the volume of white matter in the frontal lobe and the age curve, indicating a high correlation between the volume of white matter and age [94]. Although neurons exhibit age-related synaptic decline and altered expression of neurotransmitter receptors, the total number of neurons does not change with aging. Therefore, age-dependent cognitive decline cannot be fully explained by age-dependent changes in neurons. Furthermore, white matter is crucial for transmitting electrical signals across different brain areas, and therefore, white matter malfunction can lead to serious neurobehavioral and cognitive impairments [95].

Diffusion tensor imaging (DTI) studies have found white matter alterations in aging brains, such as atrophy in white matter [96], disruption of white matter tracts [97], impairment in white matter integrity [98, 99], increased inflammation [100, 101], loss of myelination [102], and vessel impairments [103]. Further studies have shown that the risk of blood vessels increased with aging by 2.7% at the age of 36 years and 24.3% at the age of 69 years, and a higher vascular risk was associated with smaller brain size and white matter hyperintensities at the age of 70 years [104].

Stenosis caused by hyaluronic fibrosis of arterioles and small blood vessels is the main vascular change in white matter [105], which in turn leads to chronic cerebral ischemia and increases the incidence of brain lesions such as white matter lesions (WML), infarction, and cerebral microbleeds in older adults [106].

Some researchers indicate that abnormal alterations in white matter are due to hypoperfusion causing ischemia of white matter [107]. Toxin levels may rise in aged white matter, causing a decline in the vascular density [108], impairing autoregulation in the cerebral vascular system [109], and aggravating the hypoperfusion of white matter in elderly individuals. The major blood supply of WM is via the long perforating branch artery of the cerebral surface and cerebral pia mater vascular network and the terminal branch of the choroidal artery and vistriatum artery from the subependymal artery. There is limited or no anastomosis between the two vascular branches, rendering the periventricular white matter in the arterial junction area more vulnerable to systemic or focal local cerebral blood flow reduction [110] and hypoperfusion. Tortuous arterioles are well known to form in aged WM [111] and are assumed to have a curling shape in the cavity where the brain parenchyma is lost. Tortuous arteries decrease blood flow to WM because of an increase in vessel length and a loss of kinetic energy in the curling arterioles. In addition, excess collagen deposits in the walls of veins and venules in the deep WM of aging brains [111], which may lead to reduced WM blood flow by constriction of the vessel lumen. Thickened venous walls resulting from collagen deposition also delay the elimination of toxins via the bloodstream, accordingly impairing perivascular drainage [112]. Consistently, a model of chronic cerebral hypoperfusion caused by ligation 30 days later found axonal damage, diffuse demyelination and glial cell changes, and other white matter damage [113].

Aging is related to changes in the vascular system, including arteriolosclerosis and large artery stiffening [114, 115]. Among elderly individuals, there is a high probability of arteriolosclerosis in small blood vessels, and some researchers believe that abnormal alterations in white matter are due to dysfunction of the blood-brain barrier (BBB) as changes in small blood vessels may lead to damage to the BBB, chronic leakage of fluid and macromolecules in white matter [116], and increased concentrations of cerebrospinal fluid albumin and IgG values [117]. In addition, albumin extravasation is widespread in the aging brain and increases in white matter [118]. Moreover, BBB permeability is increased in normal-appearing white matter in patients, and its presence in normal-appearing white matter is consistent, playing a causal role in disease pathophysiology [119]. In addition, changes in blood vessels and endothelial cells are related to aging and can be accelerated by different vascular risk factors.

Changes in vascular function may be seriously affected by genetic and epigenetic factors. Smoking and chronic diseases can cause abnormal alterations in white matter [120], which has an impact on blood vessels, further leading to WML. Hypertension can promote microatheromatosis; lead to lipohyalinosis in media, the thickening of vessel walls [121], stenosis in the lumen of the small perforating arteries and arterioles stenosis, or the occlusion of vessels to different degrees until the lumen is completely blocked; and then induce ischemia, rendering white matter vulnerable to cerebral ischemia [122]. In addition to the potential effect on arteriosclerosis, subsequent hemodynamic changes may lead to reduced cerebral blood perfusion [123]. Chronic hypertension and hypotension alter cerebral blood flow (CBF) autoregulation and may affect the autoregulatory range. "Autoregulation" usually refers to the adaptation of CBF to acute and chronic changes in arterial blood pressure and perfusion pressure. Pressure autoregulation maintains fairly stable perfusion within the mean system pressure range of 60–150 mmHg. Hypertension can also disrupt the BBB and cause changes in white matter through brain edema, astrocyte activation, destructive enzymes, or other toxins that cross the walls of damaged blood vessels [124]. In addition to hypertension, diabetes can affect WML and change the transfer of glucose and insulin through the BBB, thereby affecting local metabolism and microcirculation, further changing the membrane permeability, reducing local blood flow, and accelerating white matter ischemia [125, 126]. In addition, hypertension and diabetes lay a foundation for altered neurovascular coupling and a regional decline in vasomotor capacity; these factors all reshape the sigmoidal physiologic curve of the adaptive system into "a straight line" and result in WML.

Abnormal alterations in white matter are highly heritable, and some polymorphisms have been found in various candidate genes, such as apolipoprotein E (epsilon 4±), methylenetetrahydrofolate reductase (cytosine677thymine), and angiotensinogen (Met235Thr), which are related to white matter damage [127, 128]. RNA microarray and pathway analyses indicated that multiple genes exhibit altered RNA transcription (immune regulation, cell cycle, apoptosis, proteolysis, ion transport, cell structure, electron transport, and metabolism), and white matter has no lesions (WM[C]) represented areas with a complex molecular phenotype [129, 130]. In addition, some studies have found that 241 specific genes were expressed for WM[C] associated with inflammation, oxidative stress, detoxification, and hormonal responses, and also including genes related to brain repair, long-term potentiation, and axon guidance. Genes correlated with oligodendrocyte proliferation, axon repair, long-term potentiation, and neurotransmission are also included [131]. These results provide an explanation for ischemia, BBB dysfunction, systemic oxidative stress, and inflammation in the WML. However, in meta-analyses, some results indicate that there is no convincing evidence of an association between white matter damage and candidate gene polymorphisms [132].

In pathological aging, age-related white matter hypoperfusion may lead to white matter ischemia, which causes white matter damage. Some researchers believe that impaired perivascular drainage throughout the brain is a result from venous collagenosis, tortuosity lesions, and thickened arteriolar walls in aged WM, which may cause the decreased clearance of Aβ [133]. In turn, the reduction in Aβ clearance causes elevated Aβ deposition in and around vascular walls, narrowing or even blocking the lumen [134] and further microvascular pathology and subsequent hypoperfusion. Declined Aβ clearance also aggravates age-related reductions in angiogenesis possibly due to the combination of Aβ to vascular endothelial growth factor (VEGF) [135], which is a growth factor vital for angiogenesis. AD patients exhibit greater BBB disruption [136] and an elevated expression of receptor for advanced glycation end products (RAGE) [137], which is essential for the influx of peripheral Aβ into the brain. Additionally, a decreased expression of the efflux receptor for Aβ and lipoprotein receptor-related protein (LRP)-1 and increased influx receptor RAGE have been reported in AD patients [138], which could be expected to enhance the intracellular accumulation of Aβ. These characteristics may partially explain WML in AD. In addition, researchers believe that there is a significant loss of cholinergic innervation in cerebral blood vessels in the brains of AD patients. The lack of cholinergic receptors in penetrating vessels [139] may also lead to cerebral hypoperfusion [140].

10.2.2 Demyelination

As previously mentioned, DTI studies have found white matter alterations in aging brains; in this section, we discuss myelin loss in white matter.

Neuroimaging evidence supports the loss of the myelin sheath during aging. DTI studies show that fractional anisotropy (FA) decrease and mean diffusivity (MD) increase in elderly individuals [95, 141], suggesting impaired WM integrity. A greater decrease in FA in distal regions of the corpus callosum was also found compared to that in medial regions with aging [142]. Both the age-related loss of forebrain white matter volume and the decrease in FA in subcortical white matter tracts are related to cognitive decline.

At the microscopic level, the myelin sheath degenerates in appearance and divides into layers during normal aging [143], which is consistent with the changes in the thickness, density, and function of the myelin sheath in elderly rhesus monkeys with aging [144]. In elderly individuals, the myelin sheath is damaged, including the division and expansion of the myelin sheath and the complete degeneration of axons and the myelin sheath. In addition, age-related myelin histopathology is closely related to the decrease in FA and the decrease in corpus callosum compound action potentials. Studies have also shown that the partial loss of myelin, axons, and oligodendritic cells and mildly reactive astrocyte proliferation gradually occur during the aging process [105]. In addition, Sugiyama found that in old rats, there was an increase in myelin division, myelin balloon formation, and axon separation [145]. Similar to the myelin sheath, white matter axon function [146], conduction efficiency, and threshold [147] were impaired. Hampton showed that older animals are more susceptible to axon damage than young animals and exhibit lower remyelination efficiency [148]. These pathological changes may be the basis of age-related cognitive decline due to the loss of effective neural communication.

Why does demyelination occur with brain aging? Some researchers propose that demyelination occurs due to the inflammatory reaction and abnormal protein decomposition destroying the balance of lipids and proteins, which are the main components of myelin, leading to the loss of the myelin sheath and functional abnormalities. For example, ganglioside, which plays an important role in brain aging, decreases after the age of 70 years [149]. Additionally, an experiment in which myelin lipids were replaced with radionuclides revealed that the metabolic rate of cerebroside and ganglioside in elderly adults was significantly higher than that in young adults. In addition to lipids, the contents of myelinogenic oligodendrocyte-specific protein (MOSP) and 2–3-cyclic nucleotide phosphodiesterase (CNPase) increased with aging [150]. The overexpression of CNPase, which maintains the myelin sheath and axons, in the brains of elderly individuals can lead to structural loosening and vacuoles of the myelin sheath. Mice without CNPase show age-dependent glioma [151]. Moreover, a considerable number of CNPase and MOSP hydrolytic fragments were found in the aged brain, which was consistent with the abnormal proteasome system and the abnormal increase in calpain-1 activity in the aging brain [150].

Oligodendrocytes produce myelin sheaths in the central nervous system and help with myelin regeneration. Oligodendrocyte progenitor cells (OPCs) exist in the normal adult brain and can differentiate into mature oligodendrocytes. Damage to OPC recruitment or differentiation is also closely related to a decrease in myelination. Studies have shown that the degradation of oligodendrocytes and OPCs increases with aging [152], leading to increased myelin degradation, reduced remyelination, myelin component fluctuations, and the eventual destruction of white matter integrity during normal aging [153]. In addition, research evidence indicates that the transcription factors SOX2, Oligo2, Oligo1, SOX10, Hes5, Id2, and Id4 are related to OPC recruitment, which changes in aged demyelinating mice, and the growth factors platelet-derived growth factor (PDGF), fibroblast growth factor (FGF), and the transforming growth factor-β (TGF-β) regulated by OPC recruitment-related transcription factors are downregulated in aged mice [154].

Lactate is crucial for the normal function of oligodendrocytes: In the process of myelination, oligodendrocytes may use lactate from the blood and/or astrocytes as a source of energy and the material to generate fatty acids for myelin synthesis, and after myelination, mature oligodendrocytes release lactate through myelin. Additionally, the maintenance of myelin requires the decomposition of fatty acids, during which lactate may be generated. Thus, oligodendrocytes may support axon metabolism by providing lactate as a nutrient to neighboring neurons [155]. Consistent with this study, Lee and their colleagues [156] found that lactate was inhibited to be transported from oligodendrocytes to neurons through knockdown of monocarboxylate transporter 1 (MCT1), which would lead to abnormal axon morphology and neuronal death, revealing a crucial role of oligodendrocyte-derived lactate for maintenance of axonal viability. In addition, a decrease in lactate content was also found in the hippocampus of aging accelerated mice [147].

Astrocytes in white matter play an important role in myelination as they promote myelination by clearing extracellular ions and neurotransmitters and releasing promyelination factors [157, 158]. The function of activated astrocytes under pathological conditions is complex and depends on their activated phenotypes. These harmful effects include toxicity to oligodendrocyte lineage cells, the release of inflammatory cytokines and free radicals, and the recruitment and activation of peripheral immune cells into the central nervous system (CNS). On the contrary, the beneficial functions include the clearance of inhibitory myelin debris and the release of neurotrophic factors. Studies have demonstrated that astrocytes isolated from aged brains display a proinflammatory phenotype with an increased production of inflammatory factors and free radicals [159]. On the contrary, the release of glutathione and trophic factors (e.g., basic fibroblast growth factor, glial cell-derived neurotrophic factor, etc.), which are protective against white matter oligodendrocytes and axons, is reduced in aged astrocytes [160]. Aging brains exhibit a decrease in the ability of astrocytes to remove toxic molecules, including α-synuclein [161] and glutamate [162], from the extracellular medium. Notably, a study documented an increase in astrocyte activation in the brains of elderly individuals with a significant decrease in myelin expression [163]. Therefore, the activation of astrocytes is closely related to changes in white matter in brain aging.

Some researchers also speculate that microglia, which are the resident immune cells in white matter, maintain white matter homeostasis under physiological conditions [164]. In normal aging, the density of microglia in white matter increases, while gray matter remains stable. Additionally, the age-related decrease in myelin protein is associated with increased microglial activation [163]. In monkeys, myelin repair is associated with microglial activity [165]. However, opinions regarding the effect of microglia on the structure and function of white matter differ respectively. Some studies found that microglia are in a state of promoting inflammation during the aging process, leading to brain defects [166]. However, the downregulation of neurotoxic pathways and upregulation of neuroprotective pathways in aged microglia have also been demonstrated [167]. Therefore, the effect of age-related and region-dependent microglial changes on the structure or function of white matter remains to be confirmed.

Myelin sheath loss is also found in the setting of neurodegenerative diseases. Decreased FA and increased MD were also found in mild cognitive impairment (MCI) and early phases of AD [168–170]. Moreover, AD showed atrophy in specific fiber bundles, such as cingulate tract related to memory [171, 172], fornix [173], hippocampal para-white matter bundle [174], knee of corpus callosum [175] intracortical projection fiber bundle [176], and edge WM bundles [177]. In addition, there are significant differences in the volume of regional white matter between nondementia aging and AD [178]. The correlation between Aβ42, total tau, phospho-tau, with subcortical axonal impairments has also been reported [179]. Moreover, the changes in glial cells may be the basis of white matter damage in AD patients. Compared with age-matched healthy controls, AD patients showed higher microglial density [180], increased the production of RAGE, proinflammatory cytokines, and chemokines, thus aggravating white matter damage [181]. In addition to microglia, the activation of astrocytes in the AD brain is increased [182]. Elevated production of apolipoprotein E from astrocytes is associated with Aβ accumulation [183], leading to greater white matter damage. Moreover, abnormal WML is also a risk factor for dementia, depression, and stroke [184], and the abnormal lesions are mainly related to the loss of axons and myelin sheaths [185].

Although the understanding of the physiological and pathological changes in WM with aging is still unclear, white matter is an effective treatment target for many brain injuries and disease states.

10.3 Neurovascular Unit Injury

In 2001, the American Institute of Neurological Diseases and Stroke formally established the concept of the "neurovascular unit" as follows: the neurovascular unit (NVU) is the basic structure and functional unit of the nervous system and emphasizes the importance of the interaction among neurons, glial cells, and cerebral vessels, reflecting the relationship among brain cells and the cerebrovascular system, neural activity, and cerebral blood flow [186]. The NVU is composed of neurons, vascular cells (endothelial cells, pericytes, and vascular smooth muscle cells), glial cells (astrocytes, microglia, and oligodendrocytes), and brain-specific extracellular matrix through the BBB to strictly control the passage of materials in and out of the brain [187]. BBB integrity is crucial for proper synaptic function, information processing, and neuronal connection. The loss of BBB integrity leads to increased vascular permeability and allows toxic blood-derived molecules, cells, and microbial agents to enter the brain, which is associated with reduced CBF, impaired hemodynamic responses, and inflammatory and immune responses, which can initiate multiple pathways of neurodegeneration [188–190]. Neuroimaging studies have illustrated early BBB dysfunction in neurodegenerative disorders, which is also supported by biofluid biomarker data and is consistently found in post-mortem tissue [191]. The brain needs to consume much energy from oxygen, glucose, and nutrients to maintain its activity, while the brain's energy reserves are scarce. Irreversible brain damage and even brain death can result from a few minutes of complete interrupted blood supply in the brain, such as occluded cerebral artery in stroke or after pump failure in cardiac arrest. Moreover, if the flow is not completely stopped but is reduced or not well matched to the energy demands of the tissue, more subtle brain alterations ensue and cause chronic brain damage in vulnerable areas associated with cognitive impairment, eventually leading to cognitive impairment [192]. This finding proves that nerve cells and blood vessels must interact and the neurovascular unit plays an important role in normal brain function and various neurological diseases.

Numerous studies have found that blood vessels in the brain are associated with aging. During aging, the vascular morphology changes. Vermeer found a higher incidence of cerebral infarcts in elderly individuals [193]. According to Badji and Benetos's study, compared to youth, the degree of arterial and venous stiffness increases in elderly individuals [194, 195]. In addition, numerous animal studies have shown that the capillary diameter [196], density [197], surface area [196], and unit volume length [198] are decreased in elderly individuals. A similar phenomenon has been found in humans, but it is regionally specific in humans [133, 197, 199, 200]. In addition, other human studies described that the likelihood of arteriole tortuosity increases with aging [201]. This change in blood vessels may be due to an imbalance between mitochondrial reactive oxygen species (ROS) and oxidative stress during aging [202]. This imbalance is caused by an increase in ROS, which destroys nitric oxide (NO) signaling, reduces Cavelin 1 expression, activates MMP-2 and MMP-9, and destroys the tight connection between cells, thus increasing the permeability of the BBB and finally leading to the entry of neurotoxic substances, such as Aβ, into the ischemic area. Additionally, this change can lead to endothelial cell damage in blood vessels and an increase in inflammatory signals [203, 204]. Additionally, during the aging process, the levels of endothelin-1, angiotensin II, prostaglandins, thromboxane, and other prostanoids increase. In turn, this increase can lead to the production of mitochondrial reactive oxygen species, accelerated vascular senescence, and microvascular lesion, causing inflammation [205–207]. This process can be intensified by the activation of the transcription factor NF-κB. NF-κB induces inflammatory cytokines that cause vascular endothelial damage [208], limiting the vascular responsiveness to dilatory and constrictive vasoactive substances [209]. Some studies have shown that the transcription factor JUND can protect cells from damage caused by ROS but becomes less active with aging [210]. Various factors and mechanisms lead to chronic inflammation in the aging vasculature, causing damage that cells struggle to repair, reducing vascular endothelial growth factor and antioxidant enzyme activity, and causing vascular aging [211, 212].

In addition, the blood circulation system connects peripheral blood to the brain, which is altered in neurodegenerative diseases. Maier and their colleagues found that the concentration of cytokines in the peripheral blood of patients with AD was increased and the complement system was activated [213, 214]. Genetic evidence suggests that the complement pathway plays a key role in AD [215]. In healthy individuals, Aβ peptides are degraded within or cleared from the brain via the BBB [216], but Wang et al. found that Aβ in the peripheral blood of AD patients was significantly higher than that in healthy individuals [217, 218]. Huang et al. found that the plasma levels of soluble intercellular adhesion molecules (ICAM-1) and vascular cell adhesion molecules (VCAM-1) in AD patients were higher than those in the control group, indicating vascular inflammation in AD [219]. Furthermore, Vestweber et al. found that soluble platelet endothelial cell adhesion molecule 1 (PECAM-1) accumulates to a higher level in the plasma of AD patients, further destroying the BBB [219]. The feasibility of using peripheral blood as a biomarker for the detection of AD has been fully demonstrated.

In neurovascular units, neural-vascular coupling represents a system of communication among neural, glial, and vascular structures, with astrocytes playing the most typical role among glial cells [220]. Astrocytes are not only neuro-supportive but also diverse morphologically and functionally. Throughout the brain, astrocytes are centrally positioned to dynamically mediate interactions between neurons and the cerebral vasculature, play key roles in BBB maintenance and neurovascular coupling, and control the exchange of substances between the blood and the brain. Astrocytes also play a key role in the production, maturation, and regulation of the NVU. Astrocyte dysfunction may lead to BBB impairment and neurovascular disorders [221, 222]. However, with aging, the BBB maintained by astrocytes breaks down and becomes more permeable, subsequently leading to fewer and thinner endothelial cells with fewer mitochondria in aged human brain tissue and causing vascular aging [223–226]. Astrocytes promote communication between the synapse and local vasculature mainly through Ca^{2+} signaling, while Ca^{2+} activates phospholipases which produce arachidonic acid. Arachidonic acid is converted into potent vasomodulators and vasodilating prostaglandins and regulates the blood vessel diameter and blood flow. However, with aging, disturbance in calcium homeostasis in nerve cells and a decrease in the calcium regulatory capacity occur [227, 228], accelerating vascular aging and microvascular damage through a series of reactions and causing inflammation [229, 230]. Another significant aspect of vascular-astrocyte signaling is aquaporin channels that ship water between cells. Astrocyte end-feet express high levels of aquaporin 4 (AQP4) channels which interface with vascular aquaporins [231]. This interface is lost with aging [232]. Normally, astrocytes are at rest. However, with aging, brain damage, and other neurodegenerative diseases, "good" "resting astrocytes" become "reactive astrocytes" with abnormal behavior. Reactive astrocytes are divided into the following two types: A1 and A2. A1 loses the ability to promote neuronal survival, outgrowth, synaptogenesis, and phagocytosis and induces the death of neurons and oligodendrocytes. A2 secretes substances that support neuron growth and keep neurons healthy and alive [233]. During aging, the expression of A2 is increased. In older rats, astrocytes exhibit a neurotoxic form of reactive astrogliosis and upregulate the expression of inflammatory genes which are characteristic of reactive astrogliosis and genes involved in synapse elimination pathways, but signal transducer and activator of transcription proteins (STAT3) is downregulated in aged rat brain tissue [234]. In addition, age-related metabolic alterations have been observed in astrocytes [235].

Neural vascular unit changes before the clinical pathology of the BBB in endothelial cells and radial glial cells, basement membrane and downstream activation lead to neuronal dysfunction, nerve inflammation, and nerve degeneration. These changes can all be used as neurodegenerative disease biomarkers for the early detection of cognitive impairment. Previous studies have shown that the destruction of the integrity of the BBB is reversible.

References

1. Burke SN, Barnes CA (2006) Neural plasticity in the ageing brain. Nature Rev Neurosci 7(1):30–40
2. Crook T, Bartus RT, Ferris SH, Whitehouse P, Cohen GD, Gershon S (1986) Age associated memory impairment: proposed diagnostic criteria and measures of clinical change — report of a national institute of mental health work group. Dev Neuropsychol 2(4):261–276
3. Richards M, Touchon J, Ledesert B, Richie K (1999) Cognitive decline in ageing: are AAMI and AACD distinct entities? International Journal of Geriatric Psychiatry 14(7):534–540
4. Swanson HL (1999) What develops in working memory? A life span perspective. Dev Psychol 35(4):986–1000
5. Salthouse TA (1996) The processing-speed theory of adult age differences in cognition. Psychol Rev 103(3):403–428
6. VanWormer LA, Bireta TJ, Surprenant AM, Neath I (2012) The effect of perceptual cues on inhibiting irrelevant information in older adults using a list-learning method. Exp Aging Res 38(3):279–294
7. Schneider BA, Daneman M, Murphy DR, See SK (2000) Listening to discourse in distracting settings: the effects of aging. Psychol Aging 15(1):110–125
8. West RL (1996) An application of prefrontal cortex function theory to cognitive aging. Psychol Bull 120(2):272–292
9. Grady CL, Maisog JM, Horwitz B, Ungerleider LG, Mentis MJ, Salerno JA et al (1994) Age-related

changes in cortical blood flow activation during visual processing of faces and location. J Neurosci 14(3 Pt 2):1450–1462
10. Cabeza R (2002) Hemispheric asymmetry reduction in older adults: the HAROLD model. Psychol Aging 17(1):85–100
11. Small SA, Tsai WY, DeLaPaz R, Mayeux R, Stern Y (2002) Imaging hippocampal function across the human life span: is memory decline normal or not? Ann Neurol 51(3):290–295
12. Morrison JH, Hof PR (1997) Life and death of neurons in the aging brain. Science 278(5337):412–419
13. Erraji-Benchekroun L, Underwood MD, Arango V, Galfalvy H, Pavlidis P, Smyrniotopoulos P et al (2005) Molecular aging in human prefrontal cortex is selective and continuous throughout adult life. Biol Psychiatry 57(5):549–558
14. Yankner BA, Lu T, Loerch P (2008) The aging brain. Annu Rev Pathol 3:41–66
15. Andrews-Hanna JR, Snyder AZ, Vincent JL, Lustig C, Head D, Raichle ME et al (2007) Disruption of large-scale brain systems in advanced aging. Neuron 56(5):924–935
16. Grillo FW (1999) Long live the axon. Parallels between ageing and pathology from a presynaptic point of view. J Chem Neuroanat 76(Pt A):28–34
17. Peters A, Sethares C, Luebke JI (2008) Synapses are lost during aging in the primate prefrontal cortex. Neuroscience 152(4):970–981
18. Krashes MJ, DasGupta S, Vreede A, White B, Armstrong JD, Waddell S (2009) A neural circuit mechanism integrating motivational state with memory expression in drosophila. Cell 139(2):416–427
19. Loerch PM, Lu T, Dakin KA, Vann JM, Isaacs A, Geula C et al (2008) Evolution of the aging brain transcriptome and synaptic regulation. PLoS One 3(10):e3329
20. Fraser HB, Khaitovich P, Plotkin JB, Paabo S, Eisen MB (2005) Aging and gene expression in the primate brain. PLoS Biol 3(9):e274
21. Jiang CH, Tsien JZ, Schultz PG, Hu Y (2001) The effects of aging on gene expression in the hypothalamus and cortex of mice. Proc Natl Acad Sci U S A 98(4):1930–1934
22. Zhang X, Liu H, Wu J, Zhang X, Wang Y (2009) Metabonomic alterations in hippocampus, temporal and prefrontal cortex with age in rats. Neurochem Int 54(8):481–487
23. Burke SN, Barnes CA (2010) Senescent synapses and hippocampal circuit dynamics. Trends Neurosci 33(3):153–161
24. Burke SN, Barnes CA (2006) Neural plasticity in the ageing brain. Nat Rev Neurosci 7(1):30–40
25. McEwen BS (1997) Possible mechanisms for atrophy of the human hippocampus. Mol Psychiatry 2(3):255–262
26. Jha SK, Jha NK, Kumar D, Sharma R, Shrivastava A, Ambasta RK et al (2017) Stress-induced synaptic dysfunction and neurotransmitter release in Alzheimer's disease: can neurotransmitters and neuromodulators be potential therapeutic targets? J Alzheimers Dis 57(4):1017–1039
27. Lee SH, Lutz D, Mossalam M, Bolshakov VY, Frotscher M, Shen J (2017) Presenilins regulate synaptic plasticity and mitochondrial calcium homeostasis in the hippocampal mossy fiber pathway. Mol Neurodegener 12(1):48
28. Zhang Y, Li P, Feng J, Wu M (2016) Dysfunction of NMDA receptors in Alzheimer's disease. Neurological Sciences 37(7):1039–1047
29. Song Y, Hu M, Zhang J, Teng ZQ, Chen C (2019) A novel mechanism of synaptic and cognitive impairments mediated via microRNA-30b in Alzheimer's disease. EBioMedicine 39:409–421
30. Chen Y, Fu AKY, Ip NY (2019) Synaptic dysfunction in Alzheimer's disease: mechanisms and therapeutic strategies. Pharmacol Ther 195:186–198
31. Shah K, Rossie S (2018) Tale of the good and the bad Cdk5: remodeling of the actin cytoskeleton in the brain. Mol Neurobiol 55(4):3426–3438
32. Fischer A, Sananbenesi F, Wang X, Dobbin M, Tsai LH (2007) Recovery of learning and memory is associated with chromatin remodelling. Nature 447(7141):178–182
33. Franceschi C, Garagnani P, Parini P, Giuliani C, Santoro A (2018) Inflammaging: a new immune-metabolic viewpoint for age-related diseases. Nat Rev Endocrinol 14(10):576–590
34. Pelvig DP, Pakkenberg H, Stark AK, Pakkenberg B (2008) Neocortical glial cell numbers in human brains. Neurobiol Aging 29(11):1754–1762
35. Fabricius K, Jacobsen JS, Pakkenberg B (2013) Effect of age on neocortical brain cells in 90+ year old human females--a cell counting study. Neurobiol Aging 34(1):91–99
36. Grosche A, Grosche J, Tackenberg M, Scheller D, Gerstner G, Gumprecht A et al (2013) Versatile and simple approach to determine astrocyte territories in mouse neocortex and hippocampus. PLoS One 8(7):e69143
37. Lalo U, Palygin O, North RA, Verkhratsky A, Pankratov Y (2011) Age-dependent remodelling of ionotropic signalling in cortical astroglia. Aging Cell 10(3):392–402
38. Kress BT, Iliff JJ, Xia M, Wang M, Wei HS, Zeppenfeld D et al (2014) Impairment of paravascular clearance pathways in the aging brain. Ann Neurol 76(6):845–861
39. Ogura K, Ogawa M, Yoshida M (1994) Effects of ageing on microglia in the normal rat brain: immunohistochemical observations. Neuroreport 5(10):1224–1226
40. Sheffield LG, Berman NE (1998) Microglial expression of MHC class II increases in normal aging of nonhuman primates. Neurobiol Aging 19(1):47–55
41. Perry VH, Matyszak MK, Fearn S (1993) Altered antigen expression of microglia in the aged rodent CNS. Glia 7(1):60–67
42. VanGuilder HD, Bixler GV, Brucklacher RM, Farley JA, Yan H, Warrington JP et al (2011) Concurrent

hippocampal induction of MHC II pathway components and glial activation with advanced aging is not correlated with cognitive impairment. J Neuroinflammation 8:138
43. Davies DS, Ma J, Jegathees T, Goldsbury C (2017) Microglia show altered morphology and reduced arborization in human brain during aging and Alzheimer's disease. Brain Pathol 27(6):795–808
44. Streit WJ, Sammons NW, Kuhns AJ, Sparks DL (2004) Dystrophic microglia in the aging human brain. Glia 45(2):208–212
45. Wang C, Yue H, Hu Z, Shen Y, Ma J, Li J et al (2020) Microglia mediate forgetting via complement-dependent synaptic elimination. Science 367(6478):688–694
46. Bartsch T, Wulff P (2015) The hippocampus in aging and disease: from plasticity to vulnerability. Neuroscience 309:1–16
47. Martin SJ, Grimwood PD, Morris RGM (2000) Synaptic plasticity and memory: an evaluation of the hypothesis. Annu Rev Neurosci 23(1):649–711
48. Citri A, Malenka RC (2008) Synaptic plasticity: multiple forms, functions, and mechanisms. Neuropsychopharmacology 33(1):18–41
49. Dong Z, Bai Y, Wu X, Li H, Gong B, Howland JG et al (2013) Hippocampal long-term depression mediates spatial reversal learning in the Morris water maze. Neuropharmacology 64(Complete):65–73
50. Bliss TVP, Lømo T (1973) Long-lasting potentiation of synaptic transmission in the dentate area of the anaesthetized rabbit following stimulation of the perforant path. J Physiol 232(2):331–356
51. Lynch GS, Dunwiddie T, Gribkoff V (1977) Heterosynaptic depression: a postsynaptic correlate of long-term potentiation. Nature 266(5604):737–739
52. Herring BE, Nicoll RA (2016) Long-term potentiation: from CaMKII to AMPA receptor trafficking. Annu Rev Physiol 78(1):351–365
53. Delorenzo RJ, Sun DA, Deshpande LS (2005) Cellular mechanisms underlying acquired epilepsy: the calcium hypothesis of the induction and maintainance of epilepsy. Pharmacol Ther 105(3):229–266
54. Arundine M, Tymianski M (2003) Molecular mechanisms of calcium-dependent neurodegeneration in excitotoxicity. Cell Calcium 34(4–5):325–337
55. Disterhoft JF, Thompson LT, Moyer JR Jr, Mogul DJ (1996) Calcium-dependent afterhyperpolarization and learning in young and aging hippocampus. Life Sci 59(5–6):413–420
56. Wu WW, Oh MM, Disterhoft JF (2002) Age-related biophysical alterations of hippocampal pyramidal neurons: implications for learning and memory. Ageing Res Rev 1(2):181–207
57. Ge Y, Dong Z, Bagot RC, Howland JG, Phillips AG, Wong TP et al (2010) Hippocampal long-term depression is required for the consolidation of spatial memory. Proc Natl Acad Sci USA 107(38):16697–16702
58. Tombaugh GC, Rowe WB, Chow AR, Michael TH, Rose GM (2002) Theta-frequency synaptic potentiation in CA1 in vitro distinguishes cognitively impaired from unimpaired aged Fischer 344 rats. J Neurosci 22(22):9932–9940
59. Sankar R, Shin D, Mazarati AM, Liu H, Katsumori H, Lezama R et al (2000) Epileptogenesis after status epilepticus reflects age- and model-dependent plasticity. Ann Neurol 48(4):580–589
60. Rex CS, Kramar EA, Colgin LL, Lin B, Gall CM, Lynch G (2005) Long-term potentiation is impaired in middle-aged rats: regional specificity and reversal by adenosine receptor antagonists. J Neurosci 25(25):5956–5966
61. Barnes CA (1979) Memory deficits associated with senescence: a neurophysiological and behavioral study in the rat. J Comp Physiol Psychol 93(1):74–104
62. Diogenes MJ, Costenla AR, Lopes LV, Jeronimo-Santos A, Sousa VC, Fontinha BM et al (2011) Enhancement of LTP in aged rats is dependent on endogenous BDNF. Neuropsychopharmacology 36(9):1823–1836
63. Pinho J, Vale R, Batalha VL, Costenla AR, Dias R, Rombo D et al (2017) Enhanced LTP in aged rats: detrimental or compensatory? Neuropharmacology 114:12–19
64. Bloss EB, Hunter RG, Waters EM, Munoz C, Bernard K, McEwen BS (2008) Behavioral and biological effects of chronic S18986, a positive AMPA receptor modulator, during aging. Exp Neurol 210(1):109–117
65. Radin DP, Zhong S, Purcell R, Lippa A (2016) Acute ampakine treatment ameliorates age-related deficits in long-term potentiation. Biomed Pharmacother 84:806–809
66. Jurado S (2017) AMPA receptor trafficking in natural and pathological aging. Front Mol Neurosci 10:446
67. Silva A, Stevens C, Tonegawa S, Wang Y (1992) Deficient hippocampal long-term potentiation in alpha-calcium-calmodulin kinase II mutant mice. Science 257(5067):201–206
68. Jouvenceau A, Dutar P, Billard JM (1998) Alteration of NMDA receptor-mediated synaptic responses in CA1 area of the aged rat hippocampus: contribution of GABAergic and cholinergic deficits. Hippocampus 8(6):627–637
69. Serra M, Ghiani CA, Foddi MC, Motzo C, Biggio G (1994) NMDA receptor function is enhanced in the hippocampus of aged rats. Neurochem Res 19(4):483–487
70. Ingram DK, Garofalo P, Spangler EL, Mantione CR, Odano I, London ED (1992) Reduced density of NMDA receptors and increased sensitivity to dizocilpine-induced learning impairment in aged rats. Brain Res 580(1–2):273–280
71. Topic B, Willuhn I, Palomero-Gallagher N, Zilles K, Huston JP, Hasenohrl RU (2007) Impaired maze performance in aged rats is accompanied by increased

density of NMDA, 5-HT1A, and alpha-adrenoceptor binding in hippocampus. Hippocampus 17(1):68–77
72. Stein V, House DR, Bredt DS, Nicoll RA (2003) Postsynaptic density-95 mimics and occludes hippocampal long-term potentiation and enhances long-term depression. J Neurosci 23(13):5503–5506
73. Ehrlich I, Klein M, Rumpel S, Malinow R (2007) PSD-95 is required for activity-driven synapse stabilization. Proc Natl Acad Sci USA 104(10):4176–4181
74. Elias GM, Funke L, Stein V, Grant SG, Bredt DS, Nicoll RA (2006) Synapse-specific and developmentally regulated targeting of AMPA receptors by a family of MAGUK scaffolding proteins. Neuron 52(2):307–320
75. Bradley SA, Steinert JR (2016) Nitric oxide-mediated posttranslational modifications: impacts at the synapse. Oxidative Med Cell Longev 2016:5681036
76. Nagura H, Doi T, Fujiyoshi Y (2016) Characterization of physiological phenotypes of dentate gyrus synapses of PDZ1/2 domain-deficient PSD-95-knockin mice. Eur J Neurosci 43(5):618–625
77. Squire LR (2009) Memory and brain systems: 1969–2009. J Neurosci 29(41):12711–12716
78. Yang K, Broussard JI, Levine AT, Jenson D, Arenkiel BR, Dani JA (2017) Dopamine receptor activity participates in hippocampal synaptic plasticity associated with novel object recognition. Eur J Neurosci 45(1):138–146
79. Li YC, Gao WJ (2011) GSK-3beta activity and hyperdopamine-dependent behaviors. Neurosci Biobehav Rev 35(3):645–654
80. Furini CR, Myskiw JC, Schmidt BE, Marcondes LA, Izquierdo I (2014) D1 and D5 dopamine receptors participate on the consolidation of two different memories. Behav Brain Res 271:212–217
81. Vila-Luna S, Cabrera-Isidoro S, Vila-Luna L, Juarez-Diaz I, Bata-Garcia JL, Alvarez-Cervera FJ et al (2012) Chronic caffeine consumption prevents cognitive decline from young to middle age in rats, and is associated with increased length, branching, and spine density of basal dendrites in CA1 hippocampal neurons. Neuroscience 202:384–395
82. Carvalho AL, Caldeira MV, Santos SD, Duarte CB (2008) Role of the brain-derived neurotrophic factor at glutamatergic synapses. Br J Pharmacol 153(Suppl 1):S310–S324
83. Fried PJ, Schilberg L, Brem AK, Saxena S, Wong B, Cypess AM et al (2017) Humans with Type-2 diabetes show abnormal long-term potentiation-like cortical plasticity associated with verbal learning deficits. J Alzheimers Dis 55(1):89–100
84. Clark JK, Furgerson M, Crystal JD, Fechheimer M, Furukawa R, Wagner JJ (2015) Alterations in synaptic plasticity coincide with deficits in spatial working memory in presymptomatic 3xTg-AD mice. Neurobiol Learn Mem 125:152–162
85. Varga E, Juhasz G, Bozso Z, Penke B, Fulop L, Szegedi V (2015) Amyloid-beta1-42 disrupts synaptic plasticity by altering glutamate recycling at the synapse. J Alzheimers Dis 45(2):449–456
86. Yuan Xiang P, Janc O, Grochowska KM, Kreutz MR, Reymann KG (2016) Dopamine agonists rescue Abeta-induced LTP impairment by Src-family tyrosine kinases. Neurobiol Aging 40:98–102
87. Goetzl EJ, Kapogiannis D, Schwartz JB, Lobach IV, Goetzl L, Abner EL et al (2016) Decreased synaptic proteins in neuronal exosomes of frontotemporal dementia and Alzheimer's disease. FASEB J 30(12):4141–4148
88. Turrigiano G (2012) Homeostatic synaptic plasticity: local and global mechanisms for stabilizing neuronal function. Cold Spring Harb Perspect Biol 4(1):a005736
89. Hobbiss AF, Ramiro-Cortes Y, Israely I (2018) Homeostatic plasticity scales dendritic spine volumes and changes the threshold and specificity of Hebbian plasticity. iScience 8:161–174
90. Miller KD, MacKay DJC (1994) The role of constraints in Hebbian learning. Neural Comput 6:100–126
91. Driscoll I, Davatzikos C, An Y, Wu X, Shen D, Kraut M et al (2009) Longitudinal pattern of regional brain volume change differentiates normal aging from MCI. Neurology 72(22):1906–1913
92. Seidler RD, Bernard JA, Burutolu TB, Fling BW, Gordon MT, Gwin JT et al (2010) Motor control and aging: links to age-related brain structural, functional, and biochemical effects. Neurosci Biobehav Rev 34(5):721–733
93. Liu H, Wang L, Geng Z, Zhu Q, Song Z, Chang R et al (2016) A voxel-based morphometric study of age- and sex-related changes in white matter volume in the normal aging brain. Neuropsychiatr Dis Treat 12:453–465
94. Bartzokis G, Cummings JL, Sultzer D, Henderson VW, Nuechterlein KH, Mintz J (2003) White matter structural integrity in healthy aging adults and patients with Alzheimer disease: a magnetic resonance imaging study. Arch Neurol 60(3):393–398
95. Bennett IJ, Madden DJ (2014) Disconnected aging: cerebral white matter integrity and age-related differences in cognition. Neuroscience 276:187–205
96. Lemaitre H, Crivello F, Grassiot B, Alperovitch A, Tzourio C, Mazoyer B (2005) Age- and sex-related effects on the neuroanatomy of healthy elderly. NeuroImage 26(3):900–911
97. Shenkin SD, Bastin ME, Macgillivray TJ, Deary IJ, Starr JM, Rivers CS et al (2005) Cognitive correlates of cerebral white matter lesions and water diffusion tensor parameters in community-dwelling older people. Cerebrovasc Dis 20(5):310–318
98. Zakhari N, Taccone MS, Torres CH, Chakraborty S, Sinclair J, Woulfe J et al (2019) Prospective comparative diagnostic accuracy evaluation of dynamic contrast-enhanced (DCE) vs. dynamic susceptibility contrast (DSC) MR perfusion in differentiating tumor recurrence from radiation necrosis in

treated high-grade gliomas. J Magn Reson Imaging 50(2):573–582
99. Caligiuri ME, Perrotta P, Augimeri A, Rocca F, Quattrone A, Cherubini A (2015) Automatic detection of White matter hyperintensities in healthy aging and pathology using magnetic resonance imaging: a review. Neuroinformatics 13(3):261–276
100. Sloane JA, Hollander W, Moss MB, Rosene DL, Abraham CR (1999) Increased microglial activation and protein nitration in white matter of the aging monkey. Neurobiol Aging 20(4):395–405
101. De Biase D, Piegari G, Prisco F, Cimmino I, Pirozzi C, Raso GM et al (2020) Autophagy and NLRP3 inflammasome crosstalk in neuroinflammation in aged bovine brains. J Cell Physiol 235(6):5394–5403
102. Taubert M, Roggenhofer E, Melie-Garcia L, Muller S, Lehmann N, Preisig M et al (2020) Converging patterns of aging-associated brain volume loss and tissue microstructure differences. Neurobiol Aging 88:108–118
103. Sure VN, SSVP S, Sperling JA, Evans WR, Merdzo I, Mostany R et al (2018) A novel high-throughput assay for respiration in isolated brain microvessels reveals impaired mitochondrial function in the aged mice. GeroScience 40(4):365–375
104. Lane CA, Barnes J, Nicholas JM, Sudre CH, Cash DM, Malone IB et al (2019) Associations between vascular risk across adulthood and brain pathology in late life: evidence from a British birth cohort. JAMA Neurol 77:175–183
105. Brun A, EL, E. (1986) A white matter disorder in dementia of the Alzheimer type: A pathoanatomical study. Ann Neurol 19(3):253–262
106. Salat DH, Buckner RL, Snyder AZ, Greve DN, Desikan RSR, Busa E et al (2004) Thinning of the cerebral cortex in aging. Cereb Cortex 7:7
107. Tomimoto H, Ihara M, Wakita H, Ohtani R, Lin JX, Akiguchi I et al (2003) Chronic cerebral hypoperfusion induces white matter lesions and loss of oligodendroglia with DNA fragmentation in the rat. Acta Neuropathologica 106(6):527–534
108. Klein AW, Michel ME (1977) A morphometric study of the neocortex of young adult and old maze-differentiated rats. Mech Ageing Dev 6(6):441–452
109. van Beek AH, Claassen JA, Olde Rikkert MG, Jansen RW (2008) Cerebral autoregulation: an overview of current concepts and methodology with special focus on the elderly. J Cereb Blood Flow Metab 28(6):1071–1085
110. Moody DM, Bell MA, Challa VR (1990) Features of the cerebral vascular pattern that predict vulnerability to perfusion or oxygenation deficiency: an anatomic study. AJNR Am J Neuroradiol 11(3):431–439
111. Brown WR, Moody DM, Challa VR, Thore CR, Anstrom JA (2002) Venous collagenosis and arteriolar tortuosity in leukoaraiosis. J Neurol Sci 203(10):159–163
112. HMA A, Wolfson L, Moscufo GCRG, KaplanRF WWB (2016) Cardiovascular risk factors and small vessel disease of the brain: blood pressure, white matter lesions, and functional decline in older persons. J Cereb Blood Flow Metab 36(1):132–142
113. Irving EA, Bentley DL, Parsons AA (2001) Assessment of white matter injury following prolonged focal cerebral ischaemia in the rat. Acta Neuropathol 102(6):627–635
114. Czuwala JP (1991) McDonald's blood flow in arteries: theoretical, experimental and clinical principles. Echocardiography 8(3):405–406
115. Lakatta EG, Levy D (2003) Arterial and cardiac aging: major shareholders in cardiovascular disease enterprises: part I: aging arteries: a "set up" for vascular disease. Circulation 107(1):139–146
116. Pantoni L (2002) Pathophysiology of age-related cerebral white matter changes. Cerebrovasc Dis 13(Suppl 2):7–10
117. Wallin A, Sjögren M, Edman A, Blennow K, Regland B (2000) Symptoms, vascular risk factors and blood-brain barrier function in relation to CT White-matter changes in dementia. Eur Neurol 44(4):229–235
118. Wharton SB, Baxter L, Sears JC, Matthews FE, Ince PG, Gelsthorpe C et al (2010) Alterations of the blood-brain barrier in cerebral white matter lesions in the ageing brain. Neurosci Lett 486(3):246–251
119. Topakian R, Barrick TR, Howe FA, Markus HS (2010) Blood-brain barrier permeability is increased in normal-appearing white matter in patients with lacunar stroke and leucoaraiosis. J Neurol Neurosurg Psychiatry 81(2):192–197
120. Garland EL, Gaylord SA, Boettiger CA, Howard MO (2010) Mindfulness training modifies cognitive, affective, and physiological mechanisms implicated in alcohol dependence: results of a randomized controlled pilot trial. J Psychoactive Drugs 42(2):177–192
121. Pantoni L, Garcia JH (1995) The significance of cerebral white matter abnormalities 100 years after Binswanger's report. A review. Stroke 26(7):1293–1301
122. Laurent S, Katsahian S, Fassot C, Tropeano A-I, Gautier I, Laloux B et al (2003) Aortic stiffness is an independent predictor of fatal stroke in essential hypertension. Stroke 34(5):1203–1206
123. Liao D, Cooper L, Cai J, Toole J, Bryan N, Burke G et al (1997) The prevalence and severity of white matter lesions, their relationship with age, ethnicity, gender, and cardiovascular disease risk factors: the ARIC study. Neuroepidemiology 16(3):149–162
124. Novak V (2012) Cognition and hemodynamics. Curr Cardiovasc Risk Rep 6(5):380–396
125. Manschot SM, Biessels GJ, Rutten GEHM, Kessels RPC, Gispen WH et al (2008) Peripheral and central neurologic complications in type 2 diabetes mellitus: no association in individual patients. J Neurol Sci 264:157–162
126. Novak V, Last D, Alsop DC, Abduljalil AM, Hu K, Lepicovsky L et al (2006) Cerebral blood flow velocity and periventricular white matter hyperintensities in type 2 diabetes. Diabetes Care 29(7):1529–1534

127. de Leeuw F-E (2004) Interaction between hypertension, apoE, and cerebral white matter lesions. Stroke 35(5):1057–1060
128. Henskens LHG, Kroon AA, van Boxtel MPJ, Hofman PAM, de Leeuw PW (2004) Associations of the angiotensin II type 1 receptor A1166C and the endothelial NO synthase G894T gene polymorphisms with silent subcortical white matter lesions in essential hypertension. Stroke 36(9):1869–1873
129. Simpson JE, Hosny O, Wharton SB, Heath PR, Holden H, Fernando MS et al (2009) Microarray RNA expression analysis of cerebral white matter lesions reveals changes in multiple functional pathways. Stroke 40(2):369–375
130. Tripathi A, Volsko C, Datta U, Regev K, Dutta R (2019) Expression of disease-related miRNAs in white-matter lesions of progressive multiple sclerosis brains. Ann Clin Trans Neurol 6(5):854–862
131. Xu H, Stamova B, Jickling G, Tian Y, Zhan X, Ander BP et al (2010) Distinctive RNA expression profiles in blood associated with white matter hyperintensities in brain. Stroke 41(12):2744–2749
132. Paternoster L, Chen W, Sudlow CLM (2009) Genetic determinants of white matter hyperintensities on brain scans: a systematic assessment of 19 candidate gene polymorphisms in 46 studies in 19,000 subjects. Stroke 40(6):2020–2026
133. Brown WR, Thore CR (2011) Review: cerebral microvascular pathology in ageing and neurodegeneration. Neuropathol Appl Neurobiol 37(1):56–74
134. Roher AE, Lowenson JD, Clarke S, Woods AS, Cotter RJ, Gowing E et al (1993) Beta-amyloid-(1-42) is a major component of cerebrovascular amyloid deposits: implications for the pathology of Alzheimer disease. Proc Natl Acad Sci U S A 90(22):10836–10840
135. Yang SP, Bae DG, Kang HJ, Gwag BJ, Gho YS, Chae CB (2004) Co-accumulation of vascular endothelial growth factor with beta-amyloid in the brain of patients with Alzheimer's disease. Neurobiol Aging 25(3):283–290
136. Erickson MA, Banks WA (2013) Blood-brain barrier dysfunction as a cause and consequence of Alzheimer's disease. J Cereb Blood Flow Metab 33(10):1500–1513
137. Janota C, Lemere CA, Brito MA (2016) Dissecting the contribution of vascular alterations and aging to Alzheimer's disease. Mol Neurobiol 53(6):3793–3811
138. Silverberg GD, Messier AA, Miller MC, Machan JT, Majmudar SS, Stopa EG et al (2010) Amyloid efflux transporter expression at the blood-brain barrier declines in normal aging. J Neuropathol Exp Neurol 69(10):1034–1043
139. Serrador JM, Freeman R (2017) Enhanced cholinergic activity improves cerebral blood flow during orthostatic stress. Front Neurol 8:103
140. Lim YY, Maruff P, Schindler R, Ott BR, Salloway S, Yoo DC et al (2015) Disruption of cholinergic neurotransmission exacerbates Aβ-related cognitive impairment in preclinical Alzheimer's disease. Neurobiol Aging 36(10):2709–2715
141. Sexton CE, Walhovd KB, Storsve AB, Tamnes CK, Westlye LT, Johansen-Berg H et al (2014) Accelerated changes in white matter microstructure during aging: a longitudinal diffusion tensor imaging study. J Neurosci 34(46):15425–15436
142. Sullivan EV, Rohlfing T, Pfefferbaum A (2010) Longitudinal study of callosal microstructure in the Normal adult aging brain using quantitative DTI fiber tracking. Dev Neuropsychol 35(3):233–256
143. Phillips KA, Watson CM, Bearman A, Knippenberg AR, Adams J, Ross C et al (2019) Age-related changes in myelin of axons of the corpus callosum and cognitive decline in common marmosets. Am J Primatol 81(2):e22949
144. Peters A, Sethares C (2002) Aging and the myelinated fibers in prefrontal cortex and corpus callosum of the monkey. J Comp Neurol 442(3):277–291
145. Sugiyama I, Tanaka K, Akita M, Yoshida K, Kawase T, Asou H (2002) Ultrastructural analysis of the paranodal junction of myelinated fibers in 31-month-old-rats. J Neurosci Res 70(3):309–317
146. Hinman JD, Peters A, Cabral H, Rosene DL, Hollander W, Rasband MN et al (2006) Age-related molecular reorganization at the node of Ranvier. J Comp Neurol 495(4):351–362
147. Jankelowitz SK, McNulty PA, Burke D (2007) Changes in measures of motor axon excitability with age. Clin Neurophysiol 118(6):1397–1404
148. Hampton DW, Innes N, Merkler D, Zhao C, Franklin RJ, Chandran S (2012) Focal immune-mediated white matter demyelination reveals an age-associated increase in axonal vulnerability and decreased remyelination efficiency. Am J Pathol 180(5):1897–1905
149. Posse de Chaves E, Sipione S (2010) Sphingolipids and gangliosides of the nervous system in membrane function and dysfunction. FEBS Lett 584(9):1748–1759
150. Sloane JA, Hinman JD, Lubonia M, Hollander W, Abraham CR (2002) Age-dependent myelin degeneration and proteolysis of oligodendrocyte proteins is associated with the activation of calpain-1 in the rhesus monkey & nbsp. J Neurochem 84(1):157–168
151. Rasband MN, Tayler J, Kaga Y, Yang Y, Lappe-Siefke C, Nave K-A et al (2005) CNP is required for maintenance of axon-glia interactions at nodes of Ranvier in the CNS. Glia 50(1):86–90
152. Kohama SG, Rosene DL, Sherman LS (2012) Age-related changes in human and non-human primate white matter: from myelination disturbances to cognitive decline. Age (Dordr) 34(5):1093–1110
153. Frischer JM, Weigand SD, Guo Y, Kale N, Parisi JE, Pirko I et al (2015) Clinical and pathological insights into the dynamic nature of the white matter multiple sclerosis plaque. Ann Neurol 78(5):710–721
154. Rist JM, Franklin RJ (2008) Taking ageing into account in remyelination-based therapies for multiple sclerosis. J Neurol Sci 274(1–2):64–67

155. Rinholm JE, Bergersen LH (2013) White matter lactate – does it matter? Neuroscience 276:109–116
156. Lee Y, Morrison BM, Li Y, Lengacher S, Farah MH, Hoffman PN et al (2012) Oligodendroglia metabolically support axons and contribute to neurodegeneration. Nature 487(7408):443–448
157. Lundgaard I, Osório MJ, Kress BT, Sanggaard S, Nedergaard M (2014) White matter astrocytes in health and disease. Neuroscience 276:161–173
158. Kıray H, Lindsay SL, Hosseinzadeh S, Barnett SC (2016) The multifaceted role of astrocytes in regulating myelination. Exp Neurol 283:541–549
159. García-Matas S, Gutierrez-Cuesta J, Coto-Montes A, Rubio-Acero R, Díez-Vives C, Camins A et al (2008) Dysfunction of astrocytes in senescence-accelerated mice SAMP8 reduces their neuroprotective capacity. Aging Cell 7(5):630–640
160. Khairy EY, Attia MM (2019) Protective effects of vitamin D on neurophysiologic alterations in brain aging: role of brain-derived neurotrophic factor (BDNF). Nutr Neurosci 24(8):650–659
161. Lee HJ, Kim C, Lee SJ (2010) Alpha-synuclein stimulation of astrocytes: potential role for neuroinflammation and neuroprotection. Oxidative Med Cell Longev 3(4):283–287
162. Rodriguez M, Rodriguez-Sabate C, Morales I, Sanchez A, Sabate M (2015) Parkinson's disease as a result of aging. Aging Cell 14(3):293–308
163. Bates TJ, Vonica A, Heasman J, Brivanlou AH, Bell E (2013) Coco regulates dorsoventral specification of germ layers via inhibition of TGFβ signalling. Development 140(20):4177–4181
164. Rawji KS, Yong VW (2013) The benefits and detriments of macrophages/microglia in models of multiple sclerosis. Clin Dev Immunol 2013:948976
165. Shobin E, Bowley MP, Estrada LI, Heyworth NC, Orczykowski ME, Eldridge SA et al (2017) Microglia activation and phagocytosis: relationship with aging and cognitive impairment in the rhesus monkey. Geroscience 39(2):199–220
166. Norden DM, Godbout JP (2013) Review: microglia of the aged brain: primed to be activated and resistant to regulation. Neuropathol Appl Neurobiol 39(1):19–34
167. Hickman SE, Kingery ND, Ohsumi TK, Borowsky ML, Wang LC, Means TK et al (2013) The microglial sensome revealed by direct RNA sequencing. Nat Neurosci 16(12):1896–1905
168. Teipel SJ, Grothe MJ, Filippi M, Fellgiebel A, Hauenstein K (2014) Fractional anisotropy changes in Alzheimer's disease depend on the underlying fiber tract architecture: a multiparametric DTI study using joint independent component analysis. J Alzheimers Dis 41(1):69–83
169. Kantarci K (2014) Fractional anisotropy of the fornix and hippocampal atrophy in Alzheimer's disease. Front Aging Neurosci 6:316
170. Wisse LE, Reijmer YD, ter Telgte A, Kuijf HJ, Leemans A, Luijten PR et al (2015) Hippocampal disconnection in early Alzheimer's disease: a 7 tesla MRI study. J Alzheimers Dis 45(4):1247–1256
171. Wu TC, Wilde EA, Bigler ED, Yallampalli R, McCauley SR, Troyanskaya M et al (2010) Evaluating the relationship between memory functioning and cingulum bundles in acute mild traumatic brain injury using diffusion tensor imaging. J Neurotrauma 27(2):303–307
172. Amlien IK, Fjell AM (2014) Diffusion tensor imaging of white matter degeneration in Alzheimer's disease and mild cognitive impairment. Neuroscience 276:206–215
173. Amlien IK, Fjell AM, Walhovd KB, Selnes P, Stenset V, Grambaite R et al (2013) Mild cognitive impairment: cerebrospinal fluid tau biomarker pathologic levels and longitudinal changes in white matter integrity. Radiology 266(1):295–303
174. Gold BT, Johnson NF, Powell DK, Smith CD (2012) White matter integrity and vulnerability to Alzheimer's disease: preliminary findings and future directions. Biochim Biophys Acta 1822(3):416–422
175. Zhang YZ, Chang C, Wei XE, Fu JL, Li WB (2011) Comparison of diffusion tensor image study in association fiber tracts among normal, amnestic mild cognitive impairment, and Alzheimer's patients. Neurol India 59(2):168–173
176. Teipel SJ, Grothe MJ, Filippi M, Fellgiebel A, Dyrba M, Frisoni GB et al (2014) Fractional anisotropy changes in Alzheimer's disease depend on the underlying fiber tract architecture: a multiparametric DTI study using joint independent component analysis. J Alzheimers Dis 41(1):69–83
177. Zhuang L, Sachdev PS, Trollor JN, Reppermund S, Kochan NA, Brodaty H et al (2013) Microstructural white matter changes, not hippocampal atrophy, detect early amnestic mild cognitive impairment. PLoS One 8(3):e58887
178. Salat DH, Greve DN, Pacheco JL, Quinn BT, Helmer KG, Buckner RL et al (2009) Regional white matter volume differences in nondemented aging and Alzheimer's disease. NeuroImage 44(4):1247–1258
179. Skillbäck T, Zetterberg H, Blennow K, Mattsson N (2013) Cerebrospinal fluid biomarkers for Alzheimer disease and subcortical axonal damage in 5,542 clinical samples. Alzheimers Res Ther 5(5):47
180. Flanary BE, Sammons NW, Nguyen C, Walker D, Streit WJ (2007) Evidence that aging and amyloid promote microglial cell senescence. Rejuvenation Res 10(1):61–74
181. Fang F, Lue LF, Yan S, Xu H, Luddy JS, Chen D et al (2010) RAGE-dependent signaling in microglia contributes to neuroinflammation, Abeta accumulation, and impaired learning/memory in a mouse model of Alzheimer's disease. FASEB J 24(4):1043–1055
182. Furman JL, Sama DM, Gant JC, Beckett TL, Murphy MP, Bachstetter AD et al (2012) Targeting astrocytes ameliorates neurologic changes in a mouse model of Alzheimer's disease. J Neurosci 32(46):16129
183. Bien-Ly N, Gillespie AK, Walker D, Yoon SY, Huang Y (2012) Reducing human apolipoprotein E

levels attenuates age-dependent Aβ accumulation in mutant human amyloid precursor protein transgenic mice. J Neurosci 32(14):4803–4811
184. Wardlaw JM, Smith C, Dichgans M (2019) Small vessel disease: mechanisms and clinical implications. Lancet Neurol 18(7):684–696
185. Bos D, Wolters FJ, Darweesh SKL, Vernooij MW, Hofman A (2018) Cerebral small vessel disease and the risk of dementia: a systematic review and meta-analysis of population-based evidence. Alzheimers Dement 14(11):1482–1492
186. Lo EH, Dalkara T, Moskowitz MA (2003) Mechanisms, challenges and opportunities in stroke. Nat Rev Neurosci 4(5):399–415
187. Paradis A, Leblanc D, Dumais N (2016) Optimization of an in vitro human blood-brain barrier model: application to blood monocyte transmigration assays. MethodsX 3:25–34
188. Sweeney MD, Ayyadurai S, Zlokovic BV (2016) Pericytes of the neurovascular unit: key functions and signaling pathways. Nat Neurosci 19(6):771–783
189. Zlokovic BV (2011) Neurovascular pathways to neurodegeneration in Alzheimer's disease and other disorders. Nat Rev Neurosci 12(12):723–738
190. Iadecola C (2004) Neurovascular regulation in the normal brain and in Alzheimer's disease. Nat Rev Neurosci 5(5):347–360
191. Sweeney MD, Sagare AP, Zlokovic BV (2018) Blood-brain barrier breakdown in Alzheimer disease and other neurodegenerative disorders. Nat Rev Neurol 14(3):133–150
192. Iadecola C (2013) The pathobiology of vascular dementia. Neuron 80(4):844–866
193. Vermeer SE, Koudstaal PJ, Oudkerk M, Hofman A, Breteler MM (2002) Prevalence and risk factors of silent brain infarcts in the population-based Rotterdam scan study. Stroke 33(1):21–25
194. Badji A, Sabra D, Bherer L, Cohen-Adad J, Girouard H, Gauthier CJ (2019) Arterial stiffness and brain integrity: a review of MRI findings. Ageing Res Rev 53:100907
195. Benetos A, Waeber B, Izzo J, Mitchell G, Resnick L, Asmar R et al (2002) Influence of age, risk factors, and cardiovascular and renal disease on arterial stiffness: clinical applications. Am J Hypertens 15(12):1101–1108
196. Hicks P, Rolsten C, Brizzee D, Samorajski T (1983) Age-related changes in rat brain capillaries. Neurobiol Aging 4(1):69–75
197. Bell MA, Ball MJ (1981) Morphometric comparison of hippocampal microvasculature in ageing and demented people: diameters and densities. Acta Neuropathol 53(4):299–318
198. Buchweitz-Milton E, Weiss HR (1987) Perfused capillary morphometry in the senescent brain. Neurobiol Aging 8(3):271–276
199. Ambrose CT (2012) Neuroangiogenesis: a vascular basis for Alzheimer's disease and cognitive decline during aging. J Alzheimers Dis 32(3):773–788
200. Hunziker O, Abdel'Al S, Schulz U (1979) The aging human cerebral cortex: a stereological characterization of changes in the capillary net. J Gerontol 34(3):345–350
201. Thore CR, Anstrom JA, Moody DM, Challa VR, Marion MC, Brown WR (2007) Morphometric analysis of arteriolar tortuosity in human cerebral white matter of preterm, young, and aged subjects. J Neuropathol Exp Neurol 66(5):337–345
202. Camandola S, Mattson MP (2017) Brain metabolism in health, aging, and neurodegeneration. EMBO J 36(11):1474–1492
203. Camici GG, Savarese G, Akhmedov A, Luscher TF (2015) Molecular mechanism of endothelial and vascular aging: implications for cardiovascular disease. Eur Heart J 36(48):3392–3403
204. Gu Y, Zheng G, Xu M, Li Y, Chen X, Zhu W et al (2012) Caveolin-1 regulates nitric oxide-mediated matrix metalloproteinases activity and blood-brain barrier permeability in focal cerebral ischemia and reperfusion injury. J Neurochem 120(1):147–156
205. Brandes RP, Fleming I, Busse R (2005) Endothelial aging. Cardiovasc Res 66(2):286–294
206. Wiener J, Lombardi DM, Su JE, Schwartz SM (1996) Immunohistochemical and molecular characterization of the differential response of the rat mesenteric microvasculature to angiotensin-II infusion. J Vasc Res 33(3):195–208
207. Scioli MG, Bielli A, Arcuri G, Ferlosio A, Orlandi A (2014) Ageing and microvasculature. Vasc Cell 6:19
208. Michaud M, Balardy L, Moulis G, Gaudin C, Peyrot C, Vellas B et al (2013) Proinflammatory cytokines, aging, and age-related diseases. J Am Med Dir Assoc 14(12):877–882
209. Kohn JC, Lampi MC, Reinhart-King CA (2015) Age-related vascular stiffening: causes and consequences. Front Genet 6:112
210. Paneni F, Osto E, Costantino S, Mateescu B, Briand S, Coppolino G et al (2013) Deletion of the activated protein-1 transcription factor JunD induces oxidative stress and accelerates age-related endothelial dysfunction. Circulation 127(11):1229–1240, e1221
211. Ryan NA, Zwetsloot KA, Westerkamp LM, Hickner RC, Pofahl WE, Gavin TP (2006) Lower skeletal muscle capillarization and VEGF expression in aged vs. young men. J Appl Physiol (1985) 100(1):178–185
212. Cejkova J, Vejrazka M, Platenik J, Stipek S (2004) Age-related changes in superoxide dismutase, glutathione peroxidase, catalase and xanthine oxidoreductase/xanthine oxidase activities in the rabbit cornea. Exp Gerontol 39(10):1537–1543
213. Stoltzner SE, Grenfell TJ, Mori C, Wisniewski KE, Wisniewski TM, Selkoe DJ et al (2000) Temporal accrual of complement proteins in amyloid plaques in Down's syndrome with Alzheimer's disease. Am J Pathol 156(2):489–499
214. Maier M, Peng Y, Jiang L, Seabrook TJ, Carroll MC, Lemere CA (2008) Complement C3 deficiency leads to accelerated amyloid beta plaque deposition and

neurodegeneration and modulation of the microglia/macrophage phenotype in amyloid precursor protein transgenic mice. J Neurosci 28(25):6333–6341
215. Krance SH, Wu CY, Zou Y, Mao H, Toufighi S, He X et al (2019) The complement cascade in Alzheimer's disease: a systematic review and meta-analysis. Mol Psychiatry
216. Trimpin S, Deinzer ML (2007) Solvent-free MALDI-MS for the analysis of beta-amyloid peptides via the mini-ball mill approach: qualitative and quantitative advances. J Am Soc Mass Spectrom 18(8):1533–1543
217. Wang JS, Whitehead SN, Yeung KK (2018) Detection of amyloid Beta (Abeta) oligomeric composition using matrix-assisted laser desorption ionization mass spectrometry (MALDI MS). J Am Soc Mass Spectrom 29(4):786–795
218. Wang MJ, Yi S, Han JY, Park SY, Jang JW, Chun IK et al (2017) Oligomeric forms of amyloid-beta protein in plasma as a potential blood-based biomarker for Alzheimer's disease. Alzheimers Res Ther 9(1):98
219. Huang CW, Tsai MH, Chen NC, Chen WH, Lu YT, Lui CC et al (2015) Clinical significance of circulating vascular cell adhesion molecule-1 to white matter disintegrity in Alzheimer's dementia. Thromb Haemost 114(6):1230–1240
220. Iadecola C, Nedergaard M (2007) Glial regulation of the cerebral microvasculature. Nat Neurosci 10(11):1369–1376
221. Bayraktar OA, Fuentealba LC, Alvarez-Buylla A, Rowitch DH (2014) Astrocyte development and heterogeneity. Cold Spring Harb Perspect Biol 7(1):a020362
222. Matyash V, Kettenmann H (2010) Heterogeneity in astrocyte morphology and physiology. Brain Res Rev 63(1–2):2–10
223. Montagne A, Barnes SR, Sweeney MD, Halliday MR, Sagare AP, Zhao Z et al (2015) Blood-brain barrier breakdown in the aging human hippocampus. Neuron 85(2):296–302
224. Mooradian AD (1988) Blood-brain barrier transport of choline is reduced in the aged rat. Brain Res 440(2):328–332
225. Stewart PA, Magliocco M, Hayakawa K, Farrell CL, Del Maestro RF, Girvin J et al (1987) A quantitative analysis of blood-brain barrier ultrastructure in the aging human. Microvasc Res 33(2):270–282
226. Erdo F, Denes L, de Lange E (2017) Age-associated physiological and pathological changes at the blood-brain barrier: a review. J Cereb Blood Flow Metab 37(1):4–24
227. Pantoni L, Rossi R, Inzitari D, Bianchi C, Beneke M, Erkinjuntti T et al (2000) Efficacy and safety of nimodipine in subcortical vascular dementia: a subgroup analysis of the Scandinavian multi-infarct dementia trial. J Neurol Sci 175(2):124–134
228. Landfield PW, Cadwallader-Neal L (1998) Long-term treatment with calcitriol (1,25(OH)2 vit D3) retards a biomarker of hippocampal aging in rats. Neurobiol Aging 19(5):469–477
229. Attwell D, Buchan AM, Charpak S, Lauritzen M, Macvicar BA, Newman EA (2010) Glial and neuronal control of brain blood flow. Nature 468(7321):232–243
230. Takano T, Tian GF, Peng W, Lou N, Libionka W, Han X et al (2006) Astrocyte-mediated control of cerebral blood flow. Nat Neurosci 9(2):260–267
231. Hubbard JA, Hsu MS, Seldin MM, Binder DK (2015) Expression of the astrocyte Water Channel Aquaporin-4 in the mouse brain. ASN Neuro 7(5):1759091415605486
232. Duncombe J, Lennen RJ, Jansen MA, Marshall I, Wardlaw JM, Horsburgh K (2017) Ageing causes prominent neurovascular dysfunction associated with loss of astrocytic contacts and gliosis. Neuropathol Appl Neurobiol 43(6):477–491
233. Liddelow SA, Guttenplan KA, Clarke LE, Bennett FC, Bohlen CJ, Schirmer L et al (2017) Neurotoxic reactive astrocytes are induced by activated microglia. Nature 541(7638):481–487
234. De-Fraja C, Conti L, Govoni S, Battaini F, Cattaneo E (2000) STAT signalling in the mature and aging brain. Int J Dev Neurosci 18(4–5):439–446
235. Jiang T, Cadenas E (2014) Astrocytic metabolic and inflammatory changes as a function of age. Aging Cell 13(6):1059–1067

Part VI

Clinical Studies of Abnormal Aging

The Early Stage of Abnormal Aging: Cognitive Impairment

11

Huajie Shang, Chang Xu, Hui Lu, and Junying Zhang

Abstract

Cognitive impairment has become an important aspect affecting the health and quality of life of middle-aged and elderly people, which is defined as the difficulty of thought processing and leads to memory loss, difficulties making decisions, the inability to concentrate, and difficulties learning. The process of cognitive ability decline with age goes through subjective cognitive impairment (SCI) to mild cognitive impairment (MCI). There is abundant evidence supporting the link between cognitive impairment and several modifiable risk factors, such as physical activity, social activity, mental activity, higher education, and management of cardiovascular risk factors (diabetes, obesity, smoking, hypertension, and obesity). Meanwhile, these factors also provide a new perspective for the prevention of cognitive impairment and dementia.

Keywords

Subjective cognitive impairment · Mild cognitive impairment · Risk factors

11.1 The Syndrome of Cognitive Impairment

Cognition is a type of human psychological activity that refers to the psychological process by which individuals recognize and understand information. Cognitive ability includes the ability to identify oneself and one's environment, and involves perception, attention, learning, memory, thinking, and language [1]. Cognitive ability is the most important psychological aspect that supports the successful accomplishment of various activities. In clinical practice, neuropsychological scales are used to measure cognitive functions, such as general cognitive abilities and cognitive domains [2]. Additionally, these scales enable the recognition and identification of some cognitive impairments that are difficult to detect during daily life. Cognitive impairment refers to the higher cortical dysfunctions caused by brain lesions such as neurodegeneration, vascular dis-

H. Shang · C. Xu
State Key Laboratory of Cognitive Neuroscience and Learning, Faculty of Psychology, Beijing Normal University, Beijing, China

Beijing Aging Brain Rejuvenation Initiative (BABRI) Centre, Beijing Normal University, Beijing, China

H. Lu
Beijing Aging Brain Rejuvenation Initiative (BABRI) Centre, Beijing Normal University, Beijing, China

J. Zhang (✉)
Institute Of Basic Research In Clinical Medicine, China Academy Of Chinese Medical Sciences, Beijing, China

Beijing Aging Brain Rejuvenation Initiative (BABRI) Centre, Beijing Normal University, Beijing, China

Z. Zhang (ed.), *Cognitive Aging and Brain Health*, Advances in Experimental Medicine and Biology 1419, https://doi.org/10.1007/978-981-99-1627-6_11

ease, mental illness, or brain damage caused by accidents [3]. The damage is not limited to individuals of a certain age, sex, or other demographic characteristics.

Cognitive impairment is likely to occur at any stage during the lifespan development, especially in middle and old age. The progressive stages of cognitive impairment generally include normal cognitive decline with aging, asymptomatic stage, subjective cognitive impairment (SCI), mild cognitive impairment (MCI), and dementia. In 2019, there were an estimated 50 million people worldwide suffering from dementia, and this number is expected to rise to 152 million by 2050 (https://www.alzint.org/resource/world-alzheimer-report%2D%2D2019/). Alzheimer's disease (AD), a progressive neurodegenerative disease, is the most common type of dementia (60–80%) among older adults and is characterized by memory loss, language problems, and other cognitive deficits [4].

11.1.1 Subjective Cognitive Impairment

SCI is defined as cognitive impairment with only subjective, perceived impairments, but without objective deficits and appropriate concerns about memory loss [5]. At this stage, individuals experience subjective memory loss (e.g., not remembering names or the placement of objects); however, the cognitive performance, as measured by psychological state tests, remains within the normal range. SCI was hypothesized more than 20 years ago during the evolution of our understanding of AD [6]. For example, a study in 1986 estimated that the SCI phase lasted for an average 15 years before the transition to MCI in the pathological progression of AD [7]. The incidence of SCI is approximately 25–50% among older adults living in the community, and a higher prevalence is associated with older age and lower education levels[8]. With the aging of the population and the evidence of SCI indicating non-typical cognitive decline and progressive dementia in some elderly individuals, SCI has attracted increasing attention. For example, older people with SCI are twice as likely to develop dementia than those without SCI [9]. SCI affects the elderly who are concerned about memory decline and forgetfulness, lack of interest in life and activity participation, and worsens negative emotional symptoms.

11.1.2 Mild Cognitive Impairment

MCI, the intermediate stage between SCI and mild dementia, presents as both subjective and objective cognitive decline without affecting most daily life functions [10, 11]. The four clinical subtypes of MCI are amnestic MCI-single domain, amnestic MCI-multiple domains, non-amnestic MCI-single domain, and non-amnestic MCI-multiple domains. Amnestic MCI is when significant clinical memory impairment exists but has not yet met the diagnostic criteria for dementia. Non-amnestic MCI refers to functional decline not related to memory which affects attention, use of language, or visuospatial skills. Amnestic MCI has a high possibility of progressing to AD, including both single and multiple domains, while the non-amnestic subtypes have a low possibility of progressing to non-AD dementia [12, 13].

MCI has gradually become one of the most important diseases affecting the health of the elderly. The annual conversion rate of MCI to AD is 10–15%, 5–6 times higher than that of normal elderly individuals who converted to AD, and more than 32% of patients progress to AD within 5 years of an MCI diagnosis [10, 14]. The prevalence of MCI varies from region to region. The prevalence of MCI is 22.2% among people aged 71 years or older in the United States [15]. Furthermore, surveys in some countries with low and medium income found that the prevalence of MCI varied from 0.8% in China to 4.3% in India [16]. Another meta-analysis in 2018 showed that the pooled prevalence of MCI in the older Chinese population was 14.71% which covers 22 provinces [17]. The prevalence was 16.72% in clinical samples and 14.61% in nonclinical samples. According to the Beijing Brain Health Research Plan for the elderly, the incidence of MCI in the elderly residents of Beijing communities is 15.7% [18]. Additionally, while MCI is associated with a significant risk of developing dementia, not all patients with MCI will progress to dementia, as some patients will maintain mild cognitive impairment and approximately 18% of patients may convert to cognitively normal elderly individuals [19]. In recent years, a large number of basic and clinical studies have focused on predicting the transformation of MCI into AD. Although the etiology of the transformation remains unclear, this prediction is crucial for the diagnosis and early treatment of dementia.

Cognitive impairment has become an important disease affecting the health and quality of life of middle-aged and elderly adults. It might be a short-term problem or a permanent condition. The manifestations of cognitive impairment include not only memory disorders, language impairment, attention impairment, execution function impairment, and visual-spatial disorders but also emotional and behavioral disorders such as anxiety, depression, psychomotor agitation, and impulse control disorder [3]. These emotional and behavioral disorders are also the causes of disability. Cognitive impairment associated with aging has emerged as one of the major challenges in public health.

11.2 Risk and Protective Factors for Cognitive Impairment

A decrease in childhood education and exercise, reduced social interactions, increased smoking, the emergence and deterioration of hearing loss, depression, diabetes, hypertension, and obesity all potentially contribute to the generation or aggravation of cognitive impairment.

11.2.1 Cardiovascular Risk Factors

Prevention ameliorates modifiable risk factors and has a greater benefit than treatment alone. Any future treatment for cognitive impairment will not eliminate the necessity for its effective prevention. In a published study on risk factors, mid-life has been well defined as 45–65 years and later life as older than 65 years. We have used these demarcations throughout this chapter for consistency, but these risks are often related to the life course. Meanwhile, this chapter will focus on well-established cardiovascular risk factors for dementia, including diabetes, mid-life hypertension, and mid-life obesity [20].

11.2.1.1 Diabetes, Mid-Life Hypertension, and Mid-Life Obesity

Cognitive impairment usually presents in people aged over 65 years, when comorbidities occur at a relatively higher rate. At this age, age-related physical health problems and cognitive impairment usually co-occur. This co-occurrence is due to some physical problems, such as diabetes and hypertension, which increase the risk of vascular cognitive impairment; thus, mixed cognitive impairment is more likely to occur. The greater the risk of cardiovascular disease in an individual, the more likely they are to develop cognitive impairment. This relationship is probably related to a lack of elasticity and restoration, which contributes to all the issues mentioned above [21].

Among vascular risk factors, hypertension had the highest population attributable fraction (PAF), but all individuals who had PAF were below 5% [22]. Obesity is associated with prediabetes and metabolic syndrome, which is characterized by insulin resistance and high concentrations of peripheral insulin. Peripheral insulin anomalies are believed to cause a decrease in brain insulin production, which impairs amyloid clearance. An increase in inflammation and high blood glucose concentrations might also be the mechanisms of diabetes-impaired cognition.

11.2.2 Genetic Factors

Genetic factors contributing to cognitive impairment are complex. The *APOE ε4* allele is the genetic factor that prominently increases vulnerability to late-onset AD [23]. Carriers of *APOE ε4* heterozygotes have a three times higher risk of AD, and 15 times higher risk than *APOE ε3* homozygotes. *APOE ε4* alone is not the cause of AD, and testing for the allele is not clinically recommended as a diagnostic method for AD. Approximately half of young-onset familial AD cases are associated with mutations in the amyloid-β precursor protein, presenilin 1 (*PS1*), or *PS2* genes. Several genes influencing frontotemporal dementia have been verified, including the microtubule-associated protein tau and *C9ORF72*. In addition, the clinical suggestions of these certain diagnoses are not adequately clear for daily testing. Testing of patients with non-genetic dementia and at-risk relatives is not regularly conducted and should only be performed with complete informed consent after genetic counseling has been provided.

11.2.3 Lifestyle Risk Factors

Impairments in psychological and physical function are affected by exercise or social activities [24]. The emergence of these health and social challenges influences diagnosis, prognosis, the acceptance of treatment, and the necessity for health and social care in populations with cognitive impairments.

11.2.3.1 Smoking

Consistent with previous analyses, smoking had the third-highest PAF [25]. The association between smoking and cardiovascular pathology might explain cognitive impairment. However, cigarette smoke also contains neurotoxins, which may increase the risk of cognitive impairment [26]. Interventions are being implemented to reduce smoking habits in some countries, although smoking appears to have increased in the eastern Mediterranean and Africa [27].

11.2.3.2 Diet

Individuals who adopt a Mediterranean diet (low intake of meat and dairy, high intake of fruit, vegetables, and fish) have reduced plasma glucose and serum insulin concentrations, insulin resistance, markers of oxidative stress, inflammation, and vascular risk factors. A total of 447 healthy subjects, with an average age of 67 years, at high cardiovascular risk but with no substantial cognitive impairment or cardiovascular disease were randomly allocated to three dietetic groups [28]. The first group consumed the Mediterranean diet accompanied by additional virgin olive oil (1 L per week); the second group consumed a Mediterranean diet supplemented with mixed nuts (30 g per day); and the third group consumed the control diet (advised to reduce dietary fat intake). Urine testing was performed to determine whether the participants adhered to the dietary intervention. In the main analysis of combined cognitive alterations over 4 years, subjects in the intervention groups had better cognitive results than the control group. Significant differences in the number of patients developing MCI were not observed between the three groups using a subordinate analysis, and no subjects developed dementia, indicating that this intervention method might influence cognitive aging but not dementia syndrome [27].

A healthy lifestyle, such as not smoking, regularly eating fruits and vegetables, and exercise, can extend life expectancy and maintain health during aging, leading to increased attention on the cognitive effects of physical activity.

11.2.3.3 Physical Activity

Older adults who exercise are more likely to maintain better cognitive function than those who do not exercise. Observational research has revealed an inverse correlation between physical activity and the risk of dementia. Outcomes of one meta-analysis of 15 follow-up studies investigating 33,816 older people without dementia indicated that physical activity exerted a valid protective effect on cognitive impairments [29]. Another meta-analysis comprised of 16 studies involving 163,797 individuals without dementia showed that the relative risk (RR) of dementia in the highest physical activity groups was significantly lower than that in the lowest groups [30]. Physical activity benefits seniors without dementia, for example by improving balance, mood, and physical functioning and reducing mortality. Engaging in physical activity regularly in midlife is also associated with improvements in cognitive functioning. Physical activity is assumed to exert a neuroprotective effect, possibly by ameliorating the decreased levels of brain-derived neurotrophic factor cortisol and vascular abnormalities [31].

11.2.4 Other Risk Factors

Preliminary evidence on other potentially modifiable risk factors, including education, brain injury, and hearing, has been published respectively. All these factors might be important for mild cognitive impairment.

11.2.4.1 Years of Formal Education

A low educational level is proposed to be a risk factor for cognitive impairment because of a lower cognitive reserve, which allows people to retain function despite brain pathology. However, researchers have not confirmed whether education after secondary school is also a protective factor. A longitudinal study found that education was not associated with overall cognitive changes due to the non-neuropathological burden, nor was it associated with a higher neuropathological burden or a faster rate of cognitive decline [32].

11.2.4.2 Traumatic Brain Injury

The most common brain injury is a non-repetitive traumatic brain injury. The largest cohort study of traumatic brain injury confirmed that a history of traumatic brain injury (defined as >1 h loss of awareness) was not related to either a greater risk of the progression of dementia or AD nor increased plaques and tangles in the 1589 subjects who underwent an autopsy. However, traumatic brain injury was linked to the development of Parkinson's disease and Lewy body pathology [33].

A meta-analysis of seven studies that followed patients at least 1 year after traumatic brain injury found that traumatic brain injury was not related to the growth of risk of dementia. However, traumatic brain injury accelerated the risk of AD. The meta-analysis also showed that there is no significant difference in risk between single and repetitive traumatic brain injury [34].

11.2.4.3 Sleep Disorders

Sleep disorders have been the focus of many studies due to their role in the development of cognitive impairment. Sleep may facilitate the repair of injuries caused by other factors; however, considering the deficiency of adequate consistent and high-quality proof or systematic reviews, scientists still cannot include sleep in their calculations of PAF.

11.2.4.4 Hearing Loss

An acknowledgment of hearing loss as a risk factor for dementia is relatively new and has not been included in previous calculations of PAF or considered a precedent in the determination of those at risk of impaired cognition. Outcomes of cohort studies including hearing as a risk factor have usually found that even mild hearing loss increases the long-term risk of cognitive impairment and dementia in individuals who are cognitively healthy but hearing-impaired at baseline. However, 11 optimistic studies have discussed the effect of hearing loss on the risk of dementia. The results of the two studies showed no increased risk in adjusted analyses [35, 36].

The effect of hearing loss on the risk of cognitive decline in a meta-analysis is higher than the risks imposed by other single risk factors and is also relevant to a large number of people due to its extremely widespread incidence, which occurs in 32% of older subjects. The high RR and occurrence clarify the high PAF. Researchers used the prevalence of hearing loss in people older than 55 years to calculate PAF since this age was the youngest average age at which the occurrence of hearing loss was shown and increased the risk of dementia [37]. Hearing loss is then clustered with mid-life risk factors; moreover, evidence indicates that the dementia risk in later life remains increased.

The mechanism underlying cognitive decline related to peripheral hearing loss is still unclear; it is unproven whether the amendment, such as hearing aids, can inhibit or postpone the onset of dementia. Older age and microvascular pathology increase the risk of both peripheral hearing loss and dementia, and possibly then confound their relation. The hearing loss probably either increases the cognitive load of a sensitive brain which causes changes in the brain or brings depression or social disengagement and brain atrophy, which could lead to accelerated cognitive impairment [38].

11.2.4.5 Mental Disorder

Depressive symptoms may be a part of the clinical manifestation of impaired cognition, which has caused discussion as to the orientation of the following reason: whether depression is an antecedent symptom or a separate risk factor for cognitive impairment. One study with long times of follow-ups indicated a connection between the number of depressive episodes and the risks of cognitive impairment, which supports the claims that depression is an independent risk factor for impaired cognition. However, a cohort study that followed for 28 years showed that only in the 10 years preceding the onset of dementia did people with dementia had higher depressive symptoms than people without dementia. This suggests that mid-life depression is not an independent risk factor for impaired cognition. Whereas, it is still unclear whether the more severe depressive symptoms found in people who progress to dementia are an early reason for dementia. From the biological perspective, the emergence of depression probably increases the risk of cognition impairment because it influences stress hormones or neuronal growth factors [39].

References

1. Knight MM (2000) Cognitive ability and functional status. J Adv Nurs 31(6):1459–1468
2. Golden CJ (1979) Identification of specific neurological disorders using double discrimination scales derived from the standardized Luria neuropsychological battery. Int J Neurosci 10(1):51–56
3. Larson EB, Kukull WA, Katzman RL (1992) Cognitive impairment: dementia and Alzheimer's disease. Annu Rev Public Health 13:431–449
4. Nowrangi MA, Rao V, Lyketsos CG (2011) Epidemiology, assessment, and treatment of dementia. Psychiatr Clin North Am 34(2):275–294. vii
5. Montejo P, Montenegro M, Fernandez MA, Maestu F (2011) Subjective memory complaints in the elderly: prevalence and influence of temporal orientation, depression and quality of life in a population-based study in the city of Madrid. Aging Ment Health 15(1):85–96
6. Reisberg B (1986) Dementia - a systematic-approach to identifying reversible causes. Geriatrics 41(4):30
7. Reisberg B (1986) Dementia: a systemic approach to identifying reversible causes. Geriatrics 41(4):30–46
8. Caracciolo B, Gatz M, Xu W, Pedersen NL, Fratiglioni L (2012) Differential distribution of subjective and objective cognitive impairment in the population: a nation-wide twin-study. J Alzheimers Dis 29(2):393–403
9. Mitchell AJ, Kemp S, Benito-Leon J, Reuber M (2010) The influence of cognitive impairment on health-related quality of life in neurological disease. Acta Neuropsych 22(1):2–13
10. Winblad B, Palmer K, Kivipelto M, Jelic V, Fratiglioni L, Wahlund LO et al (2004) Mild cognitive impairment--beyond controversies, towards a consensus: report of the International Working Group on Mild Cognitive Impairment. J Intern Med 256(3):240–246
11. Gauthier S, Reisberg B, Zaudig M, Petersen RC, Ritchie K, Broich K et al (2006) Mild cognitive impairment. Lancet 367(9518):1262–1270
12. Petersen RC (2004) Mild cognitive impairment as a diagnostic entity. J Intern Med 256(3):183–194
13. Busse A, Hensel A, Guhne U, Angermeyer MC, Riedel-Heller SG (2006) Mild cognitive impairment: long-term course of four clinical subtypes. Neurology 67(12):2176–2185
14. Hebert LE, Scherr PA, Bienias JL, Bennett DA, Evans DA (2003) Alzheimer disease in the US population - prevalence estimates using the 2000 census. Arch Neurol 60(8):1119–1122
15. Plassman BL, Langa KM, Fisher GG, Heeringa SG, Weir DR, Ofstedal MB et al (2008) Prevalence of cognitive impairment without dementia in the United States. Ann Intern Med 148(6):427–434
16. Sosa AL, Albanese E, Stephan BCM, Dewey M, Acosta D, Ferri CP et al (2012) Prevalence, distribution, and impact of mild cognitive impairment in Latin America, China, and India: a 10/66 population-based study. PLoS Med 9(2):e1001170
17. Xue J, Li JR, Liang JM, Chen SL (2018) The prevalence of mild cognitive impairment in China: a systematic review. Aging Dis 9(4):706–715
18. Li X, Ma C, Zhang JY, Liang Y, Chen YJ, Chen KW et al (2013) Prevalence of and potential risk factors for mild cognitive impairment in community-dwelling residents of Beijing. J Am Geriatr Soc 61(12):2111–2119
19. Dickerson BC, Salat DH, Greve DN, Chua EF, Rand-Giovannetti E, Rentz DM et al (2005) Increased hippocampal activation in mild cognitive impairment compared to normal aging and AD. Neurology 65(3):404–411
20. Norton S, Matthews FE, Barnes DE, Yaffe K, Brayne C (2014) Potential for primary prevention of Alzheimer's disease: an analysis of population-based data. Lancet Neurol 13(8):788–794
21. Song X, Mitnitski A, Rockwood K (2014) Age-related deficit accumulation and the risk of late-life dementia. Alzheimers Res Ther 6(5–8):54
22. Wilson RS, Mendes de Leon C, Barnes LL, Schneider JA, Bienias JL, Evans DA et al (2002) Participation in cognitively stimulating activities and risk of incident Alzheimer disease. J Am Med Assoc 287(6):742–748

23. Ólafsdóttir M, Foldevi M, Marcusson J (2001) Dementia in primary care: why the low detection rate? Scand J Prim Health Care 19(3):194–198
24. Intzandt BB, Black SEMD, Lanctôt KLP, Herrmann NMD, Oh PMD, Middleton LEP (2015) Is cardiac rehabilitation exercise feasible for people with mild cognitive impairment? Can Geriatr J 18(2):65–72
25. Norton S, Matthews FE, Barnes DE, Yaffe K, Brayne C (2014) "Potential for primary prevention of Alzheimer's disease: an analysis of population-based data": corrections. Lancet Neurol 13(11):1070
26. Swan GE, Lessov-Schlaggar CN (2007) The effects of tobacco smoke and nicotine on cognition and the brain. Neuropsychol Rev 17(3):259–273
27. Ozcebe H, Erguder T, Balcilar M, Ursu P, Reeves A, Stuckler D et al (2018) The perspectives of politicians on tobacco control in Turkey. Eur J Pub Health 28(suppl_2):17–21
28. Valls-Pedret CM, Sala-Vila ADP, Serra-Mir MRD, Corella DDP, de la Torre RDP, Martínez-González MÁMDP et al (2015) Mediterranean diet and age-related cognitive decline: a randomized clinical trial. JAMA Intern Med 175(7):1094
29. Sofi F, Valecchi D, Bacci D, Abbate R, Gensini G, Casini A et al (2011) Physical activity and risk of cognitive decline: a meta-analysis of prospective studies. J Intern Med 269(1):107–117
30. Hamer M, Chida Y (2008) Physical activity and risk of neurodegenerative disease: a systematic review of prospective evidence. Psychol Med 39(1):3–11
31. Vaughan S, Wallis M, Polit D, Steele M, Shum D, Morris N (2014) The effects of multimodal exercise on cognitive and physical functioning and brain-derived neurotrophic factor in older women: a randomised controlled trial. Age Ageing 43(5):623–629
32. Wilson RS, Yu L, Lamar M, Schneider JA, Boyle PA, Bennett DA (2019) Education and cognitive reserve in old age. Neurology 92(10):e1041–e1050
33. Crane PK, Gibbons LE, Dams-O'Connor K, Trittschuh E, Leverenz JB, Keene CD et al (2016) Association of traumatic brain injury with late-life neurodegenerative conditions and neuropathologic findings. JAMA Neurol 73(9):1062–1069
34. Guo Z, Cupples LA, Kurz A, Auerbach SH, Volicer L, Chui H et al (2000) Head injury and the risk of AD in the MIRAGE study. Neurology 54(6):1316–1323
35. Hong T, Mitchell P, Burlutsky G, Liew G, Wang JJ (2016) Visual impairment, hearing loss and cognitive function in an older population: longitudinal findings from the Blue Mountains eye study. PLoS One 11(1):e0147646
36. Lin MY, Gutierrez PR, Stone KL, Yaffe K, Ensrud KE, Fink HA et al (2004) Vision impairment and combined vision and hearing impairment predict cognitive and functional decline in older women. J Am Geriatr Soc 52(12):1996–2002
37. Gallacher J, Ilubaera V, Ben-Shlomo Y, Bayer A, Fish M, Babisch W et al (2012) Auditory threshold, phonologic demand, and incident dementia. Neurology 79(15):1583–1590
38. Bernabei R, Bonuccelli U, Maggi S, Marengoni A, Martini A, Memo M et al (2014) Hearing loss and cognitive decline in older adults: questions and answers. Aging Clin Exp Res 26(6):567–573
39. Morkem R, Barber D, Williamson T, Patten SB (2015) A Canadian primary care sentinel surveillance network study evaluating antidepressant prescribing in Canada from 2006 to 2012. Can J Psychiatry 60(12):564–570

The Late Stage of Abnormal Aging: Dementia

12

Shudan Gao, Yun Wang, Tao Ma, and Junying Zhang

Abstract

With the growth of the aging population, more age-related diseases endanger the health of the elderly, and therefore more research attention has been put on Alzheimer's disease and dementia. Dementia does not only posing a serious threat to basic daily living in old age but also impose a greater burden on social and medical care as well as the economy. It is urgent to explore the pathogenesis of Alzheimer's disease and develop effective medicine to prevent or mitigate its onset. Currently, many related mechanisms of the pathogenesis of Alzheimer's disease have been proposed, such as beta-amyloid (A) theory, Tau protein theory, and nerve and blood vessel theory. In addition, from the perspective of improving cognitive function and controlling mental state, dementia-related therapeutic drugs were developed, such as anti-amyloid agents, amyloid vaccine, tau vaccine, and tau-aggregation inhibitor. These theories of pathogenesis and the development of drugs provide valuable experience to lift the veil of cognitive disorders in the future.

Keywords

Alzheimer's disease · Pathogenesis hypotheses · Drug therapy · Clinical trials

12.1 Hypotheses of the Pathogenesis of Dementia

Alzheimer's disease (AD) is the most common form of dementia with aging and a progressive neurodegenerative disorder in the presenium and senium. Unfortunately, the pathogenesis of AD is

S. Gao
State Key Laboratory of Cognitive Neuroscience and Learning, Faculty of Psychology, Beijing Normal University, Beijing, China

Beijing Aging Brain Rejuvenation Initiative (BABRI) Centre, Beijing Normal University, Beijing, China

School of Psychology, Shandong Normal University, Jinan, China

Y. Wang
State Key Laboratory of Cognitive Neuroscience and Learning, Faculty of Psychology, Beijing Normal University, Beijing, China

Beijing Aging Brain Rejuvenation Initiative (BABRI) Centre, Beijing Normal University, Beijing, China

T. Ma
Dongfang Hospital, Beijing University of Chinese Medicine, Beijing, China
e-mail: matao@bucm.edu.cn

J. Zhang (✉)
Institute of Basic Research in Clinical Medicine, China Academy of Chinese Medical Sciences, Beijing, China

Beijing Aging Brain Rejuvenation Initiative (BABRI) Centre, Beijing Normal University, Beijing, China

Z. Zhang (ed.), *Cognitive Aging and Brain Health*, Advances in Experimental Medicine and Biology 1419, https://doi.org/10.1007/978-981-99-1627-6_12

not yet fully understood. Some researchers consider AD as a complicated heterogeneous disease, and the incidence is related to the involvement of multiple factors, including heredity factors, neurotransmitters, immune factors, and environmental factors. Many theories have been proposed based on numerous AD research, and recently, research has mainly focused on the amyloid-β (Aβ) theory, tau protein theory, and nerve and blood vessel theory. All the theories provide unique interpretations of the pathogenesis of AD.

12.1.1 Amyloid Cascade Hypothesis

An important pathological feature of AD is the presence of a large number of senile plaques (SPs) in the brain, of which Aβ is one of the main components. Aβ is a small fragment formed by the proteolytic cleavage of a large protein named amyloid precursor protein (APP). APP is a transmembrane protein that passes through the fatty acid membrane [1]. The activation of APP leads to its cleavage, followed by a series of reactions to form oligomers, fibrils, plaques, and β-sheets, which leads to the aggregation of Aβ and the disruption of cellular communication and inflammatory responses.

Aβ, such as $A\beta_{40}$ and $A\beta_{42}$, is available in monomers, dimers, higher oligomers, and fibrillary polymers. $A\beta_{42}$ is one that accumulates early in the brain and has a high tendency to form amyloid plaques and deposits. Researchers assumed that Aβ deposition has a neurotoxic effect and causes progressive neurodegeneration. Meanwhile, the resultant changes in the functional neural circuits can lead to cognitive and behavioral symptoms of AD. Lastly, oligomers formed by Aβ also have separate neurotoxic effects and can cause damage to normal synaptic function [2].

The amyloidogenic hypothesis has been modified occasionally in the past. Initially, the hypothesis referred to amyloid plaque formation and deposition as the major cause of AD. Later, more refined hypotheses focused on increasing levels of Aβ42 with an increased ratio of Aβ42/Aβ40.

Through the increase of Aβ production, changes in Aβ42/Aβ40 ratios, increased Aβ deposition, and reduced Aβ clearance, the brain gradually accumulates Aβ, resulting in the formation and precipitation of senile plaques, resulting in increased inflammatory responses, microglial activation, cytokine release, and astrocytosis [3]. These responses are closely associated with progressive neurotic injury, neuronal deficits, and cognitive dysfunction. High concentrations of the Aβ protein also cause dendritic and axonal atrophy, followed by neuronal death. The levels of insoluble Aβ are correlated with cognitive decline. Individuals with Aβ accumulation often show symptoms of AD, such as cognitive decline.

Several genetic factors could increase Aβ production and deposition. Genetic mutation and polymorphism of presenilins (PSEN1 and PSEN2) are the most important factors that could raise the levels of Aβ [4]. PSEN1 and PSEN2 encode for presenilin 1 and 2 genes, respectively. Presenilins are the catalytic subunit of γ-secretase responsible for cleavage of APP to Aβ [5].

12.1.2 Hyperphosphorylated Tau Protein

The tau protein is one of the components of microtubule-associated protein (MAP). MAP and tubulin form microtubules. Microtubules are an important component of cytoskeletal proteins in neurons, which is involved in the transport of Aβ in the cell body and axon nutrients in the axon. The main function of tau is to stabilize microtubules, which is particularly important for neurons, since microtubules serve as transport pathways for cargo in dendrites and axons [6].

Intracellular tau-containing neurofibrillary tangles (NFTs) are an important pathological feature of AD [7]. Pathological NFTs are mainly composed of hyperphosphorylated tau proteins. In individuals with AD, the hyperphosphorylated tau proteins enter insoluble intracellular NFTs and fail to bind to the microtubules in the brain cells. These hyperphosphorylated tau proteins interact with each other and form knots called NFTs, which destroy neuronal plasticity and lead to neurodegeneration. Abnormally phosphorylated tau protein is insoluble, and also has a low affinity with microtubules, leading to abnormal cytoskeletal protein structures and affecting signal transmission within and between neurons [8].

The levels of total tau and phosphorylated tau in the cerebrospinal fluid of patients with AD were both found to be increased [9], which was also associated with a decrease in neuropsychological test scores. Increased levels of total tau and phosphorylated tau in cerebrospinal fluid are an important clinical indicator of the progression of MCI to AD [10]. Tau abnormalities are common in individuals with other neurodegenerative disorders, such as frontotemporal dementia, multiple system atrophy, and corticobasal degeneration [11].

12.1.3 The Role of Inflammation

The occurrence and development of AD are accompanied by a chronic inflammatory response that damages brain tissue and is mainly associated with microglia and astrocytes. Active microglia and astrocytes have been detected in the SP of patients with AD. Microglial cells initially reduce Aβ levels through phagocytosis, but when their activity increases, chemokines initiate a "cascade" reaction that releases inflammatory cytokines, resulting in neuronal damage [6]. Microglia express receptors for advanced glycation end products (RAGE), which binds to Aβ and increases the production of cytokines, glutamate, and nitric oxide (NO). Additionally, they enhance the neurotoxic effect and inflammatory response, which results in decreased learning and memory ability. Activated glial cells aggravate AD by releasing reactive products of the acute phase of inflammation, such as alpha 1-antichymotrypsin, beta 2-macroglobulin, and C-reactive protein. Both patients with AD and animal models exhibit the over-activation of microglia that causes neuroinflammation, leading to neuronal death. Microglia are highly specialized histological macrophages in the central nervous system (CNS), similar to the peripheral mononuclear phagocytic system in terms of function and origin [7]. Microglia in the circumference of Aβ plaques have an M1 phenotype and are activated to present an inflammatory or neurotoxic phenotype that inhibits the M2 phenotype.

12.1.4 Synaptic Pathology

Age-related synaptic deficits primarily occur in the dentate gyrus of the hippocampus [8]. In patients with AD, the degree of synaptic deficits in a specific location, such as the hippocampus, has the strongest relationship with cognitive decline. These changes are associated with *N*-methyl-D-aspartate receptors (NMDAR) activation and oxidative stress, which result in AD pathology.

NMDARs are gated cation channels, which play important roles in excitatory transmission, synaptic integration, learning, and memory in the CNS. NMDAR activation, excessive Ca^{2+} flux, and free radical generation are associated with synaptic dysfunction and tau phosphorylation [9]. Therefore, damage to NMDARs and synaptic function may be involved in the pathogenesis of AD at the synaptic level.

Meanwhile, previous researchers indicated that synaptic loss was the secondary change of protein denaturation; however, Hong et al. found that synapses could be already reduced at the beginning of AD and it may begin earlier than the formation of plaques [10].

12.1.5 Reactive Oxygen Species and Oxidative Stress

Compared to other organs, neuronal cells are more vulnerable to free radical damage due to high oxygen consumption and a lack of antioxidant enzyme activity [11]. With aging, oxidative stress in the brain is intensified, and in severe cases, the protein folding function of the endoplasmic reticulum is impaired. Protease activity and autophagy that mediate the removal function of damaged proteins are also decreased, which promotes the accumulation of Aβ deposits and tau proteins, finally leading to the occurrence of AD.

Oxidative stress is mainly caused by increased levels of reactive oxygen species (ROS) and/or reactive nitrogen species, including superoxide radical anions (O_2^-), hydrogen peroxide (H_2O_2), hydroxyl radicals (HO^-), and nitric oxide (NO) [12]. In patients with AD, neurons showed significant oxidative damage and a reduction in the number of mitochondria, which are major contributors to ROS generation. The overproduction of ROS and/or an insufficient antioxidant defense leads to oxidative stress. Before the appearance of clinical symptoms of AD and abnormal Aβ deposition, ROS production increases due to mitochondrial damage [13]. The number of mitochondria decreased significantly in AD patients. Several key enzymes are involved in oxidative metabolism, including α-ketoglutarate (α-KG) dehydrogenase. Pyruvate metabolism also shows reduced levels or activity in AD. In addition, oxidative stress has been found to be directly related to neuronal dysfunction of AD both in vitro and in vivo [14].

In the brain of AD patients, superoxide free radicals are released due to mitochondrial dysfunction, which leads to oxidative stress. The superoxide anion free radicals, NO, aldehydes, carbonyl, and nitryl derivatives have neurotoxic effects, which can cause increased calcium membrane permeability, imbalance of a variety of ions, and impaired glucose transport, which results in energy metabolism disorders and leads to neurodegeneration.

12.1.6 The Vascular Hypothesis

According to the vascular hypothesis, AD is caused by age-related chronic brain hypoperfusion [15]. This process occurs because sustained and inadequate cerebral blood flow (CBF) in elderly individuals reaches a critical threshold where brain cells no longer cope with the dwindling energy supply being delivered, thus creating a neuronal energy crisis [16]. In addition, continuous hypoperfusion is an important factor in the occurrence of AD, which reduces the energy supply of brain cells, resulting in neuronal metabolism decline, oxidative stress, cognitive decline, neurodegeneration, AD, and death.

Long-term hypoperfusion of the brain not only causes changes in many biomolecules in the brain but also systematically reduces the supply of oxygen and glucose, leading to an energy crisis in nerve cells characterized by oxidative and ischemic hypoxia. Based on previous studies on aged rats, mechanically reducing CBF in aged experimental animals for 12 months causes a range of cognitive impairments, beginning with mild memory loss and learning disabilities, followed by loss of neurons in the hippocampus and posterior cingulate cortex, gliosis, severe memory loss, and extensive atrophy of the posterior parietal cortex [17]. These findings were also confirmed in other studies [18, 19].

Small vascular lesions of varying degrees in the cerebral cortex and deep basal ganglia in the white matter have been observed in 60–90% of patients with AD. The main pathological changes include the deposition of Aβ in the middle and outer membrane of the small vascular wall in the cerebral cortex and the pia. Some patients also present blood-brain barrier (BBB) destruction and macrovascular atherosclerosis. The deposition of Aβ in the small vessel wall aggravates cerebral vasospasm, which reduces CBF and leads to an insufficient local energy supply [20].

The homeostasis of the microenvironment and metabolism in the brain relies on the support of intact vascular structures. Disorders of blood vessels can cause brain disorders and diseases. Cerebrovascular diseases and neurodegenerative diseases (AD) are characterized by changes in cerebrovascular function. Neuroimaging studies in AD patients have shown that neurovascular dysfunction is detected before neurodegeneration [21]. In addition to abnormal angiogenesis and aging of the cerebrovascular system, the faulty clearance of Aβ across the BBB can lead to neurovascular uncoupling and cause brain hypoperfusion, brain hypoxia, and neurovascular inflammation, ultimately leading to neuronal dysfunction and loss.

12.1.7 The Cholesterol Hypothesis

The brain contains 23% of the total body cholesterol content [22], and brain cholesterol is considered essential for cell maintenance, neuronal transmission, and synapse formation. Cholesterol in nerve cells is mainly derived from astrocytes. Synthetic cholesterol is delivered to nerve cells via the cholesterol transporter protein APOE, which also secretes cholesterol. Cholesterol is high in the brain, which insulates axons and is an important component of dendritic spines. Cholesterol supplementation improves cognitive performance.

Cholesterol levels in the brain are independent of peripheral tissue levels because the BBB blocks the flow of peripheral cholesterol entering the CNS. Although a high-fat/high-cholesterol diet significantly raised blood cholesterol levels, there was no change in brain cholesterol levels in animal models [23]. However, cholesterol metabolites induce increased Aβ accumulation and cognitive decline in animal models of AD on a high-fat diet. Brain cholesterol is thought to function in the brain through these cholesterol metabolites. These molecules could explain the relationship between blood and brain cholesterol levels.

The maintenance of cholesterol homeostasis is essential for normal neuronal function and brain development. In neurons, the brain-specific enzyme cholesterol 24-hydroxylase (CYP46A1) converts excess cholesterol into 24-hydroxycholesterol (24-OH), which may diffuse across the BBB [24]. When 24-OH flows from the brain into the peripheral circulation, 27-hydroxyl cholesterol (27-OH) also flows into the brain.

24S-OH levels are related to cholesterol levels in the brain. Both 24S-OH and 27-OH are considered to be physiological inhibitors of brain cholesterol biosynthesis. Several studies have shown that 27-OH may accelerate cognitive deficits in individuals with AD under conditions of hypercholesterolemia, where higher cholesterol levels result in an increased influx of 27-OH into the brain [25].

Other hypotheses indicate cholesterol is necessary for the synthesis of nerve cell membranes and the processing of APP into Aβ. When cholesterol is excessive, membrane lipid transport is reduced, and this leads to the increased production and aggregation of Aβ and decreased clearance, which leads to the occurrence of AD.

12.1.8 Hypothesis of Metal Accumulation in the Brain

Biometals play an important role in AD and may directly or indirectly influence its pathogenesis. Metal ions related to the biological function of organisms are regarded as biometals. In the CNS, biometals such as copper, zinc, and iron are important cofactors for enzymatic activity, mitochondrial function, and neuronal function. In a healthy brain, free metal ions are tightly regulated and maintained at a very low level. Biometallic ions are involved in the polymerization and toxicity of Aβ. The relationship between biological metals and AD has been extensively studied. Moreover, evidence of the dyshomeostasis of biometals in the brain from individuals with AD has been reported respectively. Biometals, especially zinc and copper, are directly coordinated by Aβ accumulation [26]. Serum levels of copper, which are not associated with ceruloplasmin, are elevated in patients with AD. In addition, the Mini-mental State Examination (MMSE) score decreased with the increase in serum copper concentration. The important role of biomaterials in Aβ production has been studied in many animal models. A small amount of Cu^{2+} added to drinking water was sufficient to induce Aβ accumulation and the subsequent formation of plaques, followed by learning disabilities and cognitive impairment [27]. During normal aging, the substantia nigra, putamen, globus pallidus, caudate nucleus, and other parts of the brain have obvious accumulation of iron. An increase in the level of iron in AD patients was first demonstrated in 1953 [28]. More recently, iron accumulation in AD has been identified by magnetic resonance imaging (MRI) and is mainly localized to certain brain areas, such as the parietal cortex, motor cortex, and hippocampus. Studies of gene mutations that affect iron metabolism suggest that the dyshomeostasis of iron plays an important role in neuronal death that occurs in neurodegenerative disorders [29]. Increased iron load promoted the generation of Aβ neurons, which exacerbated the cognitive decline of transgenic AD mice [30]. There is evidence that iron regulatory elements have a direct effect on the production of APP [31].

Iron deficiency anemia is one of the most common clinical diseases. Thus, iron deficiency may partially account for hypometabolism in AD since women with iron deficiency anemia have a higher prevalence of dementia.

Exley et al. hypothesized that Al in the brain of AD patients may be related to Aβ and Al can precipitate Aβ in vitro into fibrillar structures; in addition, Al is known to increase the Aβ burden in the brain of treated animals, possibly because Al directly or indirectly affects AB anabolic and catabolic metabolism [32]. The role of Al in AD needs to be further elucidated.

12.1.9 Hypothesis of Impaired Insulin Signaling

Hyperglycemia is another risk factor for AD. Patients with type 2 diabetes (T2D) have an increased risk of dementia, including vascular dementia (VD) and AD. The largest and most recent meta-analysis of 6184 patients with diabetes and 38,350 people without diabetes [33], showed that while the insulin levels in peripheral blood increased in patients with severe AD, but the glucose clearance capacity also decreased. The insulin receptor and glucose transporter levels in the brain were significantly decreased. The neurons of patients with AD are vulnerable to damage because the energy-stressed neurons have disrupted glucose utilization. Insulin resistance leads to neuronal energy deficiency, oxidative stress, and metabolic damage that affect synaptic plasticity.

12.1.10 Cell Cycle Hypothesis

Recently, an obstacle to the normal cell cycle inhibition mechanism in the pathogenesis of AD has been proposed, which is the cell cycle reversion obstacle theory [34]. Markers of an abnormal cell cycle have been detected in individuals with all phases of AD and mild cognitive impairment (MCI), with a large number of cells in G1-S phase. This stage of DNA replication may have been completed with tetraploid neurons, where cyclins are activated. After cultured cortical cells in the brain expressed Aβ, astrocytes were activated, which in turn acted on neurons, causing a disruption of the cell cycle of differentiated neurons. Abnormal expression of cycle-related proteins in neurons from the brain of patients with AD may be related to AD-related pathological changes in neurons.

12.1.11 Cholinergic Hypothesis

The cholinergic hypothesis was proposed by Davies and Maloney in 1976 [35]. They compared the activities of the key enzymes involved in the synthesis of neurotransmitters, including acetylcholine, γ-aminobutyric acid, dopamine, noradrenaline, and 5-hydroxytryptamine, in the brains of individuals with AD and control subjects. The activity of choline acetyltransferase in the brains of subjects with AD was substantially reduced in the amygdala and hippocampus, and the concentration of acetylcholine was decreased at synapses, even affecting synaptic function. Choline acetyltransferase is a key enzyme involved in the synthesis of acetylcholine, and its catalytic activity requires the substrates choline, acetyl-CoA, and adenosine triphosphate (ATP) [36]. This is the first time that AD has been found to be due to the disruption of the cholinergic system. This finding has also been observed in individuals with other neurological and psychiatric disorders, such as Parkinson's disease (PD) and depression.

12.2 Status and Progress of Drug Therapy

12.2.1 Current Status of Drug Therapy

12.2.1.1 Improving Cognitive Function

Cholinesterase Inhibitors

No decrease in the incidence of AD was observed when cholinesterase inhibitors were routinely used. A large amount of evidence on the use of galantamine, rivastigmine, and donepezil as treatments for AD was summarized in the 2006 review from the Cochrane collaboration [37]. At optimum doses, all cholinesterase inhibitors display a modest advantage in improving cognition compared with the placebo. Since 2006, the studies delivered have validated the cognitive advantage of cholinesterase inhibitors. Advantages in terms of global changes, as measured by a clinician with caregiver involvement, and activities of daily life living have also been observed. An updated Cochrane review on rivastigmine usage in individuals with AD found a similar but slightly smaller effect [38]. The small difference in behavioral symptoms measured using a neuropsychiatric test is not clinically significant. Therefore, this finding might be specific to people with fairly limited symptoms. Cholinesterase inhibitors are adequately and clinically valid and cost-effective; thus, National Institute for Health and Care Excellence (NICE) has advised the use of any of cholinesterase inhibitors for the management of mild-to-moderate AD. Donepezil, galantamine, rivastigmine, and memantine are used to treat AD. Cholinesterase inhibitors are comparatively well tolerated; nevertheless, adverse events occurring in patients receiving these medications include vomiting, nausea, diarrhea, vivid dreams, and leg cramps. These adverse events account for higher withdrawal rates in participants receiving cholinesterase inhibitors than in control groups receiving placebos.

Memantine

Memantine is a non-competitive modulator of the N-methyl-D-aspartate receptor that renormalizes glutamatergic neurotransmission. It inhibits excitatory amino acid neurotoxicity. The doses are generally specified to be up to 20 mg per day. A meta-analysis summarized 3 trials with over 1000 patients with moderate-to-severe AD (MMSE 3–14) and 3 unpublished papers which involve approximately 1000 patients with mild-to-moderate AD, where all trials lasted for 6 months. In the moderate-to-severe set, there was a small advantageous effect on cognition, activities of daily life living, average standards of neuropsychiatric signs, and total measurement. Beneficial effects on cognition were shown in the mild-to-moderate group, but no effects on behavior or daily activities. Two trials of memantine in patients with mild-to-moderate Lewy body dementia found that the drug improved global cognition. Based on the meta-analysis, two agreement consensus panels made cautious positive recommendations that the combination of memantine and cholinesterase inhibitors in patients with moderate to severe AD had no significant effect on global cognition, or neuropsychiatric symptoms, but significantly reduced the incidence of adverse events [39].

12.2.1.2 Control of Mental Symptoms

Neuropsychiatric symptoms are very common in patients with dementia; they are mostly related to the severity of dementia and influence almost every person with dementia. While many different symptoms still exist and regularly co-occur, numerous models of different symptoms, e.g., psychotic and affective, have been established.

Antipsychotics for Agitation

Antipsychotics were the first choice of medicines for agitation in patients with dementia. Risperidone with a daily dose of no more than 1 mg enhanced agitation and psychotic symptoms, especially when aggression was the objective symptom. Haloperidol also influences aggression but not additional symptoms of agitation. Olanzapine and quetiapine did not cure psychosis, aggression, or agitation, but aripiprazole improved the condition of agitation. Withdrawal trials of antipsychotics have not observed an effect on agitation or neuropsychiatric symptoms, except for patients with the most severe symptoms [40].

Antihypertensive Drugs

A trial of the antihypertensive agent indapamide in combination with perindopril was conducted among people older than 80 years without impaired cognition who presented with hypertension (defined as 160–200/<110 mm Hg). The trial stopped early since a reduction in stroke and mortality rates in the treatment set indicated that it was immoral to continue administering the placebo [41]. However, when these data were synthesized into a meta-analysis with the other placebo-controlled trials of antihypertensive treatment, the joint risk ratio for dementia favored treatment (HR 0.87, 95% CI 0.76–1.00).

12.2.2 Progress in AD Clinical Trials

12.2.2.1 Anti-Amyloid Agents

Decreasing Aβ Production

Given the relevance of cerebral amyloidogenesis with AD, several drugs have been developed to decrease Aβ production by inhibiting the proteases that cleave Aβ from its precursor, preventing Aβ aggregation or increasing Aβ clearance. The goals are to delay clinical progression in patients with AD. Despite the preclinical rationale, no therapy has consistently shown a benefit for patients with AD.

γ-Secretase-based approaches were the first agents to show promising disease-modifying effects on AD. These drugs were able to reduce the cerebral amyloid burden and improve memory deficits in mouse models of AD [42]. Clinical trials showed a significant improvement in cognitive deficits in individuals with mild to moderate AD [43]. However, in most cases, side effects limit the dose of the therapy and increase the risk of serious adverse events in patients with mild to moderate AD [39].

Prevention of Aβ Aggregation

The prevention of Aβ aggregation is an alternative therapeutic strategy for AD. Aβ causes severe toxic reactions in neurons. Aβ monomers participate in the formation of oligomers and precipitate toxicity. The approaches for preventing Aβ aggregation have intrigued researchers. However, only limited progress has been achieved to date in developing potential aggregation inhibitors.

Homotaurine binds to the Aβ peptide and prevents the formation of neurotoxic aggregates [44]. Several other researchers have developed potential clinical agents, which include inositol and cyclohexanol isomers, a clioquinol analogue, the polypeptide colostrinin, polyphenolic compounds, omega-3-fatty acids, and curcumin.

Increasing Aβ Clearance

Aβ degradation and clearance is an additional approach mediated by several enzymes such as natural endopeptidase neprilysin (NEP), insulin-degrading enzyme (IDE), angiotensin-converting enzyme (ACE), and plasmin. The inhibitors of receptor for RAGE could also be used to promote the degradation of Aβ. RAGE is responsible for the accumulation of Aβ in the neuronal cells and aggravates its detrimental effects.

BACE1 Inhibitor

There are several approaches evaluated and reported to reduce the formation of Aβ peptide. Inhibition enzymatic activity of β-secretase (BACE) and γ-secretase are the most crucial targets in AD treatment. Both enzymes are responsible for cleaving the APP protein at variable sites to form Aβ. β-Secretase became the favored target for Aβ-centered therapy after failures of the immune- and γ-secretase-based approaches.

BACE is a membrane-bound aspartic protease type of enzyme abundantly found in the brain and responsible for the cleavage of APP to form C99 which is a small Aβ-containing fragment, and APP BACE has two isoforms: BACE-1 and BACE-2. BACE-1 inhibition is an attractive therapeutic strategy in AD treatment and is responsible for APP cleaving and Aβ generation in the brain.

It is important to evaluate these Aβ-production inhibitors at the early stage of AD and to monitor the side effects carefully. Some Aβ-production inhibitors interfere with brain function which affects fundamental mechanisms during brain development.

12.2.2.2 Aβ Vaccine

Many studies have focused on immunotherapeutic strategies for AD since early 2000 [45, 46]. Active and passive immunotherapy are the two most important methods mentioned above. Active immunotherapy (i.e., anti-amyloid vaccine) stimulates amyloid clearance by activating the immune system. This approach would reduce the amount of not only soluble amyloid species but also the amyloid plaques in the brain. The first reported Aβ vaccine was AN1792, but this trial was halted because of serious adverse events [47]. Preclinical studies showed a significant improvement in memory deficits along with a drastic reduction in the amyloid burden in transgenic mice [48]. However, significant neuroinflammatory reactions led to brain edema and death in patients who received the amyloid vaccine. In addition, Aβ immunization may cause cerebral microbleeds and iron deposition in the choroid plexus. Dementia and cognitive dysfunction are mainly caused by abnormalities in the immune system in the brain to clear toxic proteins.

Recently, passive immunotherapy approaches with anti-amyloid antibodies have been developed [49]. This approach also focuses on improving amyloid clearance in the brain without activating the microglial system and reducing the risk of neuroinflammation. The patient's cognitive ability improved, and no serious adverse events occurred. Currently, clinical trials are underway to test the clinical efficacy and safety in patients with mild AD. In the past 6 years, several antibodies such as gantenerumab from Roche, solanezumab from Eli Lilly, and bapineuzumab from Pfizer have entered the final stage of clinical trials, although they have not been further processed [50, 51]. These clinical studies were conducted only in patients with Aβ accumulation in the brain at an early stage of the disease, as confirmed by positron emission tomography (PET)scanning. Although about 0.1% of antibodies penetrated the BBB, were retained in blood, and entered circulation, not enough of them reached their target.

12.2.2.3 Tau Vaccine

Recent studies have shown the excellent potential of antibody- and vaccine-mediated therapy for AD and related tauopathies [52]. Similar to Aβ immunotherapy, tau immunotherapy also included two types broadly: active and passive approaches. Active immunization approach induced an innate immune response through the tau vaccine, which resulted in the generation of antibodies against the administered agent. Passive approach directly injected antibodies against the specific antigen to generate the response. Several approaches worked by targeting tau proteins with specific immunogens, including targeting hyperphosphorylated tau, specific tau peptides, multimeric tau peptides, and complete tau peptides [53]. In addition, several immunotherapies were reported to exert significant effects on tauopathies related to AD.

12.2.2.4 Tau-Aggregation Inhibitor

GSK-3β is an important therapeutic target in AD due to its role in tau hyperphosphorylation and the deposition of senile plaques [54]. Several classes of GSK-3β inhibitors have been reported to exert potential effects on AD. The first reported inhibitor is lithium ions (Li^+), which might exert their actions through potassium depletion or competition with Mg^{2+} ions.

In addition, several other small-molecule drug classes were also reported, such as hymenialdisine, paullones, indirubins, and thiadiazolidinones. The GLP-1 peptide and its analogue liraglutide, which are effective treatments for type II diabetes, have also been explored and found to be effective against AD in a transgenic mouse model [55].

References

1. Kienlen-Campard P et al (2008) Amyloidogenic processing but not amyloid precursor protein (APP) intracellular C-terminal domain production requires a precisely oriented APP dimer assembled by transmembrane GXXXG motifs. J Biol Chem 283(12):7733–7744
2. Hardy JA, Higgins GA (1992) Alzheimer's disease: the amyloid cascade hypothesis. Science 256(5054):184–185
3. Weller RO et al (2000) Cerebral amyloid angiopathy: accumulation of a beta in interstitial fluid drainage pathways in Alzheimer's disease. Ann N Y Acad Sci 903:110–117
4. Tanzi RE, Bertram L (2005) Twenty years of the Alzheimer's disease amyloid hypothesis: a genetic perspective. Cell 120(4):545–555
5. De Strooper B, Iwatsubo T, Wolfe MS (2012) Presenilins and γ-secretase: structure, function, and role in Alzheimer disease. Cold Spring Harb Perspect Med 2(1):a006304
6. Akiyama H et al (2000) Inflammation and Alzheimer's disease. Neurobiol Aging 21(3):383–421
7. Prinz M et al (2014) Microglia: unique and common features with other tissue macrophages. Acta Neuropathol 128(3):319–331

8. Lister JP, Barnes CA (2009) Neurobiological changes in the hippocampus during normative aging. Arch Neurol 66(7):829–833
9. Lee M-C et al (2010) Characterisation of the expression of NMDA receptors in human astrocytes. PLoS One 5(11):e14123
10. Hong S et al (2016) Complement and microglia mediate early synapse loss in Alzheimer mouse models. Science (New York, N.Y.) 352(6286):712–716
11. Coyle JT, Puttfarcken P (1993) Oxidative stress, glutamate, and neurodegenerative disorders. Science (New York, N.Y.) 262(5134):689–695
12. Chen Z, Zhong C (2014) Oxidative stress in Alzheimer's disease. Neurosci Bull 30(2):271–281
13. Uttara B et al (2009) Oxidative stress and neurodegenerative diseases: a review of upstream and downstream antioxidant therapeutic options. Curr Neuropharmacol 7(1):65–74
14. Kamat PK et al (2016) Mechanism of oxidative stress and synapse dysfunction in the pathogenesis of Alzheimer's disease: understanding the therapeutics strategies. Mol Neurobiol 53(1):648–661
15. de la Torre JC (1994) Impaired brain microcirculation may trigger Alzheimer's disease. Neurosci Biobehav Rev 18(3):397–401
16. Kenny RA, Kalaria R, Ballard C (2002) Neurocardiovascular instability in cognitive impairment and dementia. Ann N Y Acad Sci 977:183–195
17. de la Torre JC et al (1992) Chronic cerebrovascular insufficiency induces dementia-like deficits in aged rats. Brain Res 582(2):186–195
18. Liu H-X et al (2005) Altered expression of MAP-2, GAP-43, and synaptophysin in the hippocampus of rats with chronic cerebral hypoperfusion correlates with cognitive impairment. Brain Res Mol Brain Res 139(1):169–177
19. Sopala M, Danysz W (2001) Chronic cerebral hypoperfusion in the rat enhances age-related deficits in spatial memory. J Neural Trans (Vienna, Austria : 1996) 108(12):1445–1456
20. Dickstein DL et al (2010) Role of vascular risk factors and vascular dysfunction in Alzheimer's disease. Mt Sinai J Med 77(1):82–102
21. Knopman DS, Roberts R (2010) Vascular risk factors: imaging and neuropathologic correlates. J Alzheimer's Dis 20(3):699–709
22. Ohyama Y et al (2006) Studies on the transcriptional regulation of cholesterol 24-hydroxylase (CYP46A1): marked insensitivity toward different regulatory axes. J Biol Chem 281(7):3810–3820
23. Ghribi O (2008) Potential mechanisms linking cholesterol to Alzheimer's disease-like pathology in rabbit brain, hippocampal organotypic slices, and skeletal muscle. J Alzheimer's Dis 15(4):673–684
24. Martín MG, Pfrieger F, Dotti CG (2014) Cholesterol in brain disease: sometimes determinant and frequently implicated. EMBO Rep 15(10):1036–1052
25. Zmysłowski A, Szterk A (2019) Oxysterols as a biomarker in diseases. Clin Chim Acta 491:103–113
26. Dong J et al (2003) Metal binding and oxidation of amyloid-beta within isolated senile plaque cores: Raman microscopic evidence. Biochemistry 42(10):2768–2773
27. Sparks DL, Schreurs BG (2003) Trace amounts of copper in water induce beta-amyloid plaques and learning deficits in a rabbit model of Alzheimer's disease. Proc Natl Acad Sci U S A 100(19):11065–11069
28. Goodman L (1953) Alzheimer's disease; a clinicopathologic analysis of twenty-three cases with a theory on pathogenesis. J Nerv Ment Dis 118(2):97–130
29. Lane DJR, Ayton S, Bush AI (2018) Iron and Alzheimer's disease: an update on emerging mechanisms. J Alzheimer's Dis 64(s1):S379–S395
30. Becerril-Ortega J et al (2014) Iron overload accelerates neuronal amyloid-β production and cognitive impairment in transgenic mice model of Alzheimer's disease. Neurobiol Aging 35(10):2288–2301
31. Rogers JT et al (2008) Iron and the translation of the amyloid precursor protein (APP) and ferritin mRNAs: riboregulation against neural oxidative damage in Alzheimer's disease. Biochem Soc Trans 36(Pt 6):1282–1287
32. Exley C (2005) The aluminium-amyloid cascade hypothesis and Alzheimer's disease. Subcell Biochem 38:225–234
33. Biessels GJ et al (2014) Dementia and cognitive decline in type 2 diabetes and prediabetic stages: towards targeted interventions. Lancet Diab Endocrinol 2(3):246–255
34. Yang Y, Herrup K (2007) Cell division in the CNS: protective response or lethal event in post-mitotic neurons? Biochim Biophys Acta 1772(4):457–466
35. Davies P, Maloney AJ (1976) Selective loss of central cholinergic neurons in Alzheimer's disease. Lancet (London, England) 2(8000):1403
36. Ferreira-Vieira TH et al (2016) Alzheimer's disease: targeting the cholinergic system. Curr Neuropharmacol 14(1):101–115
37. Birks J (2006) Cholinesterase inhibitors for Alzheimer's disease. Cochrane Database Syst Rev 1:CD005593
38. Birks J et al (2009) Rivastigmine for Alzheimer's disease. Cochrane Database Syst Rev 2:CD001191
39. Carlson C et al (2011) Prevalence of asymptomatic vasogenic edema in pretreatment Alzheimer's disease study cohorts from phase 3 trials of semagacestat and solanezumab. Alzheimers Dement 7(4):396–401
40. Cohen-Mansfield J et al (1999) Withdrawal of haloperidol, thioridazine, and lorazepam in the nursing home: a controlled, double-blind study. Arch Intern Med 159(15):1733–1740
41. Levi Marpillat N et al (2013) Antihypertensive classes, cognitive decline and incidence of dementia: a network meta-analysis. J Hypertens 31(6):1073–1082
42. Lahiri DK et al (2007) The experimental Alzheimer's disease drug posiphen [(+)-phenserine] lowers amyloid-beta peptide levels in cell culture and mice. J Pharmacol Exp Ther 320(1):386–396

43. Fleisher AS et al (2008) Phase 2 safety trial targeting amyloid beta production with a gamma-secretase inhibitor in Alzheimer disease. Arch Neurol 65(8):1031–1038
44. Aisen P et al (2004) Clinical data on Alzhemed (TM) after 12 months of treatment in patients with mild to moderate Alzheimer's disease. Neurobiol Aging 25: S20
45. Montoliu-Gaya L, Villegas S (2016) Abeta-immunotherapeutic strategies: a wide range of approaches for Alzheimer's disease treatment. Expert Rev Mol Med 18:e13
46. Solomon B (2001) Immunotherapeutic strategies for prevention and treatment of Alzheimer's disease. DNA Cell Biol 20(11):697–703
47. Orgogozo JM et al (2003) Subacute meningoencephalitis in a subset of patients with AD after Abeta42 immunization. Neurology 61(1):46–54
48. Janus C et al (2000) A beta peptide immunization reduces behavioural impairment and plaques in a model of Alzheimer's disease. Nature 408(6815):979–982
49. Panza F et al (2012) Immunotherapy for Alzheimer's disease: from anti-β-amyloid to tau-based immunization strategies. Immunotherapy 4(2):213–238
50. Doody RS et al (2014) Phase 3 trials of solanezumab for mild-to-moderate Alzheimer's disease. N Engl J Med 370(4):311–321
51. Moreth J, Mavoungou C, Schindowski K (2013) Passive anti-amyloid immunotherapy in Alzheimer's disease: what are the most promising targets? Immun Ageing 10(1):18
52. Kontsekova E et al (2014) First-in-man tau vaccine targeting structural determinants essential for pathological tau-tau interaction reduces tau oligomerisation and neurofibrillary degeneration in an Alzheimer's disease model. Alzheimers Res Ther 6(4):44
53. Pedersen JT, Sigurdsson EM (2015) Tau immunotherapy for Alzheimer's disease. Trends Mol Med 21(6):394–402
54. Cho JH, Johnson GV (2004) Primed phosphorylation of tau at Thr231 by glycogen synthase kinase 3beta (GSK3beta) plays a critical role in regulating tau's ability to bind and stabilize microtubules. J Neurochem 88(2):349–358
55. Hansen HH et al (2016) Long-term treatment with liraglutide, a glucagon-like Peptide-1 (GLP-1) receptor agonist, has no effect on β-amyloid plaque load in two transgenic APP/PS1 mouse models of Alzheimer's disease. PLoS One 11(7):e0158205

Part VII

Cognitive Impairment Intervention

Cognitive Training Effect and Imaging Evidence

13

Xiangwei Dai, Wu Lingli, Zaizhu Han, and He Li

Abstract

Cognitive intervention is a specific form of non-pharmacological intervention used to combat cognitive dysfunction. In this chapter, behavioral and neuroimaging studies about cognitive interventions are introduced. Regarding intervention studies, the form of intervention and the effects of the interventions have been systematically sorted out. In addition, we compared the effects of different intervention approaches, which help people with different cognitive states to choose appropriate intervention programs. With the development of imaging technology, many studies have discussed the neural mechanism of cognitive intervention training and the effects of cognitive intervention from the perspective of neuroplasticity. Behavioral studies and neural mechanism studies are used to improve the understanding of cognitive interventions for the treatment of cognitive impairment.

Keywords

Cognitive intervention · Cognitive dysfunction · Behavioral plasticity · Neural plasticity

13.1 Cognitive Training Approaches

Cognitive interventions are specific forms of non-pharmacological interventions for the direct or indirect treatment of cognitive dysfunctions [1]. The theoretical basis of cognitive intervention is the characteristics of brain plasticity [2]. In other words, the brain has the ability to modify and adapt itself to its changed reality both structurally and functionally [3]. The brain is regulated not only by internal factors such as genes but also by external factors such as learning experience, sensorimotor inputs, and physical stimulation [4, 5]. Thus, cognitive interventions, as an external factor, can cause plasticity changes in the brain to improve cognitive function.

X. Dai · H. Li (✉)
Institute of Basic Research in Clinical Medicine, China Academy of Chinese Medical Sciences, Beijing, China

Beijing Aging Brain Rejuvenation Initiative (BABRI) Centre, Beijing Normal University, Beijing, China
e-mail: daixaingwei@mail.bnu.edu.cn

W. Lingli
State Key Laboratory of Cognitive Neuroscience and Learning, Faculty of Psychology, Beijing Normal University, Beijing, China

Beijing Aging Brain Rejuvenation Initiative (BABRI) Centre, Beijing Normal University, Beijing, China

Z. Han
State Key Laboratory of Cognitive Neuroscience and Learning, Faculty of Psychology, Beijing Normal University, Beijing, China
e-mail: zzhhan@bnu.edu.cn

Z. Zhang (ed.), *Cognitive Aging and Brain Health*, Advances in Experimental Medicine and Biology 1419, https://doi.org/10.1007/978-981-99-1627-6_13

Cognitive training is one of the main cognitive intervention methods and specifically refers to the training of different cognitive domains and cognitive processes [6]. Cognitive training is believed to effectively improve subjects' cognitive function and enhance their cognitive reserve [7]. This training is subdivided into single-domain and multi-domain training based on the number of domains targeted by the training. Cognitive training can be conducted in paper-based or computer-based forms. With the rapid development of computerized technology, some cognitive training programs have been able to automatically arrange and adjust training plans suitable for improving individual trainees' performance [8]. Previous research has reported that most of cognitive domains have the characteristics of plasticity, that is, the training tasks of one cognitive domain can improve the performance of subjects on both the original and other untrained tasks in that domain. Some studies have also indicated that the effect of cognitive training is transferable and can be generated in other untrained cognitive domains. Cognitive interventions are currently considered to be an effective supplement to pharmacological therapy. On the one hand, this intervention can be applied to patients with dementia, especially those with mild to moderate dementia, especially leading to more benefits in cases of limited clinical drug treatment [9]. On the other hand, as the intervention time window for dementia and cognitive-related diseases moves forward, cognitive interventions, especially cognitive training, are targeted to individuals in the pre-dementia stage when they do not meet the drug intervention standards but do require cognitive training [10]. Therefore, cognitive training has become a scientific, safe, and effective intervention for brain health.

13.1.1 Traditional Cognitive Training Approaches

Traditional cognitive training usually adopts nonspecific cognitive interventions (e.g., handcrafting, puzzle tasks, and physical exercise) to improve the overall cognitive or social function of patients with cognitive impairment [11]. These interventions can be group-based or individualized interventions.

Handcrafting usually refers to piecing beads together, paper cutting, and other crafts. This training effectively exercises finger movements of older individuals and promotes brain function (e.g., puzzle games, scrabble games, and other training tasks). In addition, traditional Chinese games, such as Huarongdao and tangrams, are also often used in training.

Exercise interventions mainly include aerobic exercise and resistance exercise. The former mainly includes walking, running, and cycling, whereas the latter includes some strength training with training equipment. Exercise interventions are subdivided into single- and multiform interventions. Multiform mixed aerobic exercises, such as aerobic dance and traditional ethnic sports, have been widely used in clinical practice. Based on the enthusiasm of older Chinese people for traditional Chinese medicine sports (e.g., Tai Chi Quan, Baduanjin, and finger exercises), the concepts of meridian and acupoints within Chinese traditional medicine have been integrated into the process of exercise intervention. For instance, Qigong training (e.g., Baduanjin, Wuqinxi, and Liuzijue) is incorporated into the comprehensive fitness program [12].

13.1.2 A Popular Cognitive Training Approach: Computer-Based Cognitive Training

Computer-based cognitive training is a cognitive training program using computer platforms. This approach has the advantages of easy operation, low barriers to access for the patients, and flexible adjustments. Using this approach, patients can set training programs suitable for their own cognitive and physical conditions and adjust the training difficulty. Moreover, this approach utilizes vivid imagery and more engaging procedures, which is more likely to attract patients, arouse their training motivation, and sustain them with a higher level of attention. These features are helpful to improve the compliance of trainees. Currently, the usage of computer-based cognitive training is increasing rapidly [12].

Computer-based cognitive intervention mainly involves the domains of memory, executive function, attention, and processing speed, which has single- and multi-domain cognitive training. Multi-domain cognitive training is often used for patients with cognitive impairment. This training aims to reduce the speed of cognitive decline and, to some degree, to repair cognitive damage. The single-domain training was mostly studied in the 1990s while multi-domain training is still widely studied today. Compared with the former, the latter training is more effective.

In general, computer-based cognitive training significantly improves the cognitive function of older people, and it is superior to traditional cognitive stimulation training [13]. However, some inconsistent training results have been reported in different studies [14, 15]. At present, the promotion of computer-based cognitive training among the elderly in Chinese communities is still very limited, potentially due to the lack of localized computer training systems and the low ownership rate of computers in community-dwelling older adults. However, when these problems are solved, computer-based cognitive training will become increasingly popular in China. Undoubtedly, computer-based cognitive training will play a key role in the prevention of Alzheimer's disease (AD).

13.1.3 Newly Developed Cognitive Training Approach: VR-Based Cognitive Training

Older adults often suffer from attention distractions, low comprehension skills, and unfamiliar computer operations when they perform computer-based cognitive training. Virtual reality (VR) technology has been adapted for cognitive training to address these issues. This technology usually refers to a high-end computer interface that includes real-time simulation and interaction with multiple sensory channels. Through the combination of VR technology and traditional cognitive training, this approach may provide a more realistic scene for older adults and help them to have a better training experience [16].

Moreover, VR-based cognitive training can combine cognitive and motor function training. With rich scene settings and realistic behavioral feedback, this type of training can improve the ability of older subjects to perform daily life-relevant skills such as walking, balance, and basic sensory perception [17].

13.2 The Effect of Cognitive Training

13.2.1 The Effect of Different Cognitive Domains

13.2.1.1 Memory

As one of the most age-sensitized cognitive domains, episodic memory is largely affected by aging. Training for this type of memory is mostly based on some memory strategies, such as the Method of Loci (MoL), verbal memory strategy, and face-name association. In contrast, the training for working memory mainly targets basic memory processing, such as the capacity of working memory.

Memory training results in improved memory abilities in older individuals [18], and the training effects have been shown to persist for up to 5 years in some subjects [19, 20]. Furthermore, 24-h episodic memory training with adapted task difficulty improved self-reported memory and self-esteem in addition to episodic memory ability in older people [21]. Therefore, memory training in a specific domain can benefit other cognitive domains.

In addition to the improvements of behavioral performance, memory training changes brain structure and function. Eight weeks of memory training caused the training group compared to the control group to have significantly increased cortical thickness. Moreover, the degree of change of cortical thickness in the right fusiform gyrus and orbital frontal gyrus was significantly correlated with behavioral changes [22]. In addition, a similar pattern was also observed in even 1-day short-term memory training. Working memory training improved the activation level in typical memory networks [23].

Based on these studies, memory training induces neural plasticity in memory-related brain regions, especially in the prefrontal cortices. The correlation between the brain and behavioral changes suggests that changes in brain plasticity may have cognitive significance.

13.2.1.2 Executive Function

The training of executive function benefits performances of cognitive control, task switching, inhibition, and selective attention [24–27]. Like memory training, executive function training also has a long-lasting training effect. The training effect of a 5-week n-back task was still observed 18 months after the training [28]. Since executive function is closely related to working memory, the executive function training can also improves the capacity of working memory [29, 30]. Moreover, executive function is widely involved in cognitive control processes related to daily activities, thus, the training in executive function also enhances other cognitive abilities such as language, reasoning, and episodic memory [29, 30].

The changes of brain functional activity caused by executive function training are mainly concentrated in prefrontal regions. In addition to the activity changes [31], changes in cerebral blood flow were found after 8 h of short-term cognitive training [24]. Compared with the control group, the experimental group showed a significant increase in the perfusion of the frontal cortex, and such increase was related to behavioral change. Note that a significant change of gray matter density in both subject groups has not been observed, which suggests that executive function training might give rise to a distinctive change between brain function and structure.

13.2.1.3 Processing Speed

Processing speed training is mainly based on audiovisual responses and processing. In the normal population, this training indeed improves cognitive function, including increased sensitivity of vision and hearing and increased processing speed. It can also lead to better memory in the corresponding sensory domains [32–34]. The effects of processing speed training can be maintained for 6 months up to 1 year, and they could transfer to other cognitive domains, such as working memory, daily living ability, and driving performance [32–35].

Processing speed training also yielded positive outcomes in patients with mild cognitive impairment (MCI). Specifically, patients with non-amnestic MCI in a single domain achieved the greatest effect of training, and the training effect persisted for 5 years [36]. Furthermore, training improves the processing speed, attention, and working memory of patients with amnestic MCI [37].

Processing speed training has been shown to increase the integrity of white matter fibers in healthy subjects and change the activation level of the hippocampus in subjects with MCI [38, 39]. Further research on MCI found that the training enabled the protective maintenance of functional connectivity in the default mode network (DMN), which protected and reduced the functional connections of the central executive network (CEN) [37].

13.2.1.4 Visuospatial Function

Visuospatial function training is often combined with working memory and processing speed training, and studies have mostly focused on healthy older adults. For example, improvements in reasoning and episodic memory were observed after 23 hours of continuous training in the n-back and letter recognition paradigms of language materials based on visuospatial working memory tasks. In addition, these effects are transferable to flashback memory span tasks, sustained attention, and episodic memory [29]. Similarly, the training effects were retained after 3 months in another training study [40]. After 10 h of training on a 3D-object tracking task, a significant improvement in motor performance was observed for participants in the training group [41].

In terms of changes in brain plasticity, a total of 10 h of visuospatial working memory training, regardless of whether the test was performed under a high or low working memory load, indicated that brain activation was weakened mainly in the anterior cingulate, anterior parietal lobe, and hippocampus under a high load task.

However, in the low load test condition, changes mainly appeared in the dorsal lateral prefrontal cortex, anterior occipital lobe, and lingual gyrus. All of these attenuated activities were associated with cognitive performance. Although the training results suggested a whole pattern of a greater attenuation of activation, a correlation between increased activity and task performance in some areas was also observed, mainly in subcortical areas such as the basal ganglia [29].

13.2.1.5 Language

Speech training mainly aims to improve the ability and willingness of older people to use verbal coding strategies [42]. Some studies have trained individuals on reading-aloud strategies in conjunction with arithmetic and found that reading-aloud strategies improved overall cognitive performance and the changes transferred to other untrained domains [43, 44]. In addition, a higher degree of improvement in speech comprehension ability, reading breadth, number span, and reasoning ability was recorded in the trained group after learning training with language materials. Additionally, this effect was maintained for 3–6 months [30, 40, 45].

13.2.2 The Effect of Different Training Approaches

Currently, various methods are used for cognitive training, and these different approaches have also produced similar but different training effects.

In the beginning, the researchers mainly used traditional training methods, such as manual exercises and physical exercises. These traditional training methods also helped to improve cognitive ability. Some methods focused on cognitive function directly, such as 3R (reminiscence, reality, and remotivation) therapy or Method of Loci (MoL). These types of training improve cognitive performance in specifically trained areas. For instance, after 8 weeks of MoL training for memory, source memory performance showed improvement [46]. Furthermore, changes in the thickness of the right fusiform and lateral orbitofrontal cortex were positively correlated with improvements in source memory performance, suggesting that structural changes have a possible functional role. Based on these findings, traditional mental exercise may induce short-term structural changes in the aging human brain and indicate structural brain plasticity in older adults. Other studies have attempted to reduce the adverse effects of aging through manual activities. Although many studies have documented age-related decreases in the ability to learn new manual motor skills, older adults still retain the ability to learn and transfer after training in the related domains [47]. They showed similar learning abilities even when compared with younger individuals [48]. In this study, training also showed advantages in terms of brain activity. After training, brain activity related to task planning and execution decreased in temporo-parieto-frontal areas and subcortical regions. This result suggested that similar processes at the level of advanced skills enhance activation efficiency.

Physical exercise, such as aerobics, Tai Chi, and yoga, is another traditional form of training. Many studies suggest that physical training, such as a relatively high level of cardiorespiratory fitness (e.g., achieved through regular physical training), is beneficial for maintaining brain health and cognitive performance [49, 50]. It is suggested that physical training could facilitate brain health by improving white matter integrity, which is sensitive during aging [49]. The literature provides evidence that motor-cognitive exercise (including but not limited to Tai Chi and dancing) combined with motor (physical) and cognitive demands may be the most beneficial exercise to preserve or improve cognition in older adults. In these studies, training not only improved physical skills and cognitive performance but also improved some key neuroimaging indicators, such as the volume of the hippocampus and white matter integrity [51, 52]. However, only the group that underwent cognitive training displayed increased levels of brain-derived neurotrophic factor (BDNF) in the brain after 5 weeks of training compared to cognitive training [53].

In addition to traditional training methods, computer-based cognitive training is currently popular. There is some debates about the effectiveness of computerized cognitive training. For healthy adults, previous literature reviews have revealed the effectiveness of computer-based cognitive training programs for healthy older adults [10, 54, 55]. Besides, these studies also showed that training effectiveness varies across different cognitive domains. Some cognitive domains, such as memory and processing, may be more easily improved by computerized cognitive training [54], while other domains, especially executive function, do not show statistically significant overall effect sizes (Hedges' g, random effects model) compared with controls [10]. These studies provide us with a deeper insight into how computerized cognitive training improves the target cognitive domains, and this effect was retained even after 2 years [56]. Regarding the changes in the brain observed after this training approach, the results are mixed. On the one hand, computer-based cognitive training has been shown to alter the brain structure and function in older adults. On the other hand, depending on the area of interest, both increases and decreases in brain structure and function have been related to behavioral performance [57]. More importantly, even for people who have already had an abnormal cognitive decline, computer-based cognitive training can help improve cognitive abilities [58, 59].

As with the two aforementioned approaches, recent cognitive intervention training for older adults using naturalistic VR environments based on daily tasks has shown good training effects. After VR training, the results proved positive outcomes in general cognition dimensions and in some specific cognitive domains such as executive functioning, attention, and visual memory [60]. Moreover, based on the highly ecological characteristics of VR, it is possible to combine the above two cognitive training methods. A model combining more training methods may provide better training results. After a 4-week exergaming training combined with VR, participants achieved improved cognitive performances in inhibition and task switching [61].

The method, content, and timing of training modulate the training effect [62, 63]. Therefore, an increasing number of studies are investigating the effectiveness of combined cognitive training, which contains a wider range of cognitive domains and methods. Even for patients with cognitive impairment, this combined cognitive training has achieved greater improvements in cognitive and brain imaging indicators than single training [64, 65]. In a multimodal intervention study combining cognitive training, Tai Chi exercise, and group counseling, researchers used the amplitude of low-frequency fluctuations (ALFF) as an indicator of brain imaging findings. Increased ALFF was observed in the middle frontal gyrus, superior frontal gyrus, and anterior cerebellum in the experimental group, and decreased ALFF was observed in the control group. More importantly, these changes were associated with some behavioral outcomes [66]. Subsequently, an examination of the same type of intervention also revealed changes in functional connectivity. For instance, the functional connectivity between the medial prefrontal cortex and the parahippocampal gyrus was significantly increased in the intervention group, and this change was also associated with changes in cognitive performance [67]. In addition, extensive training was associated with increased white matter integrity, increased cortical thickness, and changes in activity in other areas of the brain [24, 68, 69].

In summary, although we should be cautious about any potential bias due to selective reports, current evidence suggests that at least some commercial cognitive training approaches help to promote healthy brain aging.

13.2.3 The Effect on Groups with Different Cognitive Levels

13.2.3.1 Cognitive Maintenance in Older Adults with Normal Cognitive Function

Cognitive decline occurs even in healthy older people. However, they are relatively more cognitively malleable. Therefore, both the cognitive abilities and the brains of healthy older adults can benefit from cognitive training. Meta-analyses confirm this hypothesis, and the results suggest that cognitive abilities are maintained in older adults with normal cognitive function [55]. Cognitive training has been suggested to help healthy older people participate more in brain-stimulating activities, which may stimulate cognitive plasticity and thus increase cognitive reserve. First, regarding the training effects, all cognitive interventions for the healthy older adults have achieved good training effects, i.e., older adults in the training group experienced some degree of cognitive improvement after a specific period of cognitive intervention. The interventions targeted memory, executive function, language, and other cognitive abilities [28, 30, 40, 45]. The training effect persists for a specific period of time, and the results can be maintained for up to 10 years, as tested by the Activity of Daily Living Scale [70]. In addition to the cognitive plasticity changes in the behavior of healthy older adults, relevant neural indicators such as the density/thickness of gray matter [71], white matter integrity [63], and functional connections of the brain network [72] were also changed.

13.2.3.2 Cognitive Rehabilitation for Older Adults with Cognitive Impairment

Cognitive training helps to rehabilitate cognitive impairments, such as AD, mild cognitive impairment (MCI), and subjective memory impairment (SMI), in older adults who have already experienced abnormal pathological aging compared with healthy older adults.

Patients with MCI who had more severe memory decline also show different levels of enhancement after cognitive training in comparison with normal older adults [73]. These patients can achieve some degree of memory improvement with cognitive and memory training [74–77]. Even in patients with AD, memory training still shows a positive effect [78–80]. However, it is important to note that the success of the intervention is more related to individual characteristics (e.g., baseline cognitive status and education level) than to the degree of memory impairment [18]. Also, the changes produced by memory training may generalize to other cognitive domains. Researchers administered episodic memory-based cognitive training for 24 hours with adaptive task difficulty and found significant improvements in episodic memory in older participants, as well as in self-reported memory and self-esteem [21].

For patients with MCI and SMI, cognitive training combined with integrated cognitive training, such as mindfulness, life instruction, and exercise, is also an effective method of alleviating cognitive decline [81]. Individuals who received a period of cognitive training performed better in the areas of memory and attention, and this benefit remained after 6 months. More importantly, comprehensive training reduced the rate of conversion from MCI to AD, with fewer people in the training group converting to AD than in the non-training group [82–84]. Considering training approaches, in addition to memory, computerized cognitive training has been proven to be efficacious in improving global cognition, other cognitive domains, and psychosocial functioning in people with MCI, but evidence for efficacy in people with dementia is weak and limited to trials of immersive techniques [58]. Regarding physical activity, healthy older adults are more likely to achieve cognitive improvements when involved in physical activity interventions [85]. However, moderate exercise can still improve cognitive and brain health in these patients [50].

13.3 The Brain Mechanism Underlying Cognitive Training

13.3.1 Structural Plasticity of the Brain

In studies of cognitive training for older adults, cognitive training affects brain regions through two main pathways. One is mainly associated with the cognitive aging of brain regions, and the other is associated with key brain regions affected by the aging process. One of the important aging hypotheses is the prefrontal hypothesis, which suggests that the prefrontal cortex is more susceptible to the cognitive aging process than other regions [86–88]. Most cognitive training interventions for older adults change the function and structure of the prefrontal cortex [23, 31, 42, 66]. In addition, aging affects other areas (e.g., the medial temporal lobe areas, including the hippocampus, parietal gyrus, cerebellum, and basal ganglia) in addition to the prefrontal cortex [89, 90]. These areas are also sensitive to and can be affected by the training [29, 39, 66, 89, 90]. Brain regions that support training tasks, such as the central control network associated with executive functions [31] and the parietal occipital cortex associated with training-induced visuospatial attention, are also susceptible to change [29, 31].

However, in most studies, these two affected brain pathways are not completely independent, and there are some overlap that existed in most cases. As the aim of cognitive training interventions is to improve or maintain cognitive functions affected by aging, tasks involving brain regions are closely related to important target areas of the aging brain. For instance, the hippocampus, which is not only one of the major brain regions vulnerable to the aging process but also the main area targeted by memory training, has significant effects on brain structure and function. Studies of plasticity changes in the gray matter have shown that even with comprehensive cognitive training intervention targeting a wide range of regions, changes in the thickness of the gray matter after training also did not reach the whole brain levels, and such changes were more concentrated in some related brain regions [46, 69, 91].

In terms of age differences, most studies have shown that the gray matter in older adults has the same training-related plasticity as in young individuals, and after a certain period of training, the gray matter presents positive changes [46, 69, 92]. Although older adults have less malleable brain structures than younger individuals, experience-dependent changes in the gray matter levels are more difficult to achieve, and the gray matter volume in the training group was maintained compared with the decreased gray matter volume in the control group, indicating that training still prevented the adverse effects of aging [93].

Cognitive training mainly improved white matter fiber integrity in the frontal-apex region, as evidenced by increased axial diffusivity and increased fractional anisotropy [22, 63, 68]. Even after removing the effect of baseline white matter integrity, the positive effect of cognitive training was retained. Additionally, age-comparison studies have even shown that cognitive training exerts a greater effect on white matter plasticity in older adults than in younger adults [63, 68]. Furthermore, the effect of white matter fiber integrity on training can predict migration.

Long-term migration was associated with higher white matter fiber integrity in older adults, and a 3-month follow-up was implemented at the end of the training [94].

13.3.1.1 The Plasticity of Brain Function

For brain regions affected by cognitive training, training also changes brain function. It manifests as changes in the activity pattern in the affected brain regions. In most studies, brain regions associated with the cognitive content of the training showed increased activation after cognitive training. For example, increased activation in the occipital region was observed after visual memory training, increased activation of the ventral and dorsal prefrontal lobes was detected after dual-task training, and increased activation in the frontal lobe and left caudate nucleus was observed after semantic coding strategy training [42, 95, 96]. These findings can be partially explained by the compensation hypothesis of brain plasticity, which indicates that the brain regions affected by cognitive aging are compensated by the overactivation of the current or other regions, and this increased activation implies the success of training-induced compensation [97]. In addition to this evidence, the activation of new brain regions after cognitive training also completes the compensatory hypothesis in another way [98].

However, other studies also have reported a significant degree of decreased brain activity. For example, after a total of 10 h of visuospatial and working memory training, both low and high working memory loads tests, showed decreased brain activation. Specifically, it is shown in the anterior cingulate, anterior, parietal lobe, and hippocampus in the high load; whereas in low load, the dorsolateral activity mainly appears in the prefrontal lobe, anterior occipital lobe, and the lingual gyrus [29]. To explain the decrease in brain activation, one opinion is that it might be due to the increased processing efficiency caused by training [99].

In addition to the greater changes in brain activation, increased and decreased activation of different brain regions were both reported in one study at the same time [100]. Additionally, cerebral blood flow velocity (CBF) and ALFF changed after cognitive training, including a significant increase in perfusion in the right inferior frontal cortex and an increase in ALFF in the middle frontal gyrus and anterior cerebellum [24, 66]. Moreover, these different studied metrics may represent more effective indicators than a conventional gray matter analysis. For example, some studies measured both gray matter volume and CBF and found that changes in CBF were more sensitive than changes in gray matter volume [24].

Cognitive rehabilitation not only relies on changes in individual regions but also, more importantly, requires restructuring the coordination between the various brain regions. Currently, an increasing number of studies use different methods for analyzing brain connections to explore the effects of cognitive training on the brain.

Through cognitive training, training participants exhibited changes in functional connectivity under resting conditions, and the changes in brain areas were mainly located in key networks, such as the default mode network (DMN) and the dorsal attention network (DAN), which were affected by aging in a previous study. Additionally, some other regions are affected by tasks, such as the central execution network (CEN). However, in terms of specific changes, the results are mixed.

Some studies have found that the functional connectivity of these networks is enhanced in the resting state. For example, increased functional connectivity in the DMN and DAN was observed after long-term training [69]. In another study, functional connections between the medial prefrontal and medial temporal lobes in the DMN also changed significantly after the intervention. This result showed similar increases in connections between the medial prefrontal cortex and the parahippocampal gyrus [67]. The increase in functional connections within and between networks proves the positive effect of cognitive training on the functional reorganization of brain networks.

In addition to the reported results on improved functional connectivity, studies have also presented the weakening function of brain connections. For example, after visuospatial processing speed training, functional connections in the DMN were only maintained but not enhanced, and the functional connections in the CEN were weakened [37]. After auditory perception processing training or visuomotor memory training, the functional connections between the medial parietal lobe and the inferior temporal lobe were weakened [63]. Two possible explanations for the weakening of functional connections in the network after training have been proposed. One explanation is that continuous training improved the task-related processing efficiency in certain brain regions, thereby reducing the dependence on functional connections in the network [37]. Another possible reason is that training reduces the degree of dedifferentiation of brain areas caused by aging, making the brain areas required for the task less generalized. The homogeneous activity within the network is weakened, and the functional connections of some brain networks are reduced [101]. Furthermore, the correlation between the change in functional connectivity and training migration indicated that the remodeling of network function is one of the key mechanisms of cognitive rehabilitation [63].

In general, the plasticity of brain areas affected by cognitive rehabilitation are focused on the areas of the brain associated with aging susceptibility and training task-related areas, as specifically manifested by increased brain activation, increased gray matter thickness, increased white matter integrity, and enhanced functional connections in key brain networks (such as the DMN and DAN). This evidence shows that cognitive training improves all aspects of the brain, including structure and function, which then compensate for the decline induced by aging. Notably, the changes caused by cognitive training continue during the aging process, which causes degenerative damage to brain structures and functions in the older adults. Based on this premise, cognitive training still effectively increases plasticity in the brain of older adults. Moreover, these changes in neural mechanisms are also significantly correlated with the improvement in behavioral performance, suggesting that changes in the plasticity of the elderly brain caused by cognitive training may be an important mechanism by which cognitive training improves the cognitive ability of older populations.

References

1. Bahar-Fuchs A, Clare L, Woods B (2013) Cognitive training and cognitive rehabilitation for mild to moderate Alzheimer's disease and vascular dementia. Cochrane Database Syst Rev 2013(6):CD003260
2. Park DC, Bischof GN (2013) The aging mind: neuroplasticity in response to cognitive training. Dialogues Clin Neurosci 15(1):109
3. Li T, Yao Y, Cheng Y, Xu B, Cao X, Waxman D et al (2016) Cognitive training can reduce the rate of cognitive aging: a neuroimaging cohort study. BMC Geriatr 16(1):12
4. Fisher M, Holland C, Merzenich MM, Vinogradov S (2009) Using neuroplasticity-based auditory training to improve verbal memory in schizophrenia. Am J Psychiatry 166(7):805–811
5. Hötting K, Röder BJN, Reviews B (2013) Beneficial effects of physical exercise on neuroplasticity and cognition. Neurosci Biobehav Rev 37(9):2243–2257
6. Group, C. t. C. e. c. w., & Association, C. d. C. o. N. b. o. C. M. D (2019) Consensus of Chinese experts on cognitive training. Natl Med J China 099(001):4–8
7. Mozolic JL, Long AB, Morgan AR, Rawley-Payne M, Laurienti PJJN, o. a. (2011) A cognitive training intervention improves modality-specific attention in a randomized controlled trial of healthy older adults. Neurobiol Aging 32(4):655–668
8. Tang Y, Xing Y, Zhu Z, He Y, Li F, Yang J et al (2019) The effects of 7-week cognitive training in patients with vascular cognitive impairment no dementia (the Cog-VACCINE study): A randomized controlled trial. Alzheimers Dement 15(5):605–614
9. Herrera C, Chambon C, Michel B-F, Paban V, Alescio-Lautier BJN (2012) Positive effects of computer-based cognitive training in adults with mild cognitive impairment. Neuropsychologia 50(8):1871–1881
10. Lampit A, Hallock H, Valenzuela M (2014) Computerized cognitive training in cognitively healthy older adults: a systematic review and meta-analysis of effect modifiers. PLoS Med 11(11):e1001756
11. Xu LI, Xin DU, Chen TY (2014) Main approaches to promote cognitive well-being of the elderly: a review. Chin Ment Health J 28(2):125–132
12. Wang QB, Yu S (2016) Effect of tai ji Quan on cognitive function in old adults with mild cognitive impairment. Chin J Rehabil Med 12:645–649
13. Cicerone KD, Dahlberg C, Malec JF et al (2005) Evidence-based cognitive rehabilitation: updated review of the literature from 2003 through 2008. Arch Phys Med Rehabil 86(8):1681–1692
14. Zickefoose S, Hux K, Brown J, Wulf KJB, i. (2013) Let the games begin: a preliminary study using Attention Process Training-3 and Lumosity™ brain games to remediate attention deficits following traumatic brain injury. Brain Inj 27(6):707–716
15. Shute VJ, Ventura M, Ke F (2015) The power of play: the effects of Portal 2 and Lumosity on cognitive and noncognitive skills. Comp Educ 80:58–67
16. Pingguang L, Ningguo Y, Li Y (2007) A review of international research on the application of virtual reality technology in cognitive rehabilitation. Chin J Rehab 5:13–14
17. Mirelman A, Rochester L, Maidan I, Del Din S, Alcock L, Nieuwhof F et al (2016) Addition of a non-immersive virtual reality component to treadmill training to reduce fall risk in older adults (V-TIME): a randomised controlled trial. Lancet 388(10050):1170–1182
18. Belleville S, Gilbert B, Fontaine F, Gagnon L, Ménard É, Gauthier S (2006) Improvement of episodic memory in persons with mild cognitive impairment and healthy older adults: evidence from a cognitive intervention program. Dement Geriatr Cogn Disord 22(5-6):486–499
19. Gross AL, Rebok GW, Brandt J, Tommet D, Marsiske M, Jones RN (2012) Modeling learning and memory using verbal learning tests: results from ACTIVE. J Gerontol Ser B Psychol Sci Soc Sci 68(2):153–167
20. Gross AL, Brandt J, Bandeen-Roche K, Carlson MC, Stuart EA, Marsiske M et al (2014) Do older adults use the method of loci? Results from the ACTIVE study. Exp Aging Res 40(2):140–163
21. Chambon C, Herrera C, Romaiguere P, Paban V, Alescio-Lautier B (2014) Benefits of computer-based memory and attention training in healthy older adults. Psychol Aging 29(3):731–743
22. Engvig A, Fjell AM, Westlye LT, Moberget T, Sundseth Ø, Larsen VA et al (2012) Memory training impacts short-term changes in aging white mat-

ter: a longitudinal diffusion tensor imaging study. Hum Brain Mapp 33(10):2390–2406
23. Miotto EC, Balardin JB, Savage CR, Martin MDGM, Batistuzzo MC, Amaro E et al (2014) Brain regions supporting verbal memory improvement in healthy older subjects. Arq Neuropsiquiatr 72(9):663–670
24. Mozolic JL, Hayaska S, Laurienti PJ (2010) A cognitive training intervention increases resting cerebral blood flow in healthy older adults. Front Hum Neurosci 4:16
25. Blumen H (2010) Training cognitive control in older adults with the space fortress game: the role of training instructions and basic motor ability. Front Aging Neurosci 2:145
26. Blaye A, Lucenet J, Kray J (2010) Can older adults enhance task-switching performance by verbal self-instructions? The influence of working-memory load and early learning. Front Aging Neurosci 2:147
27. Li KZH, Roudaia E, Lussier M, Bherer L, Leroux A, McKinley PA (2010) Benefits of cognitive dual-task training on balance performance in healthy older adults. J Gerontol Ser A Biol Med Sci 65A(12):1344–1352
28. Dahlin E, Nyberg L, Backman L, Neely AS (2008) Plasticity of executive functioning in young and older adults: immediate training gains, transfer, and long-term maintenance. Psychol Aging 23(4):720–730
29. Brehmer Y, Rieckmann A, Bellander M, Westerberg H, Fischer H, Bäckman L (2011) Neural correlates of training-related working-memory gains in old age. NeuroImage 58(4):1110–1120
30. Carretti B, Borella E, Zavagnin M, Beni RD (2013) Gains in language comprehension relating to working memory training in healthy older adults. Int J Geriatr Psychiatry 28(5):539–546
31. Osaka M, Yaoi K, Otsuka Y, Katsuhara M, Osaka N (2012) Practice on conflict tasks promotes executive function of working memory in the elderly. Behav Brain Res 233(1):90–98
32. Berry AS, Zanto TP, Clapp WC, Hardy JL, Delahunt PB, Mahncke HW et al (2010) The influence of perceptual training on working memory in older adults. PLoS One 5(7):e11537
33. Anderson S, White-Schwoch T, Choi HJ, Kraus N (2014) Partial maintenance of auditory-based cognitive training benefits in older adults. Neuropsychologia 62:286–296
34. Wolinsky FD, Vander Weg MW, Howren MB, Jones MP, Dotson MM (2013) A randomized controlled trial of cognitive training using a visual speed of processing intervention in middle aged and older adults. PLoS One 8(5):e61624
35. Edwards JD, Myers C, Ross LA, Roenker DL, Cissell GM, McLaughlin AM et al (2009) The longitudinal impact of cognitive speed of processing training on driving mobility. The Gerontologist 49(4):485–494
36. Valdes EG, O'Connor ML, Edwards JD (2012) The effects of cognitive speed of processing training among older adults with psychometrically-defined mild cognitive impairment. Curr Alzheimer Res 9(9):999–1009
37. Lin F, Heffner KL, Ren P, Tivarus ME, Brasch J, Chen DG et al (2016) Cognitive and neural effects of vision-based speed-of-processing training in older adults with amnestic mild cognitive impairment: a pilot study. J Am Geriatr Soc 64(6):1293–1298
38. O'Brien JL, Edwards JD, Maxfield ND, Peronto CL, Williams VA, Lister JJ (2013) Cognitive training and selective attention in the aging brain: an electrophysiological study. Clin Neurophysiol 124(11):2198–2208
39. Rosen AC, Sugiura L, Kramer JH, Whitfield-Gabrieli S, Gabrieli JD (2011) Cognitive training changes hippocampal function in mild cognitive impairment: a pilot study. J Alzheimers Dis 26(s3):349–357
40. Brehmer Y, Westerberg H, Bäckman L (2012) Working-memory training in younger and older adults: training gains, transfer, and maintenance. Front Hum Neurosci 6:63
41. Legault I, Faubert J (2012) Perceptual-cognitive training improves biological motion perception: evidence for transferability of training in healthy aging. Neuroreport 23(8):469–473
42. Kirchhoff B, Anderson B, Barch D, Jacoby L (2011) Cognitive and neural effects of semantic encoding strategy training in older adults. Cereb Cortex 22(4):788–799
43. Uchida S, Kawashima R (2008) Reading and solving arithmetic problems improves cognitive functions of normal aged people: a randomized controlled study. Age 30(1):21–29
44. Kulason K, Nouchi R, Hoshikawa Y, Noda M, Okada Y, Kawashima R (2018) The beneficial effects of cognitive training with simple calculation and Reading aloud (SCRA) in the elderly postoperative population: a pilot randomized controlled trial. Front Aging Neurosci 10:68
45. Richmond LL, Morrison AB, Chein JM, Olson IR (2011) Working memory training and transfer in older adults. Psychol Aging 26(4):813
46. Engvig A, Fjell AM, Westlye LT, Moberget T, Sundseth Ø, Larsen VA et al (2010) Effects of memory training on cortical thickness in the elderly. NeuroImage 52(4):1667–1676
47. Seidler RD (2007) Aging affects motor learning but not savings at transfer of learning. Learn Mem 14(1-2):17–21
48. Santos Monteiro T, Beets IAM, Boisgontier MP, Gooijers J, Pauwels L, Chalavi S et al (2017) Relative cortico-subcortical shift in brain activity but preserved training-induced neural modulation in older adults during bimanual motor learning. Neurobiol Aging 58:54–67
49. Tseng BY, Gundapuneedi T, Khan MA, Diaz-Arrastia R, Levine BD, Lu H et al (2013) White matter integrity in physically fit older adults. NeuroImage 82:510–516

50. Liu-Ambrose T, Barha CK, Best JR (2018) Physical activity for brain health in older adults. Appl Physiol Nutr Metab 43(11):1105–1112
51. Rehfeld K, Müller P, Aye N, Schmicker M, Dordevic M, Kaufmann J et al (2017) Dancing or fitness sport? The effects of two training programs on hippocampal plasticity and balance abilities in healthy seniors. Front Hum Neurosci 11(305). https://doi.org/10.3389/fnhum.2017.00305
52. Yue C, Zou L, Mei J, Moore D, Herold F, Muller P et al (2020) Tai chi training evokes significant changes in brain White matter network in older women. Healthcare (Basel) 8(1)
53. Ledreux A, Hakansson K, Carlsson R, Kidane M, Columbo L, Terjestam Y et al (2019) Differential effects of physical exercise, cognitive training, and mindfulness practice on serum BDNF levels in healthy older adults: a randomized controlled intervention study. J Alzheimers Dis 71(4):1245–1261
54. Shao YK, Mang J, Li PL, Wang J, Deng T, Xu ZX (2015) Computer-based cognitive programs for improvement of memory, processing speed and executive function during age-related cognitive decline: a meta-analysis. PLoS One 10(6):e0130831
55. Shah TM, Weinborn M, Verdile G, Sohrabi HR, Martins RN (2017) Enhancing cognitive functioning in healthy older adults: a systematic review of the clinical significance of commercially available computerized cognitive training in preventing cognitive decline. Neuropsychol Rev 27(1):62–80
56. Ball K, Berch DB, Helmers KF, Jobe JB, Leveck MD, Marsiske M et al (2002) Effects of cognitive training interventions with older adults: a randomized controlled trial. JAMA 288(18):2271–2281
57. Ten Brinke LF, Davis JC, Barha CK, Liu-Ambrose T (2017) Effects of computerized cognitive training on neuroimaging outcomes in older adults: a systematic review. BMC Geriatr 17(1):139
58. Hill NT, Mowszowski L, Naismith SL, Chadwick VL, Valenzuela M, Lampit A (2017) Computerized cognitive training in older adults with mild cognitive impairment or dementia: a systematic review and meta-analysis. Am J Psychiatry 174(4):329–340
59. Hu M, Wu X, Shu X, Hu H, Chen Q, Peng L et al (2021) Effects of computerised cognitive training on cognitive impairment: a meta-analysis. J Neurol 268(5):1680–1688
60. Gamito P, Oliveira J, Alves C, Santos N, Coelho C, Brito R (2020) Virtual reality-based cognitive stimulation to improve cognitive functioning in community elderly: a controlled study. Cyberpsychol Behav Soc Netw 23(3):150–156
61. Huang KT (2020) Exergaming executive functions: an immersive virtual reality-based cognitive training for adults aged 50 and older. Cyberpsychol Behav Soc Netw 23(3):143–149
62. Belleville S, Mellah S, de Boysson C, Demonet J-F, Bier B (2014) The pattern and loci of training-induced brain changes in healthy older adults are predicted by the nature of the intervention. PLoS One 9(8):e102710
63. Strenziok M, Parasuraman R, Clarke E, Cisler DS, Thompson JC, Greenwood PM (2014) Neurocognitive enhancement in older adults: comparison of three cognitive training tasks to test a hypothesis of training transfer in brain connectivity. NeuroImage 85:1027–1039
64. Feng W, Wang D, Tang L, Cheng Y, Wang G, Hu G et al (2018) Effects of different cognitive trainings on amnestic mild cognitive impairment in the elderly: a one-year longitudinal functional magnetic resonance imaging (MRI) study. Med Sci Monit 24:5517–5527
65. De Marco M, Meneghello F, Pilosio C, Rigon J, Venneri A (2018) Up-regulation of DMN connectivity in mild cognitive impairment via network based cognitive training. Curr Alzheimer Res 15(6):578–589
66. Yin S, Zhu X, Li R, Niu Y, Wang B, Zheng Z et al (2014) Intervention-induced enhancement in intrinsic brain activity in healthy older adults. Sci Rep 4:7309
67. Li R, Zhu X, Yin S, Niu Y, Zheng Z, Huang X et al (2014) Multimodal intervention in older adults improves resting-state functional connectivity between the medial prefrontal cortex and medial temporal lobe. Front Aging Neurosci 6:39
68. Lövdén M, Bodammer NC, Kühn S, Kaufmann J, Schütze H, Tempelmann C et al (2010) Experience-dependent plasticity of white-matter microstructure extends into old age. Neuropsychologia 48(13):3878–3883
69. Hampson M, Pieramico V, Esposito R, Sensi F, Cilli F, Mantini D et al (2012) Combination training in aging individuals modifies functional connectivity and cognition, and is potentially affected by dopamine-related genes. PLoS One 7(8):e43901
70. Rebok GW, Ball K, Guey LT, Jones RN, Kim HY, King JW et al (2014) Ten-year effects of the advanced cognitive training for independent and vital elderly cognitive training trial on cognition and everyday functioning in older adults. J Am Geriatr Soc 62(1):16–24
71. Lampit A, Hallock H, Suo C, Naismith SL, Valenzuela M (2015) Cognitive training-induced short-term functional and long-term structural plastic change is related to gains in global cognition in healthy older adults: a pilot study. Front Aging Neurosci 7:14
72. Wang P, Zhu XT, Qi Z, Huang S, Li HJ (2017) Neural basis of enhanced executive function in older video game players: an fMRI study. Front Aging Neurosci 9:382
73. Vermeij A, Kessels RPC, Heskamp L, Simons EMF, Dautzenberg PLJ, Claassen JAHR (2017) Prefrontal activation may predict working-memory training gain in normal aging and mild cognitive impairment. Brain Imaging Behav 11(1):141–154
74. Olchik MR, Farina J, Steibel N, Teixeira AR, Yassuda MS (2013) Memory training (MT) in mild cognitive

impairment (MCI) generates change in cognitive performance. Arch Gerontol Geriatr 56(3):442–447
75. Greenaway M, Duncan N, Smith G (2013) The memory support system for mild cognitive impairment: randomized trial of a cognitive rehabilitation intervention. Int J Geriatr Psychiatry 28(4):402–409
76. Rapp S, Brenes G, Marsh A (2002) Memory enhancement training for older adults with mild cognitive impairment: a preliminary study. Aging Ment Health 6(1):5–11
77. Simon SS, Hampstead BM, Nucci MP, Duran FLS, Fonseca LM, Martino M d GM et al (2018) Cognitive and brain activity changes after mnemonic strategy training in amnestic mild cognitive impairment: evidence from a randomized controlled trial. Front Aging Neurosci 10:342
78. Moore S, Sandman CA, McGrady K, Patrick Kesslak J (2001) Memory training improves cognitive ability in patients with dementia. Neuropsychol Rehabil 11(3-4):245–261
79. Kixmiller J (2002) Evaluation of prospective memory training for individuals with mild Alzheimer's disease. Brain Cogn 49(2):237–241
80. Cahn-Weiner DA, Malloy PF, Rebok GW, Ott BR (2003) Results of a randomized placebo-controlled study of memory training for mildly impaired Alzheimer's disease patients. Appl Neuropsychol 10(4):215–223
81. Peter J, Schumacher LV, Landerer V, Abdulkadir A, Kaller CP, Lahr J et al (2018) Biological factors contributing to the response to cognitive training in mild cognitive impairment. J Alzheimers Dis 61(1):333–345
82. Rojas GJ, Villar V, Iturry M, Harris P, Serrano CM, Herrera JA et al (2013) Efficacy of a cognitive intervention program in patients with mild cognitive impairment. Int Psychogeriatr 25(5):825–831
83. Talassi E, Guerreschi M, Feriani M, Fedi V, Bianchetti A, Trabucchi M (2007) Effectiveness of a cognitive rehabilitation program in mild dementia (MD) and mild cognitive impairment (MCI): a case control study. Arch Gerontol Geriatr 44:391–399
84. Pereira-Morales AJ, Cruz-Salinas AF, Aponte J, Pereira-Manrique F (2018) Efficacy of a computer-based cognitive training program in older people with subjective memory complaints: a randomized study. Int J Neurosci 128(1):1–9
85. Vaportzis E, Niechcial MA, Gow AJ (2019) A systematic literature review and meta-analysis of real-world interventions for cognitive ageing in healthy older adults. Ageing Res Rev 50:110–130
86. Dempster FN (1992) The rise and fall of the inhibitory mechanism: toward a unified theory of cognitive development and aging. Dev Rev 12(1):45–75
87. West RL (1996) An application of prefrontal cortex function theory to cognitive aging. Psychol Bull 120(2):272
88. Greenwood PM (2000) The frontal aging hypothesis evaluated. J Int Neuropsychol Soc 6(6):705–726
89. Kirchhoff BA, Anderson BA, Smith SE, Barch DM, Jacoby LL (2012) Cognitive training-related changes in hippocampal activity associated with recollection in older adults. NeuroImage 62(3):1956–1964
90. Voss MW, Prakash RS, Erickson KI, Basak C, Chaddock L, Kim JS et al (2010) Plasticity of brain networks in a randomized intervention trial of exercise training in older adults. Front Aging Neurosci 2:32
91. Engvig A, Fjell AM, Westlye LT, Skaane NV, Dale AM, Holland D et al (2014) Effects of cognitive training on gray matter volumes in memory clinic patients with subjective memory impairment. J Alzheimers Dis 41(3):779–791
92. Boyke J, Driemeyer J, Gaser C, Büchel C, May A (2008) Training-induced brain structure changes in the elderly. J Neurosci 28(28):7031–7035
93. Wenger E, Schaefer S, Noack H, Kühn S, Mårtensson J, Heinze H-J et al (2012) Cortical thickness changes following spatial navigation training in adulthood and aging. NeuroImage 59(4):3389–3397
94. Wolf D, Fischer FU, Fesenbeckh J, Yakushev I, Lelieveld IM, Scheurich A et al (2014) Structural integrity of the corpus callosum predicts long-term transfer of fluid intelligence-related training gains in normal aging. Hum Brain Mapp 35(1):309–318
95. Nyberg L, Sandblom J, Jones S, Neely AS, Petersson KM, Ingvar M et al (2003) Neural correlates of training-related memory improvement in adulthood and aging. Proc Natl Acad Sci 100(23):13728–13733
96. Erickson KI, Colcombe SJ, Wadhwa R, Bherer L, Peterson MS, Scalf PE et al (2007) Training-induced plasticity in older adults: effects of training on hemispheric asymmetry. Neurobiol Aging 28(2):272–283
97. Cabeza R, Dennis NA (2012) Frontal lobes and aging. In: Principles of frontal lobe function, 2nd edn. Oxford University Press, New York, pp 628–652
98. Belleville S, Clement F, Mellah S, Gilbert B, Fontaine F, Gauthier S (2011) Training-related brain plasticity in subjects at risk of developing Alzheimer's disease. Brain 134(6):1623–1634
99. Stern Y (2009) Cognitive reserve. Neuropsychologia 47(10):2015
100. Miotto EC, Balardin JB, Savage CR, Martin M d GM, Batistuzzo MC, Amaro Junior E et al (2014) Brain regions supporting verbal memory improvement in healthy older subjects. Arq Neuropsiquiatr 72(9):663–670
101. Goh JO (2011) Functional dedifferentiation and altered connectivity in older adults: neural accounts of cognitive aging. Aging Dis 2(1):30

Lifestyle Adjustment: Influential Risk Factors in Cognitive Aging

14

Chen Liu, Xiangwei Dai, Yanglan Li, and He Li

Abstract

Cognitive aging is inevitable. However, researchers have demonstrated that lifestyle adjustments can reduce the risk of cognitive impairment. A healthy diet style, the Mediterranean diet, has been proven to benefit the elderly. Oil, salt, sugar, and fat are, on the contrary, risk factors for cognitive dysfunction because of the resultant high caloric intake. Physical and mental exercises, especially cognitive training, are also beneficial for aging. At the same time, several risk factors need to be noted, such as smoking, alcohol consumption, insomnia, and excessive daytime sleeping, which are highly relative to cognitive impairment, cardiovascular diseases, and dementia.

Keywords

Cognitive aging · Lifestyle · Diet · Exercise · Sleeping

C. Liu
State Key Laboratory of Cognitive Neuroscience and Learning, Faculty of Psychology, Beijing Normal University, Beijing, China

Beijing Aging Brain Rejuvenation Initiative (BABRI) Centre, Beijing Normal University, Beijing, China
e-mail: c_liu@mail.bnu.edu.cn

X. Dai · H. Li (✉)
Institute of Basic Research in Clinical Medicine, China Academy of Chinese Medical Sciences, Beijing, China

Beijing Aging Brain Rejuvenation Initiative (BABRI) Centre, Beijing Normal University, Beijing, China
e-mail: daixaingwei@mail.bnu.edu.cn

Y. Li
Beijing Aging Brain Rejuvenation Initiative (BABRI) Centre, Beijing Normal University, Beijing, China

14.1 Contribution of Diet to Cognitive Function

14.1.1 Mediterranean Diet Benefits

Nutritional intake plays a vital role in cognition, and studies have shown that a Mediterranean diet promotes physical health and reduces the risk of cognitive impairment [1]. A healthy diet maintains body functioning and increases cortical excitability and inhibition, while also slowing aging-related cognitive impairment [2].

A "Mediterranean diet" [3] refers to the diet style of the Mediterranean coastal areas of southern Europe, such as Greece, Spain, France, and southern Italy. The diet style mainly involves foods such as vegetables, fruit, fish, cereals, beans, and olives. Vegetables and fruits should be fresh and local seasonal ingredients to avoid the loss of trace elements and antioxidants. Fish or poultry foods should be eaten twice a week. Other foods, such as cereals, typically include wheat, barley, oats, rice, barley wheat, and corn.

Z. Zhang (ed.), *Cognitive Aging and Brain Health*, Advances in Experimental Medicine and Biology 1419, https://doi.org/10.1007/978-981-99-1627-6_14

Food is processed as minimally as possible to prevent a high intake of vitamins, minerals, and fibers.

14.1.2 Omega-3 Polyunsaturated Fatty Acids (Omega-3 PUFAs) Benefits

Omega-3 PUFAs are polyunsaturated fatty acids characterized by the presence of a double bond three carbon atoms away from the terminal methyl group [4]. α-Linolenic acid (ALA), eicosapentaenoic acid (EPA), docosahexaenoic acid (DHA), and docosahexaenoic acid (DPA) are considered omega-3 PUFAs. Among these, EPA and DHA are essential for delaying cognitive aging. EPA activates neurotransmitters that accelerate the transmission and processing of the nervous system, thereby improving cognitive function [5]. Meanwhile, Omega-3 PUFAs are an essential part of nerve cell membranes and significantly contribute to brain health.

14.1.3 Negative Impacts of Oil, Salt, Sugar, and Fat on Cognition

With aging, elderly individuals experience sensory changes, including taste bud atrophy and a decrease in the number of taste buds, resulting in the loss of appetite [6]. Older people also face increased taste thresholds, decreased taste sensitivity [7], taste bias, and a preference for sweet foods.

In the elderly population, hyperglycemia and high blood viscosity predispose patients to vasculopathy [8], leading to an inadequate blood supply to the brain and resulting in cerebral white matter lesions. Studies have shown that changes in cognitive function can be predicted by white matter lesions in the brain [9]. Therefore, hyperglycemia is a potential risk factor for impairment in cognitive function. Additionally, neuronal cell damage resulting from a hyperglycemic environment can lead to cognitive decline [10].

Research has shown that a high carbohydrate caloric intake in older adults may be a dietary factor increasing the risk of mild cognitive impairment (MCI) [11]. Another risk factor for cognitive dysfunction is obesity caused by excessive fat content, which is usually coupled with disorders of glycolipid metabolism and insulin resistance [12].

14.2 Exercising Benefits

14.2.1 Physical Exercising: A Protective Factor

Studies have shown that exercise stimulates neuronal metabolism in the brain and improves synaptic plasticity, thereby improving cognitive function in the elderly [13]. Long-term physical exercise prevents cognitive decline in older people by approximately 25%, the risk of Alzheimer's disease (AD) by 37%, and mild cognitive impairment by 46% [14].

Eight Pieces of Brocade, Tai Chi, and other traditional Chinese aerobic exercises have the characteristics of low speed, easily mastered movements, and physical and mental integration and are suitable for elderly individuals. In addition, these exercises do not require special venues or equipment, rendering them suitable for large-scale promotion [15].

The hippocampus and medial prefrontal cortex (mPFC) are responsible for memory function. The hippocampus is a brain area critical for memory. Improvements in cognitive function are related to the encoding, consolidation, and extraction of memories, all of which involve the mPFC. Research by Jing and colleagues claims that Eight Pieces of Brocade may improve functional connectivity between the hippocampus and mPFC, leading to improved memory function [16]. Moreover, aerobic exercises (e.g., jogging, yoga, and swimming) can prevent a decline in brain function and improve executive control functions, such as information processing, sensory perception speed, and working memory [17].

14.2.2 Mental Exercising: A Protective Factor

Due to the plasticity of cognitive functioning, cognitive training can improve an individual's cognitive function and reasoning ability with long-lasting effects [18]. Cognitive training is an operable and side-effect-free method and can improve cognitive functions in older adults [19].

Cognitive training consists of the following three main types of training: strategy training, process training, and multimode integrated training [20]. Strategy training is defined as the ability to learn certain methods and strategies to become more skilled in using cognitive resources to perform different tasks. Cognitive skills (e.g., memory, reasoning, and problem-solving) can be improved by increasing the volume of the key brain regions responsible for the training task. Process training is defined as the constant processing of a specific cognition, familiarity with the task, and corresponding cognitive abilities promoted by process training. Certain fields such as working memory, processing speed, attention, and execution control are suitable for process training purposes. Multimode comprehensive training refers to a type of cognitive intervention combined with lifestyle management that includes multiple components such as cognitive, social, and motor skills. This comprehensive training method is the most effective training method with migration effects.

Cognitive training can also be effective at the level of the brain. Neural structure and function change according to the external environment and experience of the brain [21]. In addition, older adults have increased cerebral cortical gray matter volume and improved white matter connection integrity through cognitive training [22].

14.3 Habits and Aging

14.3.1 Smoking Habits

The number of people worldwide using tobacco products for chronic smoking, mainly in the form of cigarettes, is approximately two billion. Globally, at least four million people die each year from smoking-related diseases [23].

Smoking has always been considered harmful to the health of the general public. Several recent studies have shown that smoking is the main risk factor for fatal brain diseases and is related to other chronic diseases. One disease caused by smoking that has a major impact on health problems worldwide is a chronic obstructive pulmonary disease (COPD). Additionally, respiratory bronchiolitis, interstitial pneumonia, the accumulation of poisonous substances in the lungs, increased blood pressure, and type 2 diabetes are all related to cigarettes [24]. Moreover, smoking increases the risk of ischemic stroke, depression, cognitive impairment, and dementia, including Alzheimer's disease and vascular dementia [25]. After years of research, researchers eventually reached the conclusion that smoking increases the risk of ischemic stroke [26, 27]. More recently, other neurological diseases, e.g., Parkinson's disease (PD) and AD, have also been studied in relation to smoking [28, 29].

Data from international cohort studies indicate that at least a 1.7 times greater risk of AD is associated with smoking. As the total amount of cigarettes smoked increases, the risk of AD evidently increases. Genes, mostly APOE, are mediators that moderate the relationship among one's smoking status, exposure, and AD risk.

Research findings have proved that smoking can directly cause cognitive impairment [30, 31]. One study of young adults found that smokers were inferior to nonsmokers in the mounting of cognitive functions, including attention, impulse control, expressive vocabulary, auditory-verbal memory, information processing speed, and general intelligence. Additionally, an experiment based on a balloon analog risk task found increased levels of risk-taking in young adult smokers with behavioral measurements [32].

14.3.2 Alcohol Consumption

Many domains of functioning, such as behavior, memory, and judgment, have been found to be impaired by alcohol abuse. Additionally, one risk factor for dementia is heavy alcohol consumption [33]. Alcoholic dementia is presumed to be caused by the neurotoxic effects of alcohol, and alcohol abuse is associated with neuropathologic changes [34]. However, research findings have been inconsistent regarding the treatment of alcohol-related dementia. One study suggested that the risk of dementia increased with alcohol consumption, whereas other studies found no associations between the two factors. Multiple studies investigating the relationship between alcohol consumption levels and cognitive performance have found similar inconsistencies [35, 36].

Moderate drinking is defined within an extensive range. The result depends on the population studied and the assessment tools used [37]. Some studies reported no associations between the amount of alcohol consumption and the risk of dementia and AD and that even drinking light or moderate amounts of alcohol can be beneficial for elderly individuals. Studies have proven that light to moderate alcohol consumption reduces mortality from all causes, including coronary heart disease and cerebrovascular disease [38, 39]. The Rotterdam Study (2002) conducted by Ruitenberg indicated that a lower risk of any dementia (hazard ratio 0.58) and vascular dementia (hazard ratio 0.29) was related to light to moderate drinking, which was defined as one to three drinks per day. Another study also found that a lower risk of vascular brain and lower hippocampus and amygdala atrophy in APOE4 carriers were associated with light to moderate alcohol consumption (one drink per week to four drinks per day) [40].

While chronic alcohol abuse can still lead to neurodegeneration and volume loss-related cognitive impairment in the human hippocampus, some studies have demonstrated that alcohol consumption may matter: light to moderate alcohol intake has been proven beneficial for cognition. Dhanabalan and colleagues found that the number of granule cells and the volume of the granular cell layer in the group of chronic alcohol abusers were significantly reduced compared to those in the control group [41]. The impact of alcohol abuse may be more significant than that of aging on the structure and function of the dentate gyrus. The increased cell death or reduced number of new cells in the granular cell layer could be an explanation for the loss of neurons in the alcoholic group.

14.3.3 Social Activities

Researchers have found that healthy social relationships can protect against depression [42], coronary heart disease [43], functional decline [44], mortality [45], and many other negative outcomes. A meta-analysis conducted by Kuiper and colleagues suggests that fewer social interactions increase the risk of dementia [46]. One potential explanation is that brain structures can be influenced by social interactions, leading to more efficient recruitment of brain networks [47]. Another possible explanation relates to chronic psychological distress. Wilson and colleagues suggested that chronic psychological distress is a risk factor for AD [48]. Here, dementia can be prevented through social interactions, which may work as a buffer against stress.

14.3.4 Learning Ability

Physicians commonly advise older adults to participate in mentally and psychologically stimulating activities to reduce their risk of dementia. Studies have shown that individuals with higher levels of education have a lower risk of AD [49]. Typically, people with a shorter education history have a two to four times higher risk of developing AD than those with a longer education history [50].

Older adults at risk can benefit from other interventions, such as multi-domain interventions, which have been shown in double-blind randomized controlled trials to boost or maintain cognitive function in elderly individuals [51]. The 2-year multidomain intervention in the Finnish Geriatric Intervention Study to Prevent Cognitive Impairment and Disability (FINGER) consists of diet, exercise, cognitive training, and vascular risk monitoring. This intervention has been shown to be effective in a shorter-term randomized controlled trial [52].

Studies using animal models may shed more light on the mechanism of learning and cognitive training that reduces the risk of dementia. These studies showed that there is a greater neural complexity after exposure to an enriched environment that provides stimulation [53]. One human study using magnetic resonance spectroscopy (MRS) showed that after memory training, elderly participants would show biochemical changes in the hippocampus [54].

14.4 Sleep and Aging

14.4.1 Negative Impacts of Insomnia

Sleep quality decreases with age, and 40%-70% of older adults have sleep disorders. Studies have shown that sleep deprivation plays an important role in cognitive function [55] and can trigger impaired cognitive abilities (e.g., memory breadth, attention allocation, and executive ability) [56].

Insomnia includes the inability to fall asleep within 30 min, ease of waking up during the night, difficulty falling back asleep, poor sleep quality, excessive dreaming, and less than 5 h of total sleep. Insomnia can be adjusted to in a short time frame, and life and social functioning typically remain normal. In this case, insomnia is a normal physiological response. However, cognitive and social functioning can be affected by long-term insomnia. The relevant literature has revealed that sleep deficiency not only affects the immune system by activating inflammatory responses and leading to tissue damage [57] but also induces oxidative stress responses. Sleep deficiency can inhibit cell proliferation by reducing the levels of brain-derived neurotrophic factors, reducing the number of hippocampal neurons, and decreasing cognitive function [58].

14.4.2 Excessive Daytime Sleepiness

Excessive daytime sleepiness was reported in 15% of older adults [59], which is significantly higher than that of younger adults. Indeed, sleep is progressively impaired with aging [60]. The effect of sleep complaints on elderly individuals is relevant, as these disturbances may also decrease cognitive performance in this specific population.

Excessive daytime sleepiness might represent either a secondary phenomenon of dementia or a risk factor for cognitive impairment in geriatric subjects. Individuals who report excessive daytime sleepiness are more likely to be diagnosed with incident dementia than those who do not report daytime sleepiness [61]. Additionally, disturbances in sleep and circadian rhythm frequently impair the quality of life of individuals with AD. In a community setting, insomnia at night, agitated behavior at sunset, and excessive sleep during the daytime affect 25–40% of patients with mild-to-moderate dementia due to AD, and the intensity of these changes correlates with the severity of dementia [62].

Daytime sleepiness is related to circadian rhythms. Previous research has shown that overall sleep problems are linked to disturbed circadian rhythms. Once the rhythms have been broken, it could be a major risk factor for dementia [63]. Decreased circadian activity rhythm, amplitude, and robustness have been suggested to be linked to the incidence of dementia [64].

Furthermore, studies focused on the sleep-wake cycle have proven that circadian rhythm directly influences the level of the amyloid-β (Aβ) peptide in the brain. Sleep deprivation increases Aβ deposition and accumulation, whereas sleep extension has the opposite effect. When Aβ accumulates, wakefulness is increased and sleep patterns are altered. Sleep abnormalities were reported early than cognitive function impairment. Thus, sleep and neurodegenerative diseases may influence each other [65].

14.5 Mental Health and Cognition

Aging involves a series of processes that continue throughout life. Individuals have positive and flexible coping skills to adapt and adjust to changes during this period of their lives [66]. Psychological and sociological factors have significant influences on aging. A person's religious beliefs, social relationships, perceived health, self-efficacy, socioeconomic status, and coping skills are all positively correlated with aging.

14.5.1 Negative Impacts of Loneliness

Loneliness is a subjective and negative feeling associated with deficiencies in an individual's own experience. The definition of loneliness often depends on both external and internal factors. The absence of social networks is an external factor in the development of loneliness, and personality and psychological factors are typically internal factors.

Although one reason for loneliness is the lack of social networks, loneliness differs from social isolation. A subjectively perceived sense of social isolation and feeling disconnected from others is a type of emotional isolation, which is defined as loneliness. Therefore, dissatisfaction with social interactions is more important than the absence of social interactions with others [67].

The prevalence of loneliness among older people varies across studies for the following reasons: measures of loneliness, the study population, age groups, and sample sizes [68]. Using a single-item measure from the 2002 Health and Retirement Study (HRS) (n = 8932), Theeke found that 19.3% of noninstitutionalized or community-dwelling American adults aged over 65 years felt lonely for much of the past week [69]. Using a nationally representative sample, the American Association of Retired Persons (AARP) reported that 25% of US community-dwelling respondents over the age of 70 years scored 44 or higher on the UCLA Loneliness Scale (n = 3012) [70]. A national survey of 20,255 people in China found that 29.6% of elderly individuals aged over 60 years reported "often feeling lonely" [71].

Several neuroimaging studies have shown that different brain regions are activated when processing social information in individuals with different levels of loneliness. In a functional magnetic resonance imaging (fMRI) study, lonely individuals displayed less activation in the ventral striatum in response to positive social stimuli than individuals who were not lonely [72]. Additionally, it has been found that reduced gray matter density in the left posterior superior temporal sulcus, which is involved in the early stages of social perception, was also associated with loneliness [73].

14.5.2 Negative Impacts of Depression

Depression is a common disorder among older people that has a significant impact on their well-being and quality of aging. Depressive symptomatology also has a similar impact. Depression is not only an important indicator of psychological well-being but is also considered a significant predictor of mental and physical health as well as longevity. Longitudinal studies have shown that increased depressive symptoms are significantly correlated with increased difficulty with daily activities. Regarding mortality, older adults with depressive disorders are at increased risk, according to community-based data [74].

Studies suggest that a history of depression may increase the risk of AD. Potential biases exist because measurement tools of depression vary across different studies. In retrospective case-control studies, the greater recall of depression history in AD patients may bias the results because the experimental group was assessed more carefully than the control group [75]. In short, the relationship between depression and the risk of the subsequent development of dementia remains unclear. Hence, the relationship between a history of depression and the risk of developing dementia in later life still cannot be clarified. Researchers have raised the possibility that depression may be a prodromal symptom of AD and may appear before other illness-defining symptoms of AD. Therefore, depression is a signal rather than a risk factor for AD [76].

Ownby and colleagues utilized meta-analysis to explore the relationship between a history of depression and the risk of developing AD [75]. The results indicated that individuals with a history of depression were more likely to be diagnosed with AD later in life. This finding involved a variety of study types, including retrospective versus prospective data collection and the strictness of diagnostic criteria for AD and depression.

The neurobiological mechanisms linking depression to cognitive decline and increased risk of dementia require more investigation. It has been revealed that depression may be a prodromal symptom of AD and is influenced by AD pathology.

14.5.3 Negative Impacts of Anxiety

The study of anxiety has been particularly neglected in the context of dementia and neurodegenerative disease. In 1998, Beekman and colleagues stated that epidemiological studies of psychopathology have focused on depression and dementia but not sufficiently on anxiety [77]. A recent article by Flint (2005) supports this view [78]. Research output in this area has significantly increased from an average of 215 publications cited per year in the PsycINFO database in the early 1990s to an average of 434 per year from 2002 to 2007.

Despite the increase in research output, the controversies regarding the prevalence and nature of anxiety remain unresolved. Flint (2005) discussed the nature of anxiety in older adults and suggested qualitative differences in anxiety in younger people [78], but other researchers have ignored these differences [79]. Controversial and contradictory opinions exists regarding the comorbidity of anxiety and depression. Some researchers claim that most anxiety co-occurs with depression [78], while other scholars state that the prevalence of mixed anxiety and depression is lower than that of depression and anxiety alone [79].

The findings suggest that anxiety is associated with the development of AD in participants who are not cognitively asymptomatic. The research found that in cognitively asymptomatic participants, anxiety and AD development were associated with higher levels of cognitive impairment. Delaying the onset of AD and decreasing the threat raised by psychological symptoms could be avenues for reducing the risk of AD in the future [80]. Furthermore, a Swedish study involving 1082 non-demented twins showed that twins who experienced high levels of anxiety had a 48% higher risk of developing dementia than those who did not. In co-twin analyses, the resulting correlation between anxiety symptoms and dementia indicated genetic mediation, and this relationship was stronger in dizygotic twins than monozygotic twins [81].

Independently of age, anxiety is likely to damage the brain directly through permanent stress, or indirectly through avoidance behaviors, inactive lifestyle, and loss of cognitive reserve and resiliency in the brain. The preventive potential of anxiety therapy and the role of lifestyle factors is of interest in future research. Here, it emerges that the vicious circle of anxiety and memory loss can be potentially prevented by cognitive behavioral therapy (CBT). Because anxiety is a modifiable risk factor for future dementia, anxiety should be treated seriously at an early age. Better biomarkers and psychometric measurements are needed to more carefully investigate the relationship between anxiety and dementia in large cohort studies at the level of temporal or functional relations [82].

References

1. Gallucci M, Pallucca C, Di Battista ME, Fougère B, Grossi E (2019) Artificial neural networks help to better understand the interplay between cognition, mediterranean diet, and physical performance: clues from TRELONG study. J Alzheimers Dis 71(4):1321–1330
2. Mohajeri MH, Troesch B, Weber P (2015) Inadequate supply of vitamins and DHA in the elderly: implications for brain aging and Alzheimer-type dementia. Nutrition 31(2):261–275
3. Serra-Majem L, Roman B, Estruch R (2006) Scientific evidence of interventions using the Mediterranean diet: a systematic review 64(Suppl_1):S27–S47
4. Agrawal R, Gomez-Pinilla F (2012) 'Metabolic syndrome' in the brain: deficiency in omega-3 fatty acid exacerbates dysfunctions in insulin receptor signalling and cognition. J Physiol 590(10):2485–2499
5. Tan Z, Harris W, Beiser A, Au R, Himali J, Debette S et al (2012) Red blood cell omega-3 fatty acid levels and markers of accelerated brain aging. Neurology 78(9):658–664
6. O'Leary P (1991) Toward healthy aging: human needs and nursing response. SLACK Incorporated Thorofare, NJ
7. Sakai M, Ikeda M, Kazui H, Shigenobu K, Nishikawa T (2016) Decline of gustatory sensitivity with the

progression of Alzheimer's disease. Int Psychogeriatr 28(3):511–517
8. Kövamees O, Shemyakin A, Checa A, Wheelock CE, Lundberg JO, Östenson C-G et al (2016) Arginase inhibition improves microvascular endothelial function in patients with type 2 diabetes mellitus. J Clin Endocrinol Metab 101(11):3952–3958
9. Hensel A, Wolf H, Busse A, Arendt T, Gertz H (2005) Association between global brain volume and the rate of cognitive change in elderly humans without dementia. Dement Geriatr Cogn Disord 19(4):213–221
10. Lang BT, Yan Y, Dempsey RJ, Vemuganti R (2009) Impaired neurogenesis in adult type-2 diabetic rats. Brain Res 1258:25–33
11. Roberts RO, Roberts LA, Geda YE, Cha RH, Pankratz VS, O'Connor HM et al (2012) Relative intake of macronutrients impacts risk of mild cognitive impairment or dementia. J Alzheimers Dis 32(2):329–339
12. Shoelson SE, Lee J, Goldfine AB (2006) Inflammation and insulin resistance. J Clin Invest 116(7):1793–1801
13. Uemura K, Shimada H, Makizako H, Doi T, Yoshida D, Tsutsumimoto K et al (2013) Cognitive function affects trainability for physical performance in exercise intervention among older adults with mild cognitive impairment. Clin Interv Aging 8:97
14. Reimers C, Knapp G, Tettenborn B (2012) Impact of physical activity on cognition. Can physical activity prevent dementia? Aktuelle Neurol. 39(06):276–291
15. Antonishen K (2015) Exercise mode heterogeneity among reported studies of the qigong practice Baduanjin. J Bodyw Mov Ther 19(2):278–283
16. Tao J, Liu J, Egorova N, Chen X, Sun S, Xue X et al (2016) Increased hippocampus–medial prefrontal cortex resting-state functional connectivity and memory function after Tai Chi Chuan practice in elder adults 8:25
17. Angevaren M, Aufdemkampe G, Verhaar H, Aleman A, Vanhees L (2008) Physical activity and enhanced fitness to improve cognitive function in older people without known cognitive impairment. Cochrane Database Syst Rev (3):CD005381
18. Zehnder F, Martin M, Altgassen M, Clare L (2009) Memory training effects in old age as markers of plasticity: a meta-analysis. Restor Neurol Neurosci 27(5):507–520
19. Kelly ME, Loughrey D, Lawlor BA, Robertson IH, Walsh C, Brennan S (2014) The impact of cognitive training and mental stimulation on cognitive and everyday functioning of healthy older adults: a systematic review and meta-analysis. Ageing Res Rev 15:28–43
20. Lustig C, Shah P, Seidler R, Reuter-Lorenz PA (2009) Aging, training, and the brain: a review and future directions. Neuropsychol Rev 19(4):504–522
21. Kolb B, Whishaw IQ (1998) Brain plasticity and behavior. Annu Rev Psychol 49(1):43–64
22. Engvig A, Fjell AM, Westlye LT, Skaane NV, Dale AM, Holland D et al (2014) Effects of cognitive training on gray matter volumes in memory clinic patients with subjective memory impairment. J Alzheimers Dis 41(3):779–791
23. DeMarini DM (2004) Genotoxicity of tobacco smoke and tobacco smoke condensate: a review. Mutat Res 567(2-3):447–474
24. Kissela BM, Khoury J, Kleindorfer D, Woo D, Schneider A, Alwell K et al (2005) Epidemiology of ischemic stroke in patients with diabetes: the greater Cincinnati/Northern Kentucky Stroke Study. Diabetes Care 28(2):355–359
25. Vani G, Anbarasi K, Shyamaladevi CS (2015) Bacoside a: role in cigarette smoking induced changes in brain. Evid Based Complement Alternat Med 2015:286137
26. Fratiglioni L, Wang H-X (2000) Smoking and Parkinson's and Alzheimer's disease: review of the epidemiological studies. Behav Brain Res 113(1-2):117–120
27. Kalmijn S, Van Boxtel MP, Verschuren MW, Jolles J, Launer LJ (2002) Cigarette smoking and alcohol consumption in relation to cognitive performance in middle age. Am J Epidemiol 156(10):936–944
28. Cervilla J, Prince M, Mann A (2000) Smoking, drinking, and incident cognitive impairment: a cohort community based study included in the Gospel Oak project. J Neurol Neurosurg Psychiatry 68(5):622–626
29. Ezzati M, Lopez AD (2003) Estimates of global mortality attributable to smoking in 2000. Lancet 362(9387):847–852
30. Almeida OP, Garrido GJ, Lautenschlager NT, Hulse GK, Jamrozik K, Flicker L (2008) Smoking is associated with reduced cortical regional gray matter density in brain regions associated with incipient Alzheimer disease. Am J Geriatr Psychiatry 16(1):92–98
31. Richards M, Jarvis MJ, Thompson N, Wadsworth ME (2003) Cigarette smoking and cognitive decline in midlife: evidence from a prospective birth cohort study. Am J Public Health 93(6):994–998
32. Wallsten TS, Pleskac TJ, Lejuez CW (2005) Modeling behavior in a clinically diagnostic sequential risk-taking task. Psychol Rev 112(4):862
33. Saunders PA, Copeland JR, Dewey ME, Davidson IA, Mcwilliam C, Sharma V et al (1991) Heavy drinking as a risk factor for depression and dementia in elderly men: findings from the Liverpool longitudinal community study. Br J Psychiatry 159(2):213–216
34. Harper CG, Kril JJ (1990) Neuropathology of alcoholism. Alcohol Alcohol 25(2-3):207–216
35. Elias PK, Elias MF, D'Agostino RB, Silbershatz H, Wolf PA (1999) Alcohol consumption and cognitive performance in the Framingham Heart Study. Am J Epidemiol 150(6):580–589
36. Orgogozo JM, Dartigues JF, Lafont S, Letenneur L, Commenges D, Salamon R et al (1997) Wine consumption and dementia in the elderly: a prospective community study in the Bordeaux area. Rev Neurol (Paris) 153(3):185–192
37. Piano MR (2017) Alcohol's effects on the cardiovascular system. Alcohol Res 38(2):219

38. Berger K, Ajani UA, Kase CS, Gaziano JM, Buring JE, Glynn RJ et al (1999) Light-to-moderate alcohol consumption and the risk of stroke among US male physicians. N Engl J Med 341(21):1557–1564
39. Marmot M, Brunner E (1991) Alcohol and cardiovascular disease: the status of the U shaped curve. BMJ 303(6802):565
40. Topiwala A, Allan CL, Valkanova V, Zsoldos E, Filippini N, Sexton C et al (2017) Moderate alcohol consumption as risk factor for adverse brain outcomes and cognitive decline: longitudinal cohort study. BMJ 357:j2353
41. Dhanabalan G, Le Maître TW, Bogdanovic N, Alkass K, Druid H (2018) Hippocampal granule cell loss in human chronic alcohol abusers. Neurobiol Dis 120:63–75
42. Santini ZI, Koyanagi A, Tyrovolas S, Mason C, Haro JM (2015) The association between social relationships and depression: a systematic review. J Affect Disord 175:53–65
43. Hemingway H, Marmot M (1999) Evidence based cardiology: psychosocial factors in the aetiology and prognosis of coronary heart disease: systematic review of prospective cohort studies. BMJ 318(7196):1460
44. Avlund K, Lund R, Holstein BE, Due P, Sakari-Rantala R, Heikkinen R-L et al (2004) The impact of structural and functional characteristics of social relations as determinants of functional decline. J Gerontol B Psychol Sci Soc Sci 59(1):S44–S51
45. Holt-Lunstad J, Smith TB, Layton JB (2010) Social relationships and mortality risk: a meta-analytic review. PLoS Med 7(7):e1000316
46. Kuiper JS, Zuidersma M, Voshaar RCO, Zuidema SU, van den Heuvel ER, Stolk RP et al (2015) Social relationships and risk of dementia: a systematic review and meta-analysis of longitudinal cohort studies. Ageing Res Rev 22:39–57
47. Stern Y (2012) Cognitive reserve in ageing and Alzheimer's disease. Lancet Neurol 11(11):1006–1012
48. Wilson R, Evans D, Bienias J, De Leon CM, Schneider J, Bennett D (2003) Proneness to psychological distress is associated with risk of Alzheimer's disease. Neurology 61(11):1479–1485
49. Evans DA, Hebert LE, Beckett LA, Scherr PA, Albert MS, Chown MJ et al (1997) Education and other measures of socioeconomic status and risk of incident Alzheimer disease in a defined population of older persons. Arch Neurol 54(11):1399–1405
50. Prince M, Albanese E, Guerchet M, Prina M (2014) World Alzheimer report 2014: dementia and risk reduction: an analysis of protective and modifiable risk factors
51. Ngandu T, Lehtisalo J, Solomon A, Levälahti E, Ahtiluoto S, Antikainen R et al (2015) A 2 year multidomain intervention of diet, exercise, cognitive training, and vascular risk monitoring versus control to prevent cognitive decline in at-risk elderly people (FINGER): a randomised controlled trial. Lancet 385(9984):2255–2263
52. Dahlin E, Neely AS, Larsson A, Bäckman L, Nyberg L (2008) Transfer of learning after updating training mediated by the striatum. Science 320(5882):1510–1512
53. Diamond MC (1988) Enriching heredity: the impact of the environment on the anatomy of the brain. Free Press
54. Valenzuela MJ, Jones M, Rae WWC, Graham S, Shnier R, Sachdev P (2003) Memory training alters hippocampal neurochemistry in healthy elderly. Neuroreport 14(10):1333–1337
55. Zielinski MR, Davis JM, Fadel JR, Youngstedt SD (2013) Influence of chronic moderate sleep restriction and exercise training on anxiety, spatial memory, and associated neurobiological measures in mice. Behav Brain Res 250:74–80
56. Porter NM, Bohannon JH, Curran-Rauhut M, Buechel HM, Dowling AL, Brewer LD et al (2012) Hippocampal CA1 transcriptional profile of sleep deprivation: relation to aging and stress. PLoS One 7(7):e40128
57. Ashley NT, Sams DW, Brown AC, Dumaine JE (2016) Novel environment influences the effect of paradoxical sleep deprivation upon brain and peripheral cytokine gene expression. Neurosci Lett 615:55–59
58. Sahu S, Kauser H, Ray K, Kishore K, Kumar S, Panjwani U (2013) Caffeine and modafinil promote adult neuronal cell proliferation during 48 h of total sleep deprivation in rat dentate gyrus. Exp Neurol 248:470–481
59. Foley D, Ancoli-Israel S, Britz P, Walsh J (2004) Sleep disturbances and chronic disease in older adults: results of the 2003 National Sleep Foundation Sleep in America Survey. J Psychosom Res 56(5):497–502
60. Wolkove N, Elkholy O, Baltzan M, Palayew M (2007) Sleep and aging: 1. Sleep disorders commonly found in older people. CMAJ 176(9):1299–1304
61. Merlino G, Piani A, Gigli G, Cancelli I, Rinaldi A, Baroselli A et al (2010) Daytime sleepiness is associated with dementia and cognitive decline in older Italian adults: a population-based study. Sleep Med 11(4):372–377
62. Moran M, Lynch C, Walsh C, Coen R, Coakley D, Lawlor B (2005) Sleep disturbance in mild to moderate Alzheimer's disease. Sleep Med 6(4):347–352
63. Gehrman P, Marler M, Martin JL, Shochat T, Corey-Bloom J, Ancoli-Israel S et al (2005) The relationship between dementia severity and rest/activity circadian rhythms. Neuropsychiatr Dis Treat 1(2):155
64. Tranah GJ, Blackwell T, Stone KL, Ancoli-Israel S, Paudel ML, Ensrud KE et al (2011) Circadian activity rhythms and risk of incident dementia and mild cognitive impairment in older women. Ann Neurol 70(5):722–732
65. Ju Y-ES, Lucey BP, Holtzman DM (2014) Sleep and Alzheimer disease pathology—a bidirectional relationship. Nat Rev Neurol 10(2):115–119
66. Warnick J (1995) Listening with different ears: counseling people over sixty. QED Press

67. Hawkley LC, Burleson MH, Berntson GG, Cacioppo JT (2003) Loneliness in everyday life: cardiovascular activity, psychosocial context, and health behaviors. J Pers Soc Psychol 85(1):105
68. Ong AD, Uchino BN, Wethington E (2016) Loneliness and health in older adults: a mini-review and synthesis. Gerontology 62(4):443–449
69. Theeke LA (2009) Predictors of loneliness in US adults over age sixty-five. Arch Psychiatr Nurs 23(5):387–396
70. Wilson C, Moulton B (2010) Loneliness among older adults: a national survey of adults 45+. AARP
71. Yang K, Victor CR (2008) The prevalence of and risk factors for loneliness among older people in China. Ageing Soc 28(3):305–327
72. Cacioppo JT, Norris CJ, Decety J, Monteleone G, Nusbaum H (2009) In the eye of the beholder: individual differences in perceived social isolation predict regional brain activation to social stimuli. J Cogn Neurosci 21(1):83–92
73. Kanai R, Bahrami B, Duchaine B, Janik A, Banissy MJ, Rees G (2012) Brain structure links loneliness to social perception. Curr Biol 22(20):1975–1979
74. Bruce ML, Seeman TE, Merrill SS, Blazer DG (1994) The impact of depressive symptomatology on physical disability: MacArthur studies of successful aging. Am J Public Health 84(11):1796–1799
75. Ownby RL, Crocco E, Acevedo A, John V, Loewenstein D (2006) Depression and risk for Alzheimer disease: systematic review, meta-analysis, and metaregression analysis. Arch Gen Psychiatry 63(5):530–538
76. Jorm AF (2001) History of depression as a risk factor for dementia: an updated review. Aust N Z J Psychiatry 35(6):776–781
77. Beekman AT, Bremmer MA, Deeg DJ, Van Balkom AJ, Smit JH, De Beurs E et al (1998) Anxiety disorders in later life: a report from the Longitudinal Aging Study Amsterdam. Int J Geriatr Psychiatry 13(10):717–726
78. Flint AJ (2005) Anxiety and its disorders in late life: moving the field forward. Am J Geriatr Psychiatry 13(1):3–6
79. Wetherell JL, Sorrell JT, Thorp SR, Patterson TL (2005) Psychological interventions for late-life anxiety: a review and early lessons from the CALM study. J Geriatr Psychiatry Neurol 18(2):72–82
80. Burke SL, Cadet T, Alcide A, O'Driscoll J, Maramaldi P (2018) Psychosocial risk factors and Alzheimer's disease: the associative effect of depression, sleep disturbance, and anxiety. Aging Ment Health 22(12):1577–1584
81. Petkus AJ, Reynolds CA, Wetherell JL, Kremen WS, Pedersen NL, Gatz M et al (2016) Anxiety is associated with increased risk of dementia in older Swedish twins. Alzheimers Dement 12(4):399–406
82. Becker E, Rios CLO, Lahmann C, Ruecker G, Bauer J, Boeker M (2018) Anxiety as a risk factor of Alzheimer's disease and vascular dementia. Br J Psychiatry 213(5):654–660

GPSR Compliance

The European Union's (EU) General Product Safety Regulation (GPSR) is a set of rules that requires consumer products to be safe and our obligations to ensure this.

If you have any concerns about our products, you can contact us on ProductSafety@springernature.com

In case Publisher is established outside the EU, the EU authorized representative is:

Springer Nature Customer Service Center GmbH
Europaplatz 3
69115 Heidelberg, Germany

Batch number: 10731385

Printed by Printforce, the Netherlands